Industrial Drying Systems

J E 'Ed' Smith

Industrial Drying Systems

Guidelines for Agriculture, Food, and Wood Products

Springer

J E 'Ed' Smith
JE Smith Co
Terrell, TX, USA

ISBN 978-3-031-31862-7 ISBN 978-3-031-31863-4 (eBook)
https://doi.org/10.1007/978-3-031-31863-4

This Springer imprint is published by the registered company Springer Nature Switzerland AG
The registered company address is: Gewerbestrasse 11, 6330 Cham, Switzerland

Preface

This book is a water-drying resource and guideline written for the global wood, agricultural, and food industries, and based on many decades of my work as a mechanical engineer and a systemic fraud researcher.

This book is not structured to be an engineering text with extensive proofs based on Newtonian principles. Instead, this book points the reader to resources and solutions for the many technical, social, financial, and legal issues common in global industrial wood, agricultural, and food products and their drying systems.

The primary drying technologies focused on in this book are:

Wood Products such as lumber, timber, flitches, roundwood and poles, plywood, OSB, fiberboard, particle board, paper, and paperboard manufacturing systems.

Agricultural and Food Products such as grain, biomass, and food drying systems.

Product Drying Schedules and Drying System Equipment – Understanding the Difference

The reader must understand that all three product industries (wood, agriculture, food) have distinct challenges in which specific drying schedules (temperature, humidity, pressure, drying rates, quality control, color, etc.) must be both carefully developed and chosen for maximum product quality, health, safety, and economic yield.

The reader must also understand that the same required physical equipment to deliver the required product drying schedules may or may not exist in all three industries. However, most of the same design criteria for drying equipment exists in all three industries. **This is the reason that the reader must study all the different drying systems presented in this book to become competent in designing any one of the many drying systems discussed in the book.**

How to Study This Book The reader should first scan this study manual page by page to become familiar with how the book is organized, and to see the many subjects discussed, and the many recommended published books and technical resources listed in this manual. The table of contents located at the front of the book and the technical and word indexes at the back of the book are helpful in quickly locating specific topics, industry terms, and industry resources.

Required Time to Study This Book to Completion For people who are new to industrial drying, this book will require a minimum of 8 to 12 hours of study time to complete. For people who already have experience in industrial drying and desire to expand their working knowledge, this book will require several weeks or months to master this book because the reader will have to both acquire and study the many recommended reference books listed throughout this manual. For people who want to become world-class experts in industrial drying, expect to spend several months or years becoming familiar with the research, sciences, engineering, financial, and legal requirements of the many wood, agricultural, and food industrial drying systems discussed in this book.

Overlap and Repetition Expect to see overlap and repetition of the same subject matter and industry terms in different sections of the book. This occurs because of the many issues with water drying processes, especially wood, known to be a highly-variable, difficult-to-dry anisotropic material. Compounding the fiber drying problems are the many different classes, types, configurations, and applications of existing global industrial drying systems.

Industry Technology Overlaps Many sections of the book discuss technologies common in one industry that are applicable to many other industries. Thus, again, the reader must understand that to become competent in one drying industry, they should be aware of how similar drying technical problems are solved in other industries.

Four themes are emphasized throughout the book:

1. Energy-efficient drying technologies designed to produce minimal product degrading
2. Derivative safety, health, environmental, financial, and legal issues
3. Emphasis on durable long-lasting drying system equipment
4. Management philosophies for success in the workplace

Engineers, scientists, technologists, OEMs, constructors, service techs, insurance underwriters, attorneys, and agricultural, food, and wood product manufacturers should contact the following key organizations for purchasing additional helpful publications on water drying science, technologies, current research activities, and safety standards and codes.

The American Society of Refrigeration and Air Conditioning Engineers (ASHRAE) for purchasing a copy of the *ASHRAE Handbook*, available on CD or printed form. The ASHRAE Handbook is divided into four sections: Refrigeration, Fundamentals, HVAC System and Equipment, and HVAC Applications. Go to www.ashrae.org for ordering the latest edition of the handbook.

For wood drying, this design guideline should also be used with the latest editions of:

1. *Dry Kiln Operator's Manual*
2. *Dry Kiln Schedules for Commercial Woods: Temperate and Tropical*

Both books are published by the US Forest Products Society in Madison, Wisconsin, USA. Many of the subjects presented in the above books overlap with each other.

In addition, readers are encouraged to study the many informative articles on drying technology at: *Drying Technology*: An International Journal at www.tandfonline.com, the Taylor & Frances Online publishing group connected to the latest R&D, and engineering aspects of drying and dewatering, including mathematical modeling of drying and dryers.

Also contact the organization TAPPI located in Atlanta, Georgia, USA, for the latest drying research in the pulp and paper industry.

Also contract the Center for Advanced Research in Drying (CARD) at WPI in Worchester, MA, for current research activities (mostly pulp and paper) and past publications.

Also contact schools of forestry, wood technology, and research centers such as the University of Canterbury, Christchurch, New Zealand.

Also contact the Department of Agriculture in the country where the drying project will be located for additional help in locating current applicable technical publications.

Also contact the American Society of Agricultural and Biological Engineers (ASABE) for the latest technical articles, codes, and standards for the agricultural and food industry.

Also contact the American Society of Mechanical Engineers (ASME) and the American National Standards Institute (ANSI) for the latest publications about industrial energy and drying systems and applicable design and safety standards and codes.

Internet Videos In addition to the above resources, the reader should study the many drying equipment videos now on the internet to watch real drying equipment in operation. Simply Google the "type dryer & video" and numerous helpful videos will appear.

If the reader masters the contents of this and the above books and studies the many helpful publications available at the above organizations, they will have an exceptional understanding of successful, energy-efficient, industrial drying system design and operation practices.

Many significant social, moral, and legal issues are associated with all types of industrial drying operations. Unlike during the past century, these issues are becoming a larger part of our lives. We can no longer afford to think passively about complex environmental issues or leave them for others to solve. Each of us should become both informed and active in confronting these issues on a global scale. **This is the reason I published this design guideline.**

Due to the large scope and complexity of the safety, health, and environmental codes referenced in this book, the reader should consult with local engineers, attorneys, and industry experts familiar with the latest applicable local and national codes and standards for the country, state, or region in which the product manufacturing facility is located.

We now live in a complex, different, and extremely dangerous world. Nations around the globe are developing nuclear weapons at an alarming rate seeking economic and political power in an ongoing struggle for diminishing land and resources. Egocentric humans have already destroyed thousands of life forms and ecosystems that took millions of years to evolve.

We all should now look to the future through the eyes of rational thought founded on valid science. We can no longer allow global ignorance, complacency, and greed to control the planet. We all should read, listen, and have compassion for all life forms around the world.

If we are true to our humanity, we will welcome the future and make the world a better place for all humans and its inhabitants.

Terrell, TX, USA J E 'Ed' Smith

Contents

Chapter 1
Primary Drying Factors

The primary drying factors for a specific material are the material physical properties that determine the required total drying energy and drying rates for that material. These factors must be known before selecting a drying technology for any material.

1.1 Required Energy

For industrial water drying, the total required energy includes the required heat energy and the required electrical energy used by the entire drying system, not just the dryer proper.

1.2 Drying Rates

The drying rate for a specific material is determined by the following variables:

Three-dimensional physical description (size) of the material's thickness, width, and length.

The product's species, specific gravity, initial moisture content, final (target) moisture content, and product degrade issues including color, mold, and pest-control issues. In addition, the geographic location where the product was grown and harvested will impact how the product dries and the levels of degrading that occurs during drying.

In many materials, drying rates and fiber temperatures must be limited by the dryer to prevent damage to the material caused by fiber internal moisture content, moisture gradients, or temperatures that produce excessive fiber stresses resulting in warp, cracks, splits, collapse, and thermal degradation.

J E 'Ed' Smith, *Industrial Drying Systems*,
https://doi.org/10.1007/978-3-031-31863-4_1

Thus, in all industrial dryers, the designer must determine if the final drying system chosen is to be either energy-limited or grade-limited, or a combination of the two.

Is the drying technology low-temp or high-temp? – The differences in drying rates are significant.

Is the drying technology diffusion-flow or pressure-flow? – The differences in drying rates are significant.

These are the reasons the successful industrial drying of all wood, agricultural, and food product materials must be based on a combination of controlled research, properly applied engineering, field testing, and dryer operator training and experience.

Chapter 2
Step-by-Step How to Design an Energy-Efficient Drying System

The following tasks are typical of projects involving industrial drying systems.

2.1 Initial Project Studies

Required expertise – Study this design guideline front to back to become familiar with its structure, entire content, and the included reference materials

Select a project manager experienced in design-build projects

Determine the project's maximum required drying production capacity for the different products

Determine the products' primary drying factors: species, average specific gravity, average initial moisture content, final moisture content, product sizes, degrade issues, color issues, etc.

Investigate available drying technologies such as air drying, forced air drying, heated convection drying, dehumidification drying, high frequency (HF) drying, infra-red (IR) drying, vacuum drying, batch drying, jogging batch drying, continuous drying, drying with cogeneration, etc. to determine the total capital costs, labor costs, inventory costs, maintenance costs, and energy costs for different drying capacity requirements

Location requirements: utilities, energy supply issues, labor force, maximum economic radius, etc.

Site requirements: soils analysis, drainage issues, flooding risks, environmental risks, etc.

Determine the legal issues – construction permits, environmental permits, safety & health issues, fire & explosion prevention issues, equipment maintenance issues, possibly third-party inspections issues, financing issues, liability insurance issues, security issues, etc.

J E 'Ed' Smith, *Industrial Drying Systems*,
https://doi.org/10.1007/978-3-031-31863-4_2

About Systems of Measurements – *Different countries use different measurements of unit. Depending on which country the industrial drying is located, the system of measurement (units) will vary. – See Wikipedia "System of Measurements" for a complete history and current practices of the global systems of measurements such as Imperial and US Existing, Metric, and SI systems. – Numerous conversion tools and tables exist on the internet for converting measurements from one system to another.*

2.2 Conceptual Drying System Design Studies

2.2.1 *Phase One*

Determine the mix of products to be dried
Determine the required weekly drying production capacity for each product
Determine the minimum required energy efficiency of the drying system
Research available industrial drying technologies
Select a drying technology for the project
Develop lists of vendors and consultants
Prepare layout drawings of the available space (real estate) for the drying system
Prepare flow diagrams for the entire drying system

Mass flow rates – solids, liquids, and gases
Energy flow rates – BTU/hour, pressure, temperature, chemical energy

Prepare material inventory and storage calculations at specified points in the flowstreams
Determine the required capability of each machine center
Evaluate the interface specifications (capability) between machine centers
Determine where bottlenecks can occur
Conduct time, motion, and energy studies for each machine center
Prepare preliminary geometric production capacity models for each machine center
Conduct process time, motion, inventory, and energy studies for the entire dryingsystem
Determine the maximum possible weekly production capacity of the above drying system for specific products
Determine the energy requirements and energy efficiency of the entire drying system
Determine the environmental, safety, and health issues in the proposed drying system

2.2.2 Phase Two Modeling

Green-end systems – determine which process or machine centers ahead of the dryers require additional investigations to determine their impact on the performance of the drying system

Dry-end systems – determine which process or machine centers after the dryers require additional investigations to determine their impact on the performance of the drying system

Dryers – conduct geometric, computational fluid dynamics (CFD), and configuration models of the dryers to determine which process variables impact the performance of the dryers. – This part of the project may reveal significant changes to the physical size or configuration of the final drying system

Go back to phase one and recheck or redo all the previous assumptions and/or calculations until optimum economic designs are arrived at.

2.3 Construction

2.3.1 Purchasing Contracts

Prepare a timeline for starting and final startup for the project

Prepare a detailed list of machine centers and dryers including their design specifications

Select vendors for submitting quotes

Prepare bid specifications

Review bids and choose vendors

2.3.2 Construction and Contract Compliance

Select an owner's rep for this phase of the project

Manage the construction project to completion, final commissioning, and startup

Determine if third-party inspections are required before final payments to vendors

2.3.3 Startup, Acceptance Testing, and Recordkeeping

Document and record all process control and safety parameters

Coordinate with the business insurance carrier the final configurations of safety systems

Fine-tune all process control parameters to achieve optimum performance
Document drying production rates for each product
Document any product-degrading issues that may exist
Document the energy efficiency of the entire drying system
Store all construction drawings, parts lists, contracts, and process tuning records ina secure fire-proof vault
Final payments to vendors if no contractual performance issues exist

Chapter 3
Costs of Energy and Drying Equipment

Check for updates

3.1 Current Factors

1. The record-high demands for global wood, agriculture, and food products in the coming decades because of population growth, severe weather events, and large mass migration of millions of people.
2. The impact global warming will have on the required additional energy efficiency of all types of drying and energy systems.
3. Increasing prices for primary materials for manufacturing all types of equipment.
4. Large amounts of government money and derivative fraud, abuse, and waste flowing into the wood, agriculture, and food products industries due to the impact of global warming.
5. Continuing failures of world governments to contain or reduce the total population of people on the planet and the associated social chaos caused by human overpopulation.

3.2 Future Issues

All the above factors are all going to get far worse in coming decades. This will require objective, science-based government regulations for how all global industrial energy and drying systems are designed, maintained, and operated.

J E 'Ed' Smith, *Industrial Drying Systems*,
https://doi.org/10.1007/978-3-031-31863-4_3

Chapter 4
Safety, Health, and Fire Protection

The following information presents best practices.

4.1 Mission Statement by the CEO

Each business entity should have a mission statement by the chief executive officer that states that both safety and the environment come before production. Having a corporate mission statement sends a clear signal to all employees that the CEO and stockholders place the welfare of the public and their employees before profits.

4.2 Availability of Published Safety Codes

If a country or region adopts a published safety code, then they are legally required to enforce that published code, and if they enforce that code, then they are legally required to inform the end-user and the public what the published code requires of the end-user and the public.

Depending on the country where the project is located, all applicable enforced safety codes may be available on the Internet at no cost to the end-user. Publishing copyright and safety code laws vary between countries. Check with the local authority having jurisdiction where the project will be located. Typically, if a country adopts a published safety code, then the published code will be available to the public at no cost. If the published code is not available on the Internet, the local authority having jurisdiction should provide a free copy of the code at no cost to the end-user. The reason these laws exist is to inform the end-user of what the law requires.

J E 'Ed' Smith, *Industrial Drying Systems*,
https://doi.org/10.1007/978-3-031-31863-4_4

4.3 Health and Safety Program Resources in the United States

OSHA regulations

These can be downloaded from http://www.gpo.gov

Safety consultants and engineers

Many can be found on the Internet or through the applicable state board of engineers.
American Industrial Hygiene Association, www.aiha.org
American Society of Safety Engineers, www.asse.org
Board of Certified Safety Professionals, www.bcsp.org
National Safety Information Exchange, www.nsie.org

Resources for workplace accident, injuries, and death statistics

National Safety Council, www.nsc.org
NIOSHA, www.cdc.gov/niosh

Resources for the effects wood smoke has on the health of humans and animals

Clean Air Revival, Inc., www.burningissues.org/fact-sheet.htm
Environmental Protection Agency, www.epa.gov

Resources for design, operation, maintenance, and safety standards and codes

American National Standards Institute, www.webstore.ansi.org
American Society of Heating, Refrigeration and Air Conditioning Engineers, www.ashrae.org
American Society of Agricultural and Biological Engineers, www.asabe.org
American Society of Mechanical Engineers, www.asme.org
National Fire Protection Association, www.nfpa.org
National Safety Information Exchange, www.nsie.org
The Instrumentation, Systems, and Automation Society, www.isa.org
Global Engineering Documents, www.global.ihs.com

Employee training programs

Hire a professional safety consultant to design your safety program

Personnel safety hardware

Hardhats, safety shoes, glasses, ear plugs, electrical lockouts, etc.
Use the Internet to locate safety equipment suppliers

Safety logs

Records should be kept of safety meetings, hazards, and incidents.

Enforcement

Without enforcement, safety programs are of no value. I suggest the following:

1. Conduct daily safety meetings with workers to:

   Identify hazards
   Take corrective actions

2. Maintain records of safety actions taken

4.4 Fire and Explosion Protection

4.4.1 NFPA Resources

If your organization is not now a member of the National Fire Protection Association (NFPA), then join. The NFPA is the national authority on fire, electrical, and building safety. The NFPA publishes codes, standards, and rules developed by nationally recognized experts. Include local fire departments in your fire protection programs.

4.4.2 Fire-Rated Protection Barriers for Industrial Dryers

For (1) high-fire risk-direct fired dryers using wood as the fuel, or (2) any dryer operating above 250 °F temperatures, I recommend that a fire-rated barrier wall be located between the dryer building and the dryer control room in which the dryer instrumentation and process controls, the electrical switchgear, and the fuel-burning furnaces and heat recirculation blowers are located. Both heat supply and return air ducts should be fitted with automatic fire dampers, and safety disconnect joints designed to prevent smoke, fire, or structural damage to the control room or any ducting or equipment located inside the dryer's control room. The fire-rated wall should be located no less than 8′ from the outer adjacent side wall of the dryer building, or 10′ from the end wall of the dryer. Consult with your insurance carrier for the design and location of these smoke/fire protection barriers. Some property insurance companies or government jurisdictions may require larger fire-separation distances than those listed above. Some insurance companies or government jurisdictions may not require any fire-rated protection barriers, even for high-fire-risk dryers.

Chapter 5
Environmental Impact of Industrial Drying Systems

5.1 Introduction

The Earth's highly complex and fragile ecosystem took millions of years to evolve into its current state. Since the dawn of mankind, humans have destroyed thousands of ecosystems around the globe in the interest of personal profit. During the last two centuries, the effect of man's presence on Earth has been the massive destruction of forests and the extermination of thousands of plant and animal species.

Today, no one knows what the long-term impact of man's presence will have on the survivability of the remaining species. Scientists all around the globe agree that we are headed for a global upset in the natural order of life on Earth if the population of humans and their conduct are not both controlled and reduced. Our responsibility as caretakers of the environment will be tested during this century. Human greed and egocentricity will have to be brought under control.

If we fail to do this, our descendants will suffer the consequences. Each of us will have to rethink our perception of the environment and all its inhabitants.

5.2 Environmental Protection

Protection of the environment from the harmful effects of pollution is accomplished by:

1. Identifying the sources of pollutants
2. Identifying the types of pollutants emitted
3. Identifying the quantities of pollutants emitted
4. Limiting the levels of pollutants produced at each source

J E 'Ed' Smith, *Industrial Drying Systems*,
https://doi.org/10.1007/978-3-031-31863-4_5

Environmental discharges in industrial drying and energy systems are:

1. Air emission discharges
2. Liquid discharges
3. Solid discharges

5.2.1 Air Emissions – Sources of Significant Discharges

USEPA Air Quality Regulations for Furnaces and Dryers
Information about US federal air quality regulations can be found on the Internet. Go to www.epa.gov/ttn/atw/eparules.html. Click on National Emissions Standards for Hazardous Air Pollutants. A list of industries will appear.

Click on the following for the applicable EPA rules:

Industrial, Commercial, and Institutional Boilers and Process Heaters
Example – Plywood and Composite Wood Products

Government service personnel are available for assisting people in locating specific air quality issues for any product drying technology such as wood, agriculture, and food.

The following example discussion about wood dryers applies to the three products: wood, agricultural, and food drying.

Wood Dryers
Wood dryer air emissions are determined by the species of wood, the dryer/energy system energy usage, and the temperature the wood reaches during the drying process. Because most commercial wood dryers are heat-transfer processes, the temperature the wood reaches at the end of the drying process will approach the highest levels.

The following table demonstrates the temperatures that wood can reach during drying. During dryer control upsets, wood temperature can exceed the listed temperatures by as much as 30 °F. These upsets will increase the types and quantity of air emissions from the dryer.

Dryer system	Maximum wood temperature
Wood drying	
Air-drying yards	130 °F
Low-temp dryers	150 °F
Accelerated dryers	195 °F
High-temp dryers	270 °F
Veneer dryers	310 °F
Particle drying	310 °F
Rotary dryers	310 °F

A note of caution about all types of industrial dryer systems. – Studies have shown that the risk of fires occurs when wood fiber temperatures exceed 295 °F. – At these temperatures, any moisture in the wood boils out quickly and thermal degrading occurs rapidly. – Several direct-fired lumber kiln fires have been documented in which the lumber caught fire just as the dry-bulb temperature recorders reached 300 °F, a condition caused by burner safety control systems not being designed or maintained properly. – These fire events produce massive amounts of additional pollution above the normal levels seen in properly designed and operated drying systems.

Other than for air and solar drying, the type of heat source used with the dryer will also add to the total emissions produced by the entire drying system.

Emissions from Indirect-Fired Wood Drying Systems

An indirect-fired dryer is one in which the heat energy source does not enter the internal heating space inside the dryer. The energy source can be solar, electricity, steam, or hot liquids.

Solar-Heated Wood Drying Systems

Wood drying systems that receive their heat energy directly from the sun are the most environment friendly of all types of drying systems, producing the least amount of all types of emissions. Because of their low drying temperatures, the levels of emissions are insignificant when compared to all other types of commercial wood drying.

The most common use of solar energy is the conventional air drying yard in which stacked packages of lumber are dried by ambient air. If the lumber is dried in a kiln fitted with solar-heated hot water coils and fans, the total levels of emissions will increase due to the higher chamber temperatures and the power plant needed for the electricity used by the dryer's fans.

Electrical-Heated Wood Drying Systems

There are three main types of drying systems that use electrical energy exclusively. One is the high-frequency high frequency (HF) dryer, the infra-red (IR) dryer, and the other is the heat pump. Of the three, the heat pump is the most energy efficient. These systems are used in small- to medium-sized commercial drying systems. Heat pumps are commonly referred to as dehumidifiers due to the presence of water-condensing coils. The level of emissions produced by these designs is dictated by the temperature the wood reaches during drying, and the emissions produced by the electrical power plant needed for the electricity used by either.

Steam-Heated Wood Drying Systems

Emissions from steam-heated drying systems are produced at both the dryer and the steam-generating plant. Steam-heated wood dryers produce air emissions at dryer exhaust vents, dryer building leaks, and dryer doors. The types and levels of emissions emitted from steam-heated dryers are determined by both the rate of drying and the temperature the wood reaches during drying. The higher the temperature, the higher the levels of emissions produced. Steam boilers used for wood dryers discharge air emissions at their furnace exhaust stack. The design of the furnace and the type of fuel (wood, gas, oil) will determine the type and levels of emissions produced. Many of the emissions produced are regulated by governments due to their toxicity, their ability to cause cancer, and their impact on the environment.

Hot-Oil-Heated Wood Drying Systems
Hot-oil-heated wood drying systems will produce air emissions at dryer exhaust vents, dryer building leaks, dryer door areas, and hot-oil furnace stacks. Although the types of emissions are the same, the quantity of pollutants emitted from hot-oil-heated drying systems (dryers and hot-oil heater) is slightly less than for steam systems due to the higher thermal efficiency of hot-oil drying systems over steam drying systems.

Hot-Water-Heated Wood Drying Systems
Hot-water-heated wood drying systems will produce air emissions at dryer exhaust vents, dryer building leaks, dryer door areas, and hot-water furnace stacks. The quantity of pollutants emitted from hot-water-heated drying systems is typically less than for equivalent hot-oil drying systems, due to lower operating temperatures and higher thermal efficiency.

Hot-Air-Heated Wood Drying Systems
Hot-air-heated wood drying systems are extremely rare because of the high cost of manufacturing and maintaining the air-to-air heat exchangers suitable for industrial wood drying operations. These drying systems will produce air emissions at dryer exhaust vents, dryer building leaks, dryer door leaks, and hot-air heat exchanger discharge stacks. The quantity of pollutants emitted from hot-air-heated drying systems is strongly dependent on the design of the air-to-air heat exchanger, combustion system, and the type heating fuels used.

Emissions from Direct-Fired Wood Drying Systems
A direct-fired dryer is one in which the heat energy source enters the internal heated air steam inside the dryer. The energy source can be the combustion of wood, gas, or oil. Direct-fired wood dryers will produce emissions at dryer exhaust vents, dryer building leaks, dryer door areas, and furnace dump stacks (if any are present).

The different types and quantity of toxic compounds emitted from direct-fired wood-fueled dryers will far exceed those for all other types of direct-fired dryers.

Because of the higher net thermal efficiency of direct-fired systems over indirect-fired systems, the total annual mass of emissions produced from direct-fired systems is between 50% and 80% of the emissions from indirect-fired systems.

Direct-fired wood-fueled dryers also create the highest risk for fires than all other types of dryers. During dryer fires, large amounts of toxic compounds are released into the environment as well as being exposed to people fighting the fires.

Wood Dryers – A Global Industry in Need of Major Reform
In many countries, wood drying systems are exempted from air emissions regulations. Because of this, many dangerous toxic emissions are subjected to people near wood drying systems. This is especially true for dryer operators who are constantly

around or for plant people working downstream of wood-waste direct-fired high-temperature lumber kilns.

All heated dryers should be fitted with negative-air venting blowers and tall discharge stacks to keep the entire dryer at a slightly negative pressure to collect and then direct the toxic compounds produced during drying away from the dryers. Negative-air venting is an old technology easily applied to all types of heated wood dryers. This technology was applied to veneer dryers many decades ago. In addition to keeping the dryer proper at a negative pressure, main doors of moving-package dryers should be fitted with pressurized airlocks to prevent toxic compounds from leaking out into the areas outside the main doors.

Of all types of wood drying systems in existence today, by far, the **most toxic and most dangerous design** is the wood-waste-fueled, direct-fired, high-temp lumber dryer. In this type of dryer, the combustion products from burning wood are injected directly into the circulated air/moisture steam inside the dryer and onto the surfaces of the lumber. This situation creates significant health and safety hazards for dryer operators, plant employees, and the public in the vicinity of these dryers.

Numerous toxic compounds known to cause cancer are produced by these drying systems that produce large floating clouds of hot steam mixed with wood ash, carbon, carbon monoxide, wood tars, and other toxic chemicals. Toxic wood ash and carbon compounds from the burning wood fuel are deposited on the lumber and inside these dryers. These materials are spread across the lumber mill and during the handling of the lumber after it has been dried. When the contaminated lumber packages are broken down at the entrance to a planer mill, large clouds of toxic dust are produced.

Dryer operators are constantly exposed to dangerous risks. Leaking dryer doors or vents can produce large clouds that can drift long distances before being dispersed by winds.

Plant personnel are exposed to these toxic compounds when entering the dryer during load changes or routine cleaning or maintaining dryer equipment.

In some dryer operations burning wet sawdust, dryer operators complain of boils on their legs, ankles, and arm pits. In direct-fired dryers burning dry shavings, both dryer operators and planer machine operators complain of heavy clouds of ash and carbon around the dryers and planer mill in-feed systems. In some dryer operations, the smoke and water vapor are so thick that the visibility around the dryer is less than one foot. This increases the risk of injury or death from fork trucks.

Is it now time for all governments around the planet to step in and stop the construction of these drying systems? In my opinion, the situation has reached a level at which something needs to be done. Many wood products plants around the globe have spent large amounts of capital on these dangerous "cheap" drying systems and will surely protect their investments by using lobbyists to keep regulations out of their plants.

Objective testing is needed now to determine the total environmental impact of all types of global wood drying systems and where the entire global industry should be going. We may even discover that upper drying temperature limits should be established for the different species. The people doing these studies must be totally competent, neutral, and objective.

5.2.2 *Liquids – Sources for Significant Discharges*

Liquid discharges from wood dryers can contain chemicals classified as unsafe by the EPA. Potential sources for regulated liquid discharges are listed below:

Lumber Dryers

Dehumidifier (heat pump) liquid condensate from condensers
Expansive water (sap) released during the initial stages of drying
Spray systems for humidification
Steam heating system condensate leaks
Hot-oil-heated oil spills
Hot water/glycol-heated glycol spills

Liquid's compliance statues and laws in the United States

Federal – US CFR EPA – Contact the regional office of the EPA in your area.
State – Some states or regions may exceed federal regulations.

5.2.3 *Solids – Sources for Significant Discharges*

Solids produced from wood dryers can contain chemical compounds deemed unsafe by scientists. Wood burning furnaces generate toxic solid compounds. Disposal and handling techniques should comply with applicable local and federal regulations.

Solid's compliance statues and laws in the United States

Federal – US CFR EPA –Contact the regional office of the EPA in your area.
State – Some states or regions may exceed federal regulations.

Chapter 6
Economic Issues in Industrial Drying Systems

6.1 Introduction

The economic issues in all industrial drying systems are determined by the requirements of the final product being produced. The drying equipment, operational costs, and safety/health issues vary depending on what the final product is.

For example, Kiln-dried lumber, plywood, oriented-strand-board (OSB), particleboard, furniture, treated products, agricultural crops, and food products.

The following paragraph explains why industrial drying systems are so varied and challenging for those designing and operating these systems and for the public using these products.

All wood, agricultural, and food products are subject to challenging anisotropic fiber shrinkage problems.

Wood fibers also possess different mechanical strength and thermal properties depending on how the fibers are oriented inside a log. Geographic location and elevation also impact how wood/agricultural/food products grow, specific gravity, initial moisture content, and reactions to drying processes. The rates at which a product dries are also dependent on the physical size of the product entering a drying system. Wood especially also produces a significant amount of ash (noncombustibles) and toxic compounds when being used as a fuel. Many plants in their natural state are toxic to both wildlife and humans.

At some point in an integrated wood-products manufacturing plant, the flow of fiber is diverted from machining operations to the drying operation. After the drying is completed, the fiber may then be sent to a second machining operation. In every facility that includes drying, specific capital expenditures and manpower exist solely for removing water from the wood.

J E 'Ed' Smith, *Industrial Drying Systems*,
https://doi.org/10.1007/978-3-031-31863-4_6

6.2 Drying System Scope for Wood Products

Depending on the operation, the scope of what constitutes the drying system varies. In some cases, the sole purpose of a manufacturing facility is drying wood. In some operations, the drying system is integrated into sawing, chipping, peeling, treating, and finishing operations. In many wood-products manufacturing operations, hot steam or liquids may be produced for drying wood as well as for other uses such as hot presses, space heating, and electricity generation.

For example, in lumber manufacturing operations, the scope of the "drying system" for removing water from the fiber starts at the saw mill's green board trimmer and ends at the dry-end planer. Every piece of equipment and utility support system between the green trimmer and the dry planer is there for the specific purpose of removing water from the lumber.

6.3 Economic Scope

An **example** of the **economic scope** of a drying system (for a lumber manufacturing operation) is shown below and is typical for all types of wood drying systems.

1. The market demand for the dry finished product
2. The availability and delivered cost of fiber (logs)
3. All capital and depreciation costs for the drying system
4. The impact on the environment the drying system adds to the plant
5. The health and safety risks associated with drying
6. The total cost of labor for operating the drying system
7. The total maintenance costs for the drying system
8. The required management skills for operating the drying system
9. The total economic impact of product degrading caused by drying
10. The total electrical and heat energy costs for the drying system
11. The total cost of inventory due to drying
12. The total costs of all types of insurance due to the drying system
13. The liability risks associated with the drying system
14. Technology pitfalls common in drying systems

Economic studies for each part of the system require objective analysis of all the issues to ensure a successful project.

1. Market demand for the product
   This depends on the domestic and export markets for the manufactured products including all sales costs and risks for serving the markets.
2. Fiber availability and cost

Because the availability of large quality logs is diminishing, the delivered unit costs are increasing. In many cases, the reduced quality and size of logs will be such that even with modern recovery technologies, the profitability of the plant may be at risk.

3. Capital and depreciation costs
   Standard accounting practices apply here. This includes all capital and depreciation costs for the equipment used in the drying operation. In a typical plant, the drying system includes the following:

   Green product sorting, inventory, and stacking equipment
   Green storage yards
   Lift trucks, transfer cars, and all rail, carts, and tram handling systems
   Dryers and kilns
   Steam boilers and steam systems
   Burners and furnaces for drying equipment
   Rough-dry storage and tempering systems
   Package breakdown equipment at the dry-end finishing mill
   Product moisture measuring and MIS reporting systems

4. Environmental impact
   This is the total environmental impact of all the equipment in the drying system. This includes all air pollutants, water pollutants, soils contaminations, and all **gross energy** required to manufacture, transport, construct, and install every product (steel, aluminum, copper, concrete, etc.) required of all the equipment used in the drying system.
5. Health and safety risks
   This is the total health and safety risks for the drying system.
6. Labor costs
   This includes all labor costs for operating and maintaining the entire drying system. This also includes time and motion studies of process material flowing through the system and all labor required to handle materials or operate machinery (such as fork trucks).
7. Maintenance costs
   This includes all labor, materials, and outside contract services for the items listed above.
8. Required management skills and training
   This includes the expertise, training, and management skills required for the system.
9. Degrade costs
   This includes the total direct and indirect costs of all types of products degrading due to drying, including the time and expenses for the resolution of drying-related claims.

Degrade Capacity Function (DCF)
DCF is a term in wood drying that applies to the total capacity of a drying system to waste product value and reduces profits at the plant level. Thus, DCF is an

accounting term that refers to all losses associated with product degrade. The term includes all fiber, labor, and overhead costs associated with degrade issues. DCF can vary from 0% to 100% of the product's market value.

Degrade in wood drying is categorized as **static** or **dynamic**.

Static degrade is associated with final moisture content alone.
Dynamic degrade is associated with drying rates alone.
Both types are present in every wood drying system.

Both types of degrade can involve **rebound** dynamics. The term "rebound" refers to degrade **shifts** that occur after the fiber leaves the dryer proper.

Static degrade includes value losses due solely to fiber shrinkage during drying. The term includes longitudinal, radial, and tangential shrinkage losses.

Dynamic degrade includes value losses due solely to drying rates. Dynamic losses occur significantly in the drying of hardwoods subject to checking, splits, and honeycomb when excessive drying rates are used.

Rebound degrade shifts occur after the fiber leaves the dryer proper. Moisture changes cause the fiber to either expand or shrink resulting in grade shifts.

10. Energy costs
    This includes heating fuel (gas/oil/wood) and electrical direct costs, including all processing and handling costs. If cogeneration capability exists on site, adjustments will be required according to the contract with the electric utility company.
11. Inventory costs
    This is the shared economic cost between log purchasing and sales income. This can be a significant cost if the practice of air drying is used.
12. Insurance costs
    This is the total cost for general liability, property and casualty, boiler and machinery, flood, workers compensation, and employee health care coverage.
13. Legal/financial risks
    In today's litigation environment, this can be a significant risk as well as a cost. Business liability risks involve product claims and acts of company employees. Company vehicle accidents, product defects, mold claims, property scams, frauds committed against or by company employees, etc.
14. Technology pitfalls
    Technology pitfalls occur when businesses use technological gadgets instead of common sense, experience, and proven designs to increase profits. In wood drying, the available technologies for successful drying have been in existence for

over 50 years. The engineering principles involved in designing successful wood drying systems date back over a century.

It is common to see equipment salesmen market hi-tech gadgets (electronics) to mills claiming returns on investment in months. This may be true in some cases, but in others, the same capability existed in other older proven methods.

Chapter 7
Legal Issues in Industrial Drying Systems

7.1 Disclaimer

The following list was prepared from terms and concepts commonly seen in vendor, contractor, and sales agreements, and is NOT intended to be a legal reference. All business practices should follow the advice of a licensed attorney.

7.2 Retain an Attorney

If you do not have an attorney on retainer, it would be wise to do so. After you retain your attorney, review each of the following issues with him. Have your attorney review this book before your final meeting with him. Have your attorney work with your safety consultants in preparing your operating and safety manuals for your business, including each of your manufacturing plants.

7.3 Contracts, Assignments, and Warranties

Contracts for purchasing and sales should be reviewed.

Assignments for vendors, service people, consultants, and engineers should be reviewed. Warranties should be reviewed prior to purchasing equipment.

J E 'Ed' Smith, *Industrial Drying Systems*,
https://doi.org/10.1007/978-3-031-31863-4_7

7.4 Qualification/Assignment/Ethical Issues for Consultants and Engineers

Definition of a consultant (Webster) – "To ask the advice or opinion of"
Definition of an engineer (Webster) – "One trained in engineering"
Definition of a licensed engineer – One who holds a license to practice engineering

Defining Assignments for Consultants and Engineers
General consulting agreements are usually verbal. If losses are caused by the consultant's advice, recovery for damages may not be possible.

Assignments for engineers should be specific to the scope of work and put in writing.

Ethical and Legal Requirements of Engineers
The primary role of all engineers is to protect the public. Engineers are also required to comply with all applicable local, national, and international design codes. Engineers should also protect their reputations and themselves from making mistakes that can get them into trouble.

Errors and Omissions Insurance for Engineers
All engineers should contact their insurance carrier and their client's insurance carrier in writing before getting involved in any project that poses an obvious risk to the public at large, their client, or themselves. The claimed risk should be objective, accurately described, and without personal bias. Consult an attorney while drafting the letter. The letter should be sent by registered mail, return-receipt-required, so the engineer has legal proof the letters were both mailed and received.

One example of a significant legal risk is low-cost wood-fueled, direct-fired lumber kilns that are known by most people working in the wood-products industry and government agencies to create significant health and safety risks for plant personnel, cause numerous dangerous kiln fires, and produce numerous toxic and cancer-causing emissions.

Failure to Take Extra Care
Engineers can be sued in civil courts if they only do just what codes, standards, and guidelines require and not take additional precautions to prevent property losses or injuries to the public.

Legal Risks of Using Engineer-Wannabes and Technically Unqualified Consultants
Errors can occur that could lead to major losses, fires, explosions, injuries, deaths, plant closings, and lawsuits. Both engineers and property owners can be held liable for following the advice of technically unqualified consultants. Relationships with all consultants should be at arm's length to avoid risky conflicts of interests and minimize personal liability.

Legal Issues When Retaining Consultants, Engineers, Licensed Engineers, and Experts

1. People who work in an industry often think of themselves as an expert in that industry.
2. Any person can market themselves as consultants to any industry.
3. Any engineer can market themselves as consultants to any industry.
4. A consultant cannot market themselves as engineers unless they have a degree in engineering.
5. In many states, a consultant cannot provide engineering services without having engineering licenses. Check the laws for the state where the project is located. If a person represents themselves to the public as being an engineer, they are required by law to hold an engineering license in the state where the person engages in engineering practice. Laws are rapidly changing on this issue.
6. An expert is a person who has special skills or knowledge above that held by the average person.
7. In some states, experts are required to have a professional license in their field of practice before being allowed to testify in court.
8. Check with the state board of engineers for applicable laws for the state in which the project is to be located in. Licensed engineers may be required.
9. In some states, an engineer cannot investigate an insurance claim or loss for another party unless they are licensed in that state.
10. Some states require private investigator licenses for people investigating property losses.

7.5 Qualifications of the Designer

When is a licensed engineer required? This depends on the project. Examples are given below:

Furnaces

> Check with your attorney and the state fire marshal.
> National design and testing standards apply.

Buildings and their foundations

> If over a certain size, a licensed structural engineer may have to either design or seal the drawings. Check with the state board of engineers.

Fire protection systems

> A fire-protection license is required in most states.
> NFPA standards are required in the United States.

Waste treatment systems

> Check with the local government environmental regional office.

Lumber kiln structures

This depends on the size, location, and the state. It is always best to have a resident licensed structural engineer review structural designs.

Environmental control equipment and testing

Check with the local government environmental regional office.

7.6 Qualifications of the Equipment Manufacturer

This depends on the project. Examples are given below.

Furnaces and burners – check with the state fire marshal, your insurance carrier, and your attorney.
National standards apply – NFPA, ASME, FM, CSA, etc.
Fire protection systems – NFPA standards apply.
Waste treatment systems – National and regional standards apply.
Lumber kiln structures – This depends on the location and the state. It is always best to have a licensed structural engineer review the building designs.
Environmental control equipment – Designs must be approved by previous testing. National standards apply.

7.7 Qualifications of the Installer

This depends on the project.
Check with the manufacturer of the equipment.

7.8 Qualifications of Machine Operators

All machinery operators should be adequately trained for safe operations.

7.9 Commissioning Equipment

Final configurations should comply with all applicable purchasing contracts, codes, and standards. The people doing the commissioning should be qualified.

7.10 Third-Party Issues

Third-party requirements vary by the type of project and physical location.

Project financial institutions may require third-party inspections.
Insurance companies may require third-party inspections.
Government agencies may require third-party inspections.

7.11 Keeping Secure Complete Records on Equipment Involves the Following

Delegation of a responsible party for record keeping which is usually either the end-user's maintenance or engineering manager.

Fire-Proof Records Vaults

I recommend this for every facility, both large and small.

Machinery Identification Numbers

Each machine center should have a plant ID number.

Equipment Safety Manuals

Get these from the equipment manufacturers.

Operator Manuals and Procedures

Get these from the equipment manufacturers.

List of Applicable Codes and Standards for Machinery

These can be ordered from Global Engineering Documents. Discuss this with the original equipment manufacturer, plant safety managers, and the project engineer.

Parts Lists and Construction Drawings

Get these from the equipment manufacturers or design firms and keep them in a safe vault.

Electrical Schematics and Process Controls Documentation

Get these from the equipment manufacturers or design firms and keep them in a safe vault.

7.12 Preventive Maintenance and Inspection Programs

Follow the recommendations of reputable established equipment manufacturers.

Use software programs that link machine centers to purchasing and maintenance costs.

Inspections and Process Instrument Calibration Programs
These involve:

Inspections for boiler and machinery insurance coverage

Federally mandated inspections

Pressure vessels

State-mandated inspections

Pressure vessels
Fire-protection systems

Process controls and instrumentation calibration

Field data sheets records kept in a secure filing system
Contact the ISA at www.isa.org for published standards

7.13 Concepts in Moral and Legal Negligence

A negligence scale encompasses total care to gross negligence.

Total Care

Total care adopts acceptable risk management practices:
Hiring competent people to manage safety programs
Identifying safety risks on an ongoing basis
Implementing checks and balances
Ongoing monitoring by upper management

Ignorance

The sequential effect of ignorance on the ego:
Ignorance displaces the ego from reality testing, creating false beliefs.
False beliefs lead to errors in judgment.
Errors in judgment lead to mistakes.
Mistakes lead to personal and economic losses.
Losses lead to projection (placing blame on others).
Projections lead to heightened chaos.
Heightened chaos leads to additional mistakes

Omission (Webster) – An Act of Neglect, or to Leave Undone

A singular incidence of conduct

Oversight (Webster) – To Not See (Perception), or Consider (Analysis)

A singular incidence of conduct

Accident (Webster) – An Incident Occurring by Chance or Unintentionally

A singular incidence of conduct

Wanton Conduct (Webster) – Immoral Conduct

Conduct that occurs in patterns

Reckless Conduct (Webster) – Lacking Caution

Conduct that occurs in patterns

Occasional Disregard for the Safety and Welfare of Others – The factors are:

Intent
Ignorance
Oversights
Omissions

Gross Negligence – Total Disregard for the Safety and Welfare of Others. The factors are:

Intent
Ignorance
Oversights and omissions
Recklessness
Ongoing pattern of neglect
Ongoing refusal to take care

Consequences of Gross Negligence – The consequences are:

High probability of death, injuries, business losses, and failures
High probability of civil lawsuits
Possible criminal charges and imprisonment

Chapter 8
Industrial Drying System Design, Operation, and Maintenance

8.1 Introduction

Industrial wood, agricultural, and food products drying systems involve numerous complex engineering principles that are rarely fully understood by even those who design these systems.

Especially in the forest products industry, most of the drying equipment in operation today was designed by people who have very little formal engineering training. Most final designs are based on past observations of fabricators, consultants, and contractors who learned by trial and error what met their customer's expectations.

In far too many cases, equipment designs are **pirated** from established original equipment manufacturers (OEMs). Many OEMs' business philosophy is to "**copy, cut, and polish**" their competitors' designs and then wait for feedback from the field about the success or failure of the pirated designs.

Low-end manufacturers reduce the size of bearings, shafts, motors, and structural items and then polish and promote their products as being leading edge and superior to their competitors' designs. These companies rarely retain or employ graduate or licensed engineers for improving their products.

The time lag between a new design and its operation can be from one hour to several months. Often, performance and reliability problems do not surface until after the consultant, salesman, and contract technicians have left the plant. In many cases, pirated designs require complete replacement to keep the plant in operation.

The industrial drying industry is no different than other industries in which machinery plays a large part in the profitability of manufacturing plants. People often identify themselves too much with the complex inner workings of plants and then project their personal needs, desires, and logic into the physical environment they work in.

Additionally, because all people are human, they all make mistakes. All people can become trapped by their faulty perceptions of what is or is not reasonable, suffer

J E 'Ed' Smith, *Industrial Drying Systems*,
https://doi.org/10.1007/978-3-031-31863-4_8

from inflated egos, or hesitate to ask for assistance from others for fear of having their mistakes exposed. The fallout from this human trait is that past mistakes are often kept secret, and solutions are not explored. If not corrected, this very common kind of conduct by people can lead to consequences that can be both costly and dangerous.

For energy and drying systems, which are by far the most difficult processes to understand, the use of defective plant equipment that does not comply with accepted engineering and safety standards can lead to large economic losses, plant shutdowns, injuries, death, and lawsuits.

This book was written to introduce the reader to basic engineering principles encountered in safe, energy-efficient, successful industrial drying operations. Although extensive engineering skills and experience are needed to master each of the technologies discussed in this book, I have avoided presenting them to make this book more of a technical reference than an engineering text.

8.2 Legal Duty of the Engineer and the Owner

It is the **legal duty of the engineer and the owner** to study technical books such as this one and implement high-level management programs that verify on an ongoing basis that all industrial drying systems meet all applicable industry, safety, and environmental codes and standards.

Internet Resources for Studying Business Ethics, Moral Attitudes, Social Responsibility, and Deviant Social Conduct Such as Psychopathy and Narcissism.

There are numerous well-researched, science-based videos on the Internet and YouTube that discuss these issues in detail.

I suggest the following three YouTube videos as a starting point to learn about the science and research behind moral reasoning and the high costs to society when people do bad things:

"The Neuroscience of Real-Life Monsters: Psychopaths, CEOs and Politicians" by Octhuio Choi, MD, PhD.

"Living in the Future's Past" by Jeff Bridges.

"The Reproducibility Crisis" by Dr. Dorothy Bishop, and Dr. Sabine Hossenfelder.

Chapter 9
Drying System Definition Review

9.1 Introduction

The definition of **drying system** depends on the industry (wood, agriculture, food) and addresses what is and is not required for a drying process to be present in a specific manufacturing facility.

The following example is for wood-products manufacturing and can be used as a model for other industrial drying processes.

A **wood-products** drying system is everything including required support processes from the moment a tree is harvested to the moment a manufactured dry finished product is graded.

One Example Is in Lumber Manufacturing Moments after a tree is cut down and the limbs are removed, the log starts to lose moisture. Some wood species require end coatings to then be applied to the cut surfaces and ends of the logs to prevent costly end checks from occurring. Then, the logs continue to dry until it enters the sawing operation. Then the green sawed boards continue to dry until they enter the lumber kilns. Then the kiln-dried lumber continues to either loose or regain moisture until it enters the dry-end in-line moisture measuring system.

Thus, in lumber manufacturing, the drying system is anything that requires machinery and labor to control the wood fiber moisture content from the moment a tree is harvested to the moment the dry lumber is graded for product quality. See the following example.

Successful Lumber Drying Systems Require the Following Steps:

Managing log inventories to minimize wide variances in average log moisture content

Precision sawing of green board thickness (a critical variable)

Trimming green sawed boards for defects and length

J E 'Ed' Smith, *Industrial Drying Systems*,
https://doi.org/10.1007/978-3-031-31863-4_9

Transporting green trimmed boards to a green sorter
Sorting green boards by species, thickness, widths, lengths, and grade into separate bundles
Transporting the sorted green bundles of lumber to an inventory-storage area
Managing the sorted green bundle inventory area
Transporting sorted green bundles of lumber to a green stacker
Stacking the green lumber into packages with drying stickers
Transporting the sorted green stacked packages to a green inventory area
Managing the sorted green package inventory area
Transporting the green lumber packages to the infeed area of the kilns
Loading the kilns with green stacked stickered packages
Drying the green lumber with appropriate conditions for a target moisture content
Adjusting kiln operations to control final lumber moisture content
Unloading the dried stacked stickered lumber packages from the kiln
Transporting the dried lumber packages to the dry inventory storage area
Managing the sorted (weather-protected) dried lumber package inventory area
Transporting the dried stacked stickered lumber packages to an un-stacker
Separating the dried lumber boards from the kiln stickers
Transporting layers of boards to the infeed of a dry planer operation
Collecting the kiln stacking stickers into bundles
Transporting the kiln sticker bundles back to the green lumber stacker
Feeding the kiln sticker bundles into the green stacker operation
Measuring individual board moisture content at the dry planer facility
Reporting dry-end board moisture content to kiln operators and mill management
Generating point vs. batch MIS reports for the impact drying has on product market value
Adjusting drying system operations to minimize variances in dry-end board moisture content
Adoption of redry handling systems for certain operations

Another way to **define** a wood-products drying system is **any additional** equipment, support utilities, and operating and maintenance costs and labor **to convert** a manufacturing process from a finished green-wood product to a finished dry-wood product.

Physical Sizes of Wood Drying Systems

The physical size of what is defined to be a wood drying system can vary from a small test kiln (4′ × 4′ × 6′), to a large industrial convection-type continuous softwood dryer (40′ wide × 40′ high × 400′ long) plus the entire green end system plus the entire dry-end system all of which could require up to 10 acres of real estate. If air drying is involved, the total drying system could involve over 10 acres of real estate.

Finally, every manufacturing plant should understand and constantly track the **separate financial impacts** that the type of drying system and the type logs entering the plant have on product market value. In softwood lumber manufacturing plants, this can be accomplished by statistical regression-analysis software at the dry-end in-line moisture sensor and the grading station.

Chapter 10
Green-End Systems

10.1 Introduction

The green end of an industrial system is everything upstream of the dryers. The following is an example of what the green end of an industrial drying system is, and the impact it has on the drying process.

The green end of a drying system is defined to be every piece of equipment located downstream of the final green machining process (the green trimmer) and ahead of the entrance doors of the dryers.

The layout, design, operation, and maintenance of green-end equipment will determine the success or failure of drying operations more than any other part of the drying system.

In every drying operation, primary thermal drying dynamics determines the minimum required green-end equipment for quality drying.

Next, the class of dryer system must be selected.

The three classes of drying systems are batch, jogging-batch, and continuous. The features of these three drying systems will be discussed in later sections of this book.

Finally, the type of fiber handling system must be chosen.

10.2 An Example of Green-End Systems (Wood)

This involves fork trucks, straddle carriers, roll cases, and track (rail) systems with kiln carts and trams which require stable foundations under the rails during all types of weather conditions.

The following are typical of green-end operations:

Lumber Drying Operations

J E 'Ed' Smith, *Industrial Drying Systems*,
https://doi.org/10.1007/978-3-031-31863-4_10

Antistain Dip Tanks
Dip tanks may be used to treat green lumber leaving a sawmill. Each board is submerged in a vat solution containing antistain treatment chemicals. The dip tank is in a section of conveyor chain carrying the green lumber to the green product inventory area.

Green Sorters
Lumber drying rates, drying temperature schedules, and postdrying equalization and conditioning schemes are strongly affected by species, grade, board thickness, width, and lengths. Thus, adequate product sorting equipment and procedures are required for quality drying. In one green lumber sorting operation, there may be as many as 24 bays for receiving sorted materials. Very large multispecies operations require very large numbers of green product sorting bays.

Green Sorter-to-Green-Stacker Handling and Inventory Systems
Buggies, trams, and in-line conveyors have been used for handling green sorted materials (in bundles) ahead of the green stacker.

Unscramblers
Unscramblers are inclined chains with lugs that unscramble bundles of lumber and then feed individual boards onto an in-feed chain to a green stacker.

Green Stackers
Lumber stacking can be done manually or by an automatic stacker. Most lumber mills today use automatic stackers. Automatic stackers will either be a solid-course or air-spacing type depending on the species and size of lumber.

The final height of the package leaving the green stacker varies from 3′ to 16′ depending on the operation. The width of the package leaving the stacker varies from 4′ to 12′ depending on the operation. The length of the packages leaving the stacker is determined by the length of the lumber being stacked.

Thus, sizes of packages leaving green lumber stackers can typically vary from as small as 4′ wide by 3′ high by 8′ long to as large as 12′ wide by 16′ high by 24′ long. That is a 48:1 ratio of package volume from the largest to the smallest.

The variable (**package size**) has a **very strong impact** on how **energy efficient** a lumber drying system is and how **profitable** a lumber manufacturing operation is. This will be discussed later in this book.

The level of **precision** in green lumber stacking refers to how square and consistent the packages are when leaving the green stacker. High precision lumber stacking is mandatory for successful lumber drying operations.

Some large multispecies operations may require more than one green stacker.

Green Package Sticker Handling Systems
Stickers can either be manually placed into stacked packages or by automatic stick placers. Most modern green stackers use automatic sticker placers and handling systems.

Green Package-Handling Equipment

Once the green packages with stickers leave the stacker, either rail-type transfer cars or mobile package trucks are used to transport the packages to the green package storage area or straight to the dryers.

The **maximum-allowable (safe) speed** of all green-package-handling equipment (conveyors, transfer cars, straddle carriers, fork tucks, etc.) should be determined only by engineers familiar with package-handling equipment designs to – prevent dangerous accidental spilling and toppling of heavy packages.

Stacker-to-Dryer Green Storage Area

This is the storage area located between the green stacker and the dryers. The required inventory in this area is dependent on the number of products, type, and size of the drying operation.

Green-End Wood Particle Handling Systems

Green wood particle materials include bark, chips, and saw dust. These mechanical handling systems involve mechanical conveyors, drag chains, screens, classifiers, diversion gates, storage bins, feeder bins, gravimetric flow meters, moisture sensors, preheat recovery systems, chippers, grinders, storage bins, surge bins, bin unloaders, and feeder rams, chains, or screws. Pneumatic handling systems include both low-pressure and high-pressure blowpipe systems. Also included are the many electrical and hydraulic power and control systems to make these systems operate smoothly.

Tree bark is highly abrasive because of sand and dirt accumulated during tree harvesting and log storage activities. Ells in bark-handling blowpipe systems must be fitted with replaceable liners. Liner wear rate in these ells is a strong function of air handling stream velocity.

Large Timber and Pole Green-End Systems

These systems involve specialized end trimmers and handling equipment designed to handle individual heavy items such as large timbers and poles. Hydraulic boom trucks, small cranes, and fork loader trucks are used.

Due to the numerous different sizes and weights of timbers and poles, the green-end equipment used in timber and pole drying operations varies widely.

Small timber stackers are designed like lumber stackers, but with shorter stacking forks, and thus producing narrow package widths. Large timber stackers are typically custom designed for the specific extra weight subjected to the stacker forks.

Green Timber Kilns

Although some timbers are dried in fork-truck-fed batch side loader type kilns, many are dried in long double-track batch kilns. These designs are discussed later in this book.

Green Pole Drying Systems are usually long double-track batch designs and require long pole trams fitted with outer vertical safety side posts to prevent spillage. Special attention is required to determine exactly what the door-to-door length of the kilns should be. Whole door baffles are required for sealing off air leaks at the ends of these kilns. These designs are discussed later in this book.

Chapter 11
Solid-Wood Industrial Drying Systems

11.1 Global Drying Industry Standards: Why We All Need Them

The Unregulated Wood Drying Industry

The notorious history of the global wood drying industry is a case study in technological jet lag. Furthermore, the entire global wood products industry has historically lagged the rest of the world in both evaluating where it came from and where it is headed. Even in the year 2023, after centuries of achievements in science and engineering around the globe, the global wood drying industry is still mired in myths, misinformation, disinformation, significant financial frauds, and endless drying and energy problems.

Even in the year 2023, defective hazardous toxic drying systems are being installed in manufacturing facilities all around the globe. These mistakes are due to erroneous beliefs, misinformation, and hype from equipment salesmen, pseudo drying experts, consultants, and management who have little or no formal education in engineering.

In many lumber manufacturing plants in operation today, one can still see the same mistakes being made again and again. The costs to the industry and our planet from these mistakes will be with our children for decades to come. There are numerous social issues in wood drying that require our attention. We need to understand today, not tomorrow, our mistakes and what we need to do now to lead the global wood products industry in the right direction.

We all need to understand that we now live in the age of global warming and major social upheavals will soon be upon us all if we do not prevent them now.

J E 'Ed' Smith, *Industrial Drying Systems*,
https://doi.org/10.1007/978-3-031-31863-4_11

11.1.1 Health and Safety Issues

The drying of wood is not without serious effects on the health and safety of humans and wildlife. Wood produces numerous toxic and carcinogenic compounds during both burning and drying. These issues should not be taken lightly by industry, the public, or our governments.

11.1.2 Protection of the Environment and Our Natural Resources

Scientists around the globe have been warning us for decades that the population of humans on the planet is dangerously high, and the demand for wood fiber is growing at an unsustainable level. They have repeatedly warned us that the time is near when many billions of humans will be confronted with limited resources and high costs of fossil fuels and wood fiber. Furthermore, we have been repeatedly warned that these events will cause massive environmental, social, and political problems around the globe. If we do not collectively confront these issues with viable solutions, it is a certainty that the nations of the world will be constantly fighting economic and military wars over the control of the planet's diminishing natural resources.

11.1.3 Advancing the Wood-Drying Process Through Industry Standards

Although improvements in dryer designs have occurred during the last century, most commercial wood dryers today are still perceived by many people as just big hot boxes with fans. One can walk inside a new lumber kiln today and see little significant difference from a kiln designed 50 years ago. And although kiln fans are more efficient today, and computer technology is now affordable, lumber kilns are still perceived by most people in the forest products industry as they were a century ago.

Up until engineers developed modern computers, and the long-delayed acceptance of this technology by the wood drying industry, milestones in dryer designs have been few and far between. Also true is that during the last 50 years, the entire wood drying industry has been plagued with hype from aggressive dryer salesmen and "experts" about modern computer technology and what it can do when applied to drying operations. Some of these claims are, from a scientific viewpoint, totally preposterous.

One example of this is the claim that computerized multi-zone kilns use 45% less energy to dry lumber. Another outrageous claim is that it is possible to "feel" the moisture content of lumber inside a dryer by waving your hands through the steam leaks around the door of a dryer. Claims like these raise serious questions about the motives of people who would make such claims. These days, like in the past, there is certainly no shortage of junk science and hype floating around in the wood products industry.

11.1.4 Junk Science: What Is It?

Those who have experience in the US adversarial legal system are aware of the conduct of experts retained by trial lawyers and insurance companies. These experts, most of which are prostitutes who make their living catering to their client's needs, have developed quite a reputation in the insurance community and the courts for promoting "scientific" theories that are in fact carefully contrived junk. This is where the term "junk science" comes from.

In the forest products industry, there has never been a shortage of junk science or hype, especially in lumber drying. There are also considerable faulty interpretations of junk lumber drying claims by people who do not have the analytical skills or experience to decipher junk from true science. The entire global wood drying industry has a need for an agency that has the duty of sifting through the hype, and scientifically and objectively sorting out the junk.

11.1.5 The Need for Global Dryer Standards

Today, wood dryer designers and manufacturers are not bound by any industry standards that prevent the installation of junk equipment. Because of this, one can go into many wood product operations today and see disastrous conditions around drying and energy-conversion processes.

Equipment engineering and safety codes developed decades ago are frequently disregarded. When safety interlocks fail to operate, plant electricians often bypass them in the interest of production. When process controllers fail, they are often removed or bypassed. In one plant that I recently visited, every safety interlock and every temperature controller on two lumber kilns and their burners had been removed to avoid spending money on repairs. Back in the 1980s, I visited a pole-treating plant in the USA south that had every safety interlock bypassed on every gas burner on their kilns. This same plant had a package boiler explode because the low-water safety switch had been bypassed.

When new drying systems are purchased, gullible buyers listen to salesmen with claims of golden opportunities only available with their equipment. Then when the purchasing agent gets involved in the project, the minimum engineering specifications, if any exists, are often discarded and the purchasing agent looks across the table with one thing on his mind. Who has the lowest price?

The entire global wood dryer industry needs universal standards for the key elements of safe successful drying to prevent the costly errors and abuses by unethical equipment salesmen and pseudo drying experts who belong in the used-car business or politics.

Having minimum standards would stop the abuses that have occurred throughout the history of the industry. I propose the following as a start.

1. Minimum safety standards that go beyond US OSHA and NFPA standards.
2. Energy efficiency guidelines for reducing electricity, fuel usage, and pollution.
3. Guidelines for dryer length, width, and door height so that the correct equipment is purchased.
4. Guidelines for fan performance and their application to dryers.
5. Guidelines for heating and energy recovery systems.
6. Minimum guidelines for process control systems.
7. Minimum structural design standards to minimize failures and foundation settlements.

11.2 Modeling the Drying Process

Up until 1965, few universities or research centers had computer systems capable of modeling the complex heat- and mass-transfer processes involved in the drying of wood. Prior to 1965 most drying models were mathematical equations derived from the known laws of physics, heat transfer, and fluid mechanics.

In 1967, mathematicians solved the Navier–Stokes field equation in fluid mechanics. This milestone in fluids mechanics, combined with the high speed of computer processors, set the stage for a new branch of science known as computational fluid dynamics (CFD). Using CFD, multi-dimensional models were developed that could predict the motion of fluids in complex three-dimensional fields. Once this technology evolved, the same mathematics and computation technology was used in solving heat-transfer problems, stress analysis in solids, soils analysis, electrical field theory, and a host of other engineering applications.

Today, there are numerous engineering models for combustion processes, drying processes, materials stress analysis, building designs, electrical field theory, physics, automobile accident reconstruction, fire and explosion modeling, weather prediction, global weather patterns, ocean movement patterns, and medical research. Models are the use of mathematical principles with the known laws of physics. There are numerous simple models and there are also numerous complex models.

Computational fluid dynamics (CFD) and finite element analysis (FEA) are examples of rigorous complex models in which multiple simultaneous equations are solved by computers. To reach a solution, millions of calculations may be required, something that only computers can do efficiently and accurately.

Models can be classified as:

Basic research models
System application models
Equipment design models

11.2.1 Basic Research Models

Structural properties of wood, agricultural, and food products during drying
Moisture and energy dynamics in drying
Moisture (mass and energy) dynamics during drying

- Low-temperature drying
  - Above fiber saturation point
  - Below fiber saturation point
- High-temperature drying
  - Above fiber saturation point
  - Below fiber saturation point

11.2.2 System Application Models

When researchers test a drying theory, it is usually done in a small test kiln under conditions significantly different than commercial drying systems. Often the conclusions that come out of these projects are both erroneous and misleading. In many cases, these experiments are both designed and run by people who know very little about commercial dryers. Following the experimental drying tests, the data is fitted or compared to the model by regression or statistical analysis. After this a final report is published. The report may then be distributed to people who know very little about commercial dryers, and from this, erroneous conclusions are formed. Upper management people in any product manufacturing plants may then act on false conclusions leading to very large economic losses. During the last 50 years I have seen numerous examples of this.

11.2.3 Equipment Design Models

Modern computer technology has totally changed the landscape of design engineering. Starting in the 1960s, engineers started developing design software that literally affected the careers of millions of engineers and draftsmen around the world. Today there are numerous powerful desktop design software programs that can now do what it used to take an entire staff of engineers and draftsmen to do.

Foundation design models for buildings
Structural analysis models for buildings

Pressure vessel and piping stress analysis models
Piping flow analysis and design models

Fluid dynamics models
Heat transfer models

11.2.4 Examples of Drying and Equipment Models

Because computer models have always fascinated me, starting in the early 1970s, I started developing my own models for lumber drying and equipment analysis. Some of them were:

Lumber package-drying models
Airflow models
Heat system heat-transfer models
Finite-element degrade-limiting models

The package drying models were developed for parametric studies of different dryer configurations. The airflow models were for establishing the minimum required airflow CFM through sticker openings for quality drying. The heat transfer models were used for designing dryer heating systems. And finally, I spent several years working on finite-element degrade-limiting models. These models looked at establishing drying rates for specific levels of strain induced in hardwoods.

11.2.4.1 Finite-Element Model of 1″ Northern White Oak

Optimum drying rate vs. wood moisture content

The graph below is from a finite-element model that I developed in 1977 for northern red oak while, the subject of my master's thesis.

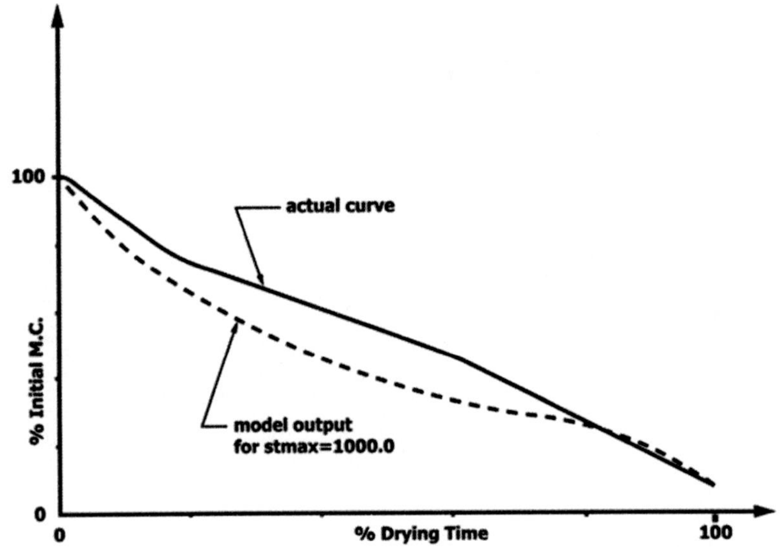

MODEL OUTPUT VERSUS EXPERIMENTAL DECAY CURVE (NO. RED OAK)

11.2.4.2 Finite-Element Package Model for Air/Heat Delivery in Lumber Kilns

K105 proprietary model – total drying time vs. package entering air temperatures

The following graph is from one of several finite-element models that I developed in 1977 specifically for **upgrading** air/heat-delivery systems in lumber kilns. The model looks at: package width, board thickness, sticker thickness, air velocity, initial moisture content, package center moisture content, package exterior moisture content, wood-specific gravity, and dry-bulb and wet-bulb temperatures. The model adjusts for fan reversals. The model also calculates the energy requirements for drying. The surface temperature and drying rate of each board in the package can be watched during the drying cycle.

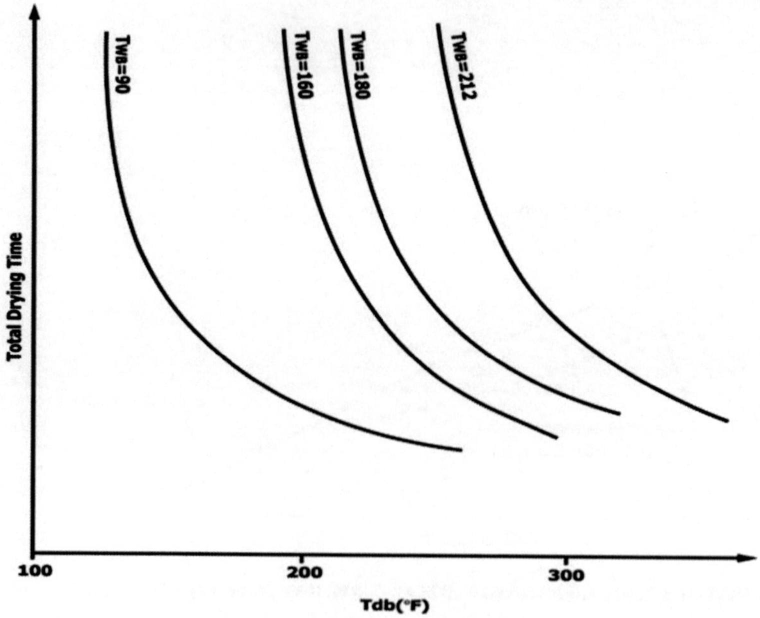

11.2.4.3 1975 Regression Model of a Steam Heating System in a Lumber Kiln

Heat transfer rate vs. (steam temperature – approaching air temperature)

The following regression curve was developed from data collected on a high-temp southern pine kiln in 1975. The data was used for converting the kiln over to 60 psig steam instead of 125 psig, with no increase in drying time. The curve shows the heat transfer rate/degree of temperature difference between the steam and approaching air temperature.

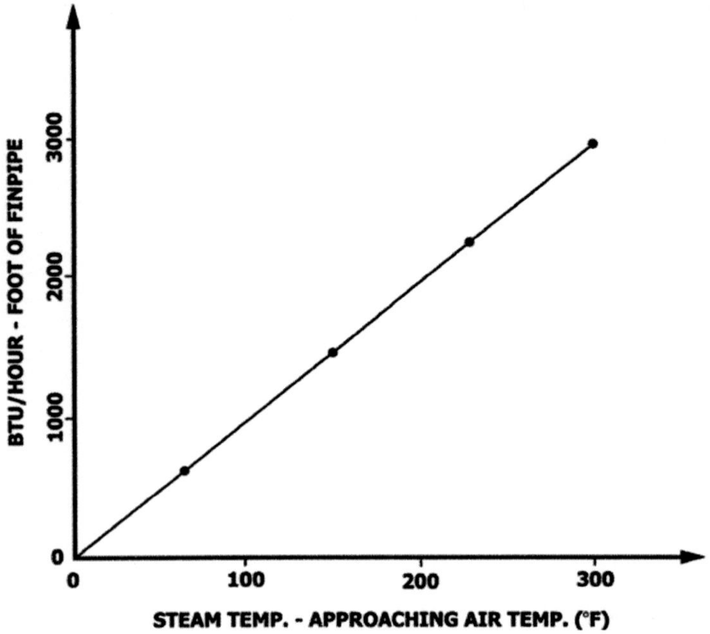

11.3 Drying Challenges, Dryer Types, Classes, and Terms in Drying

The following is a summary overview of the wood drying industry today and what the future challenges will be. This same type of analysis can easily be applied to all types of industrial drying systems (wood, agricultural, food, plastics, pharmaceutical, minerals, etc.).

11.3.1 Objectives in Wood Drying

1. To achieve dimensional stability of the final product
2. To meet the strength requirements of the final product
3. To improve finishing properties and appearance
4. To set the pitch to prevent seepage to the surface of the wood
5. To improve gluing processes used in beams, furniture, and plywood manufacturing
6. For chemical-impregnation processes to prevent decay and damage by insects
7. For weight reduction to minimize handling and shipping costs
8. To destroy fungus, mold, and insects
9. For use as a fuel in furnaces

11.3.2 Common Methods Used in Wood Drying

Convection Dryers in which heated air moves over the surface of moist wood
Electromagnetic Radiation Dryers in which electromagnetic waves are used to either heat the surface or interior of wood
Hydrostatic Presses in which wood fibers are subjected to high static pressure to displace free water from wood fibers
Vacuum Chambers in which the boiling point of water is lowered to increase drying rates

11.3.3 Major Obstacles in Wood Drying

1. Controlling the final moisture content to a level acceptable for its end use
2. Minimizing degrade caused by the drying process

11.3.4 Most Common Mistakes Made in Wood Drying

1. Buying inferior equipment
2. Failure to understand how wood dries
3. Failure to understand the causes of drying degrade
4. Failure to understand the total cost of degrade
5. Using unqualified people for designing equipment
6. Failure to adequately train equipment operators
7. Having inadequate dryer capacity
8. Failure to maintain dryer equipment

11.3.5 Future Challenges in Wood Drying

1. Declining size and quality of logs
2. Increasing environmental, safety, and health issues
3. Increasing competition in the domestic and global marketplaces
4. Growing litigious nature of wood-product consumers

11.3.6 Current Dryer Technologies

The following matrix lists different types of wood drying technologies found in wood-products manufacturing operations today. Although not every type of wood drying system is included, this gives an overview of what is out there.

Wood Dryer Technology Matrix

TYPE of FIBER		**CLASS of DRYER (3)**	
	BATCH	**JOGGING-BATCH**	**CONTINUOUS**
PARTICLES	Presses	Presses	Rotary
			Presses
			Suspension
			Bed
			Flash tube
			Furnaces
Sheets	Hot presses		Veneer
Small boards			High Frequency
Packages	Air drying yards		
	Pre-dryers		
	Dehumidifiers		
	Kilns	Kilns	Kilns
	E&C chambers	E&C chambers	E&C chambers
	High frequency		
	Storage and tempering		

11.3.7 Industry Code Assignments for Wood Dryers

In the coming decades, the wood products industry will be confronted with the task of sorting out the many issues in equipment design. Severe global demands on the planet will force all nations to set minimum standards for energy efficiency, pollution controls, and safety. To do this, we need a system of equipment design codes for all types of wood dryers.

Proposed code assignments (by the author) for wood dryers.

Temperature range

(1) 33–120 F (low-temperature drying)
(2) 121–180 F (conventional-temperature drying)
(3) 181–211 F (accelerated-temperature drying)
(4) 212–50 F (high-temperature drying)
(5) Above 250 F (hyper-temperature drying)

Operating pressure

(Z) Above 15.7 psia operating pressure (presses)
(Y) 12.7–15.7 psia operating pressure (atmospheric dryers)

(X) Below 12.7 psia operating pressure (vacuum dryers)
Type heating system
(N) Indirect-fired heating
(D) Direct-fired heating
Class of dryer
(B) Batch dryers
(JB) Jogging-batch dryers
(C) Continuous dryers
Type of fiber entering the dryer
(P) Particles
(V) Sheets
(I) Individual boards
(L) Packages of boards, timbers, poles
Stage of drying
(G) Drying above fiber saturation point
(F) Drying below fiber saturation point
(E) Equalization and conditioning
(S) Storage and tempering

Example: An indirect-fired, high-temperature, batch dryer operating at 14.7 psia for drying timbers above fiber saturation point only would have the code 4YNBLG.

11.3.8 Predominant Types of Convection Wood Dryer Systems

The following is an overview of predominant types of wood drying systems with definitions.

Pre-dryer is associated with removing free water only and degrading control

A. Atmospheric systems

1. Open space

    Air-drying yards (no applied heat or fans)
    Air-drying sheds with fans (no applied heat)

2. Closed space (insulated chamber)

    Heated pre-dryers
    Dehumidifier heat pumps

B. Below-atmospheric systems

1. Vacuum only dryers
2. Vacuum dryers with heating systems
3. Vacuum dryers with dehumidification systems

Kiln is associated with removing both free and bound water in a closed insulated chamber with degrade control

A. Atmospheric designs

1. Direct fired kilns heated by:

   (a) Oil and gas burners
   (b) Wood burners
   (c) Radio frequency

2. Indirect fired kilns

   (a) Steam-heated
   (b) Hot-water-heated
   (c) Hot-oil-heated
   (d) Hot-air-heated
   (e) Solar-heated
   (f) Heat pumps

B. Below-atmospheric designs

1. Vacuum chambers
2. Vacuum chambers with heating systems
3. Vacuum chambers with heat pumps

Equalizing and Conditioning chamber (E&C) is associated with applied heat and humidity in a closed insulated chamber for final moisture content and degrade control

A. Equalization of moisture content in:

1. Hardwoods
2. Softwoods

B. Conditioning for stress relief in:

1. Hardwoods
2. Softwoods

Storage facility is associated with kiln-dried inventory storage and tempering.

A. Open-storage space

1. Unprotected storage (not recommended)
2. Shed storage for sun, rain, and snow protection

B. Closed-storage space (buildings)

1. Temperature and humidity not controlled, no fans, no building insulation
2. Temperature and humidity controlled inside insulated buildings
3. Temperature, humidity, and air velocity controlled inside insulated buildings

11.3.9 Classes, Types, Configurations, and Applications of Wood Drying Systems

Three Main Classes of Kilns and Dryers
Batch, jogging-batch, and continuous

Common Types of Lumber Drying Systems
Convection-type kilns – by far the most prevalent design
Dehumidification kilns – used mostly for hardwoods
Pre-dryers – typically for hardwoods
Air drying yards – typically for hardwoods
High frequency dryers for small wood product operations

Lumber Kiln and Package-Handling Configurations and Terms
Numerous possible configurations, not all of which are in existence. The species and production requirements of each mill will dictate the optimum configuration. These are not all the possible or existing types of lumber kilns.

Lumber Kiln Applications
Batch Kilns – Used with All Species of Wood

Track

Softwoods – common design, 1–8 track designs have been built
Hardwoods – rare

Side-loader – often called a package kiln

Softwoods – common in treating plants and small lumber mills
Hardwoods – most common design, structural corrosion can be a problem

Sidewinder – small lumber operations, low cost of construction
Can be either a track or side loader type kiln

Softwoods – common outside the USA
Hardwoods – now rare in the USA

Moving-Package Kilns – Only Used with Easy-to-Dry, None-Refractory Species of Wood

Counter-flow – high labor and front-end costs, energy efficiency subject to design
Counter-flow and bidirectional have the same meaning.
Lumber – common with thin lumber boards
Poles and large timbers – difficult but possible, requires long end sections

Parallel-flow – energy efficiency subject to design
Parallel and unidirectional have the same meaning
Lumber – common with thin lumber boards
Poles and large timbers – difficult but possible, requires long end sections

"Progressive Kilns" A misused technical term in drying, none in the USA, and is also a common term used today in Europe, Australia, and Africa. The term "progressive" was originally used for some primitive US track kilns built prior to WW2.

The following discussion possibly can explain how the term progressive is misused today in industrial drying systems:

The English word progressive can be used as either an adjective or a noun. When used as an adjective it means "happening or developing gradually or in stages, proceeding step by step." When used in grammar, it means "a progressive tense or aspect" (Webster).

In drying processes (batch, jogging-batch, or continuous), one must specify whether the action occurring is either the drying equipment (kiln) or the drying medium.

EXAMPLE: Progressive drying can exist in any batch kiln by progressive changes in drying temperature schedules. Progressive drying can also exist in multi-zone jogging-batch kilns by both zone controls and temperature schedules in each zone. In continuous kilns, like jogging-batch kilns, progressive drying can also exist by both zone controls and temperature schedules in each zone.

The "progressive kiln debate" is only one of many examples of how different languages can create misinformation and myths in technologies when there exist no universal global industry standards. This is a much larger problem than many people are aware of.

Superheated Steam Drying

A very old but usually over-hyped drying technology in which a product is dried in a superheated steam environment. The benefits are high thermal efficiency and high fiber equilibrium moisture content that minimized fiber stresses and thus reduces fiber damage during drying. Conditions in some indirect-fired high-temperature softwood dryers can simulate superheated steam drying during the early stages of a drying cycle when the venting loading is high.

Common Track Kiln Configurations, Operations, and Terms

Single-Track Kiln Operations

Package flow: 1-way, 2-way, shuttle action, shotgun, batch, jogging-batch, continuous

Package-handling methods: fork trucks, transfer cars, cable and winches, pusher devices, package jogging devices, roller beds, rail systems, carts, trams

Infeed and outfeed section lengths: close-coupled, short-coupled, kiln-length, kiln-length + end clearance, shotgun

Kiln carts and tram handling methods: fork trucks, recirculation systems, transfer cars

Drying methods: full batch kiln drying, partial (staged) batch kiln drying, scheduled single-package batch kiln drying, jogging-batch kiln drying, continuous kiln drying, constant-rising-temperature (CRT), staged or heat-limited drying, or progressive drying (an old misleading and misused term)

Multi-Track Kiln Operations

Number of tracks: 2, 3, 4, 5, and up to 8 – Note: I personally know of only one 8-track kiln ever built.
Package flow: 1-way, 2-way, shuttle action, shotgun, parallel, counter-flow, jogging-batch, continuous
Package-handling methods: fork trucks, transfer cars, cable and winches, pusher devices, package jogging devices, roller beds, rail systems, carts, trams
Infeed and outfeed section lengths: close-coupled, short-coupled, kiln-length, kiln-length + end clearance, shotgun
Kiln carts and tram handling methods: fork trucks, recirculation systems, transfer cars
Drying methods: full batch kiln drying, partial batch kiln drying, partial (staged) batch kiln drying, scheduled single-package batch kiln drying, jogging-batch kiln drying, continuous kiln drying, constant-rising-temperature (CRT) drying, staged or heat-limited drying

Kiln Term	**Configurations and Definitions (Partial List)**
Batch	A single "product" placed in a kiln for processing to a final condition.
Bi-directional	Products move in opposite (counter-flow) directions through the kiln. Multi-track kilns can include different configurations of bi-directional package flow.
Continuous	A smooth motion method that does not include stopping or jogging actions but may include changes in speed.
Drying capacity	The average weekly drying capacity of a lumber drying operation Expressed as board feet (sales)/week.
Jogging-batch	A repetitive jogging method of feeding packages into a kiln involving a resting **period** (for loading packages onto tracks) and a timed **stroke** of moving packages through the kiln. Hydraulic rams or mechanical drives can be used.
Jogging-action	To start from a stop or repeat a starting action.
Jog period	The elapsed time between jogs.
Rest-time	The resting time of packages between stop and start of package movements.
Kiln capacity	The maximum amount of lumber that a kiln can hold at one time. Expressed as actual cubic feet or board feet (sales).
Load	The actual amount of lumber that has been placed inside a kiln for drying. Expressed as actual cubic feet or board feet (sales).
Path	In kilns, the same meaning as track (single-path, dual-path, triple-path, etc.).
Shuttle	A back-and-forth package handling method, common in double-track kilns.
Shotgun kiln	A track kiln with only one main door, both loaded and unloaded from one end.

Sister kilns	A combination of kilns that serve different or overlapping duties. Example: pre-driers, main drying kilns, E&C chambers, pre-heat-vent-recovery chambers, cooling sheds, dry-storage buildings – used for product processing, energy-recovery, or inventory (product storage).
Unidirectional	All products move in the same (parallel) direction through the kiln.
ZIGZAG	A track kiln in which the heated air stream zigs and zags through lumber packages to recover heat energy of evaporation, reduce wood fiber stresses, and flows counter to the movement of the packages through the kiln.
BDK	**A kiln in which a non-moving batch of lumber is dried to completion**
SLBDK	A side-loader batch dry kiln
TBDK	A track-type batch dry kiln
CDK	**A continuous dry kiln**
STCDK	A single-track continuous dry kiln
DTPFCDK	A double-track parallel-flow continuous dry kiln
TTPFCDK	A triple-track parallel-flow continuous dry kiln
QTPFCDK	A quad-track parallel-flow continuous dry kiln
FTPFCDK	A five-track parallel-flow continuous dry kiln
6TPFCDK	A six-track parallel-flow continuous dry kiln
CFCDK	**A counter-flow continuous dry kiln**
DTCFCDK	A double-track counter-flow continuous dry kiln
TTCFCDK	A triple-track counter-flow continuous dry kiln
QTCFCDK	A quad-track counter-flow continuous dry kiln
FTCFCDK	A five-track counter-flow continuous dry kiln
6TCFCDK	A 6-track counter-flow continuous dry kiln
JBDK	**A jogging-batch dry kiln**
STJBDK	A single-track jogging-batch dry kiln
	Parallel-flow kilns
DTPFJBDK	A double-track parallel-flow jogging-batch dry kiln
TTPFJBDK	A triple-track parallel-flow jogging-batch dry kiln
QTPFJBDK	A quad-track parallel-flow jogging-batch dry kiln
FTPFJBDK	A five-track parallel-flow jogging-batch dry kiln
6TPFJBDK	A six-track parallel-flow jogging-batch dry kiln
	Counter-flow kilns
DTCFJBDK	A double-track counter-flow jogging-batch dry kiln
TTCFJBDK	A triple-track counter-flow jogging-batch dry kiln
QTCFJBDK	A quad-track counter-flow jogging-batch dry kiln
FTCFJBDK	A five-track counter-flow jogging-batch dry kiln
6TCFJBDK	A six-track counter-flow jogging-batch dry kiln
	Counter-flow zigzag-air-flow kilns
DTCFZZAFJBDK	A double-track counter-flow zigzag air flow jogging batch dry kiln
STCAFCDK	A single-track counter-air-flow continuous dry kiln in which the lumber packages move sideways through the kiln, and the heated air stream moves counter to the movement of the lumber packages – common in Europe
STZZCAFCDK	A single-track zigzag counter-air-flow continuous dry kiln in which the lumber packages move endwise through the kiln while the heated air stream moves counter to the direction of package flow and in a zig-zag pattern through the packages
	Sidewinder batch kilns
STSWBDK	A single-track sidewinder batch dry kiln
DTSWBDK	A double-track sidewinder batch dry kiln
SLSWBDK	A side-loader sidewinder batch dry kiln – fans can be located either at the ends of the packages or facing the packages

Heat Supply Systems
Direct-Fired Kilns

Burner systems

Natural Gas – common
Fuel oils – typically small operations
Wood – fire hazards exist, will pose health issues for plant personnel

Wet – requires frequent downtime to clean out ash
Dry – high particulate discharge, ash will build up on kiln fans

Combination gas/oil/wood – rare, expensive

Heat supply duct

Over-head – most common design
Under-ground – dictated by type of soil and water table height

Heat return duct

Single-side of kiln – most common design, lowest cost
Dual-sides of kiln – rare, used to reduce duct size, horsepower, and certain heat-recovery systems that may be involved

Zone control

Single-zone – most common type
Multi-zone – rare, expensive

Indirect-Fired Kilns

Steam – most common method used in the USA, makeup water PH can be a problem

Supply pressure can range from as low as 10 psig to as high as 150 psig.

Hot liquids (water or oil) – common in Europe, high energy efficiency
Hot-air – rare due to the high cost of the air-to-air heat exchanger
Heat pump – used in dehumidification drying of hardwoods, high thermal energy efficiency but uses expensive electricity, not practical in large drying systems

Internal Fan Systems
Fan locations (3): above, below, or at the same (sidewinder) level as the lumber packages
Fan drives (2): internal or external motors with either line-shaft or cross-shaft drives
Variable-frequency drives are required for reducing electrical-energy usage

Only high-efficiency axial fan designs should be used
All internal fans must be precision balanced (aerodynamically and mass) before installation

Types and Typical Applications of Industrial Drying Systems

Type	Hobby	Typical applications Small	Medium	Large
Convection-type dryers				
Lumber, timber, poles	x	x	x	x
Veneer, sheets	x	x	x	x
Tray	x	x		
Dehumidification dryers	x	x	x	
Air drying yards	x	x	x	x
Forced air drying		x	x	
Pre-dryers		x	x	x
Radiation dryers		x	x	x
Vacuum dryers	x	x		
High frequency dryers	x	x		
HF + vacuum board dryers	x	x		
HF + convection veneer dryers		x		
Conduction dryers				x
Hydrostatic press dryers		x	x	x
Rotary dryers			x	x
Flash-tube dryers			x	x
Belt dryers		x	x	x
Screw bed dryers		x	x	x
Fluidized bed dyers			x	x

Predominant Types of Radiation Wood Dryer Systems

Heating environments in which wood particles are heated by radiation

A. Grate/bed furnaces
B. Suspension furnaces

Predominant Types of High Frequency Wood Dryer Systems

High frequency dryers

A. Atmospheric systems
B. Vacuum systems

Predominant Types of Hydrostatic Presses

High-pressure devices

A. Hydraulic rams for squeezing free water out of hog fuels
B. Mechanical rollers for squeezing free water out of mats and sheets

Predominant Designs and Applications of Wood-Drying Equipment
Particle Drying Applications

1. Rotary dryers
2. Suspension dryers (furnaces, flash tubes)
3. Mechanical conveyor dryers (screws, screen belts, chains)

Sheet Drying Applications

1. Veneer dryers (plywood and furniture manufacturing)
2. Hot presses (plywood and furniture manufacturing)

Individual Board Drying Applications

1. High frequency conveyor dryers

Package Drying Systems

A. **Pole-drying applications**

   1. **Batch drying equipment**

      A. Single-pass dryer designs (final target moisture content, without E&C)

         1. Multiple-package batch dryer designs

            a. Track kilns for poles

               1. Single-track designs (rare)
               2. Double-track designs (predominant design)

B. **Machined-wood drying applications**

   1. **Batch drying equipment**

      A. Single-pass drying designs

         Final target moisture content only (mostly softwoods)

         Final target moisture content with equalization (mostly softwoods)

         Final target moisture content with E&C (both hardwoods and softwoods)

            1. Single-package batch designs (for research and small finishing operations)

            A. Pre-dryers

            B. Kilns

            C. E&C chambers

         1. Multiple-package batch designs (for industrial operations)

            A. Pre-dryers

               1. Track pre-dryers (rare)
               2. Side-loader pre-dryers (predominant design)

A. Conventional heating systems
B. Dehumidifier heat pumps
C. Solar-heated (rare)

B. Kilns

1. Track kilns

A. Single-track (used mostly for softwoods)
B. Double-track (predominant design)
C. Triple-track (rare)

2. Side-loader kilns

A. Softwood kilns (used mostly for treating and timbers)
B. Hardwood kilns (predominant design)

3. Sidewinder kilns

A. Softwood kilns (typically for small operations)

C. Equalization and conditioning chambers

1. Track designs (applicable to softwoods)
2. Side-loader designs (common in Australia for pines)

B. Two-pass re-drying designs for track kilns (becoming rare)

High final target moisture content for re-drying high moisture content softwoods
Not used for hardwoods

2. **Jogging-batch drying applications – either parallel or counterflow track kilns**
3. **Continuous drying applications – either parallel or counterflow track kilns**

11.4 Industrial Dryer Product Loaders

Because there are **three classes of dryers** (batch, jogging-batch, and continuous), the machines for loading a product into these dryers must be designed for the specific drying operation.

However, there are **four classes of dryer product loaders** (batch, jogging-batch, cascading, and continuous).

Time, motion, economic, and safety studies must be conducted before selecting which type of dryer product loader configuration is to be used for a specific drying operation. This is especially the case with fork truck based high-production 24/7 operations during extreme weather conditions such as heavy rains, snow, and ice. These studies should include frontend costs, maintenance costs, monthly labor costs, fuels, energy, and parts costs, monthly time in use, etc., for a life expectancy of 10 years, 20 years, 30 years, etc. Follow standard accounting practices in doing these studies.

ONE: Batch dryer loaders involve mobile lift trucks, rail/tram systems, sliding belts, roll cases, hydraulics, and compressed air. These systems are designed to load a dryer for a **single batch drying operation** and then unload the same batch dryer after the batch drying cycle is complete. All batch dryers should be fitted with tight sealing doors with gaskets. Either manual or powered door operators can be used.

TWO: Jogging-batch dryer loaders are powerful automatic loaders designed to move products through a harsh dryer environment by using a repetitive push, stop, wait, push, stop, wait, push, stop, waiting scheme. In these systems, material products are loaded onto a slide, rail, tram, or roll case that traverses through a long dryer building inside which exist harsh conditions for precision machinery ball or roller bearings or electrical devices. Also included with the dryer proper are an outer-dryer infeed system and an outer-dryer outfeed system. – Small (short) units of products are loaded onto the infeed system, the loader then pushes the products a certain distance determined by drying rates, and short units of dried products are also unloaded off the outfeed section of the dryer system. To make this repetitive type of drying system operate properly, the system's controls involve programmable at-rest **periods** and pushing **strokes** both of which can exist in four different configurations.

Loader at-rest period	Loader pushing stroke
Fixed	Fixed
Fixed	Variable
Variable	Fixed
Variable	Variable

Process sensors are used to measure variables such as type product, product speed, weight, and starting inertia (acceleration), dryer residence time, temperatures, moisture content, color, etc., during the dryer operation. These measurements are then sent to a process controller programmed for the products being fed into the dryer and then automatically adjusting either the loader period or stroke or both. Because loader at-rest periods can vary from seconds to many days, and strokes can vary from inches to over 20′, and there are an unlimited possible number of combinations of these two variables, both will require specific designs dependent on the product, or products being pushed through a dryer.

All jogging-batch dryers also require entrance and exit doors specifically designed for the product species, product thicknesses, and temperature schedules used for drying the product.

In high-temperature drying of many fast-drying softwood lumber, both entrance and exit doors are often simple baffled end walls fitted with flexible rubber gaskets that rub against the outer edges and tops of the packages. These door systems may also be fitted with both an outer negative-draft vent-a-hood to collect door leakages and may also be fitted with an outer positive-draft air-seal (lock) to prevent any possible leakage to the area surrounding the dryer.

For low-temperature drying of grade-sensitive softwoods, the use of powered vertical guillotine doors is often used to close off the hot and humid interior of the dryer after a "charge of product" is pushed through both the entrance and exit doors. These configurations require careful design layouts and process controller programming of how both door end zones compensate for differences in product lengths, and how the loader feed rates automatically compensate for different drying temperature schedules and drying rates. These door and end zone designs require careful time and motion analysis before manufacturing or purchasing one of these dryers. Additionally, the long-term reliability, maintenance, and safety issues with these systems require careful analysis.

THREE: Cascading product loaders. In addition to the previous conventional jogging types of product loaders, such loader systems can also be designed for a parallel pushing action in which the motion of the products passing through the dryer simulates a true continuous flow by "**cascading or ducking-lug**" actions. In essence, as one jogging loader reaches the end of its stroke, a second retracted jogging loader picks up the load and takes over from there. Then, this action is followed by a third retracted jogging loader. Then the first loader, now fully retracted, picks up the load from there, and so on creating a continuous-like flow of products through the dryer. And the speeds at which the product moves through the dryer can be controlled by controlling the speeds at which the loaders activate. However, if these cascading types of loaders are used, attention must be focused on how the different products are both loaded onto and off the conveying systems located outside of the dryer proper – such that the product loading and unloading operations do not affect the dryer's operation.

FOUR: Continuous dryer product loaders. All continuous-dryer product loaders are, by definition, machines that move products through continuous dryers in a smooth, none-stopping, none-jerking, or erratic motion. Typically, these involve powered rail/tram systems, powered roll cases, screws, or slides. Furthermore, all continuous dryer systems must include both infeed and outfeed systems designed to automatically load and unload the main dryer product transport system without interfering with the dryer's smooth continuous motion. The doors for continuous dryers should address the same design issues discussed with jogging-batch loaders.

PARADE LOADER/DRYING SCHEMES If many different products are being loaded into the same dryer (as in a parade), process sensors and control programs must be able to track each product through the dryer, and automatically adjust drying control conditions (temperature and residence time) in each zone in the heating section of the dryer. However, there are **economic limits** to what parade control systems can do in non-batch-type drying systems. In some cases, management may have to accept a certain amount of product degrading to achieve minimum allowable production capacity. This is why all dryer product loader systems must be thoroughly evaluated before deciding on which one to use.

COMPRESSED-AIR-BLAST-LOADERS These are small (rare) product loaders that use a brief blast of compressed air to redirect or move small light-weight products across short distances.

SAFETY WARNING Because most product loader systems are extremely powerful machines, extra care must be taken to both warn and prevent plant employees from being caught between products once a loader receives a signal to move products. Sufficient warning lights, buzzers, horns, and proximity sensors are required to automatically stop the loader in the event a person enters a danger zone. And, because all these systems are extremely dangerous machines, all automatic dryer loader safety systems must both be inspected and documented on a fixed schedule. In some types of operations, statistical software may be used to notify plant management that something is not quite right with the operation, and maintenance people should now examine the loader controls or mechanical components. Finally, every one of these product loader systems must be evaluated with fault-tree analysis conducted by a process-control engineer familiar with this type of design safety analysis.

11.5 Moisture, Energy, and Drying Uniformity Terms in Wood Drying

Moisture Content of Wood

There are two methods used for defining the moisture content of wood. Both methods are expressed as a percentage.

The **dry-basis** moisture content in which the mass of water is divided by the mass of the dry wood fiber.

The **wet-basis** moisture content in which the mass of water is divided by the total weight of the water and the dry wood fiber.

The dry-basis method is typically used for machined products.

Green lumber, dry lumber, veneer, shavings, etc.

The wet-basis method is typically used for products prior to machining.

Logs, wet bark, etc.

Depending on the country or geographic location, the three different industries (wood, agriculture, and food) probably will not have uniform standards for defining moisture contents of the numerous different industry products.

Evaluating Drying Uniformity of Dryers and Drying Systems
The purpose of all drying systems is to remove water from product fibers. Knowing both the initial and final moisture contents accurately is crucial to designing successful drying systems.

The initial moisture content is the average moisture content of the product entering the dryer.
The final moisture content is the average moisture content of the product exiting the dryer.
The final moisture content distribution is a statistical graph of frequency vs moisture content.

The length of time and environmental conditions between the product dryer and the final moisture measuring system will affect both the average and standard deviations of product moisture content. Significant changes in average moisture content and standard deviations can occur between a product dryer and the final product moisture content measuring systems. Because of this, significant misinformation can exist to how any dryer is performing. Thus, all investigations of dryers should be conducted at the dryer, not at the final product moisture measuring systems.

Specific Gravity of Wood The specific gravity of wood affects the drying rate and the total energy requirement for drying. It is also a measure by which the total possible mass of water can exist in the wood and the dryer. Because the heat energy needed to evaporate a pound of water is much greater than the heat energy for heating wood fibers, the total heat energy needed to dry wood can increase dramatically when high specific gravity exists with high initial moisture contents. **The specific gravity of wood is:**

(The dry weight of wood fiber divided by its green volume)/(density of water)
The density of water is 62.43 pounds/cubic foot.

Saturated Moisture Content (SMC)

For one cubic foot of green log, there is a theoretical maximum weight of water that can exist in that one cubic foot. This water mass is determined by the specific gravity of the wood fibers in the cell walls. For a cell-wall fiber specific gravity (SGF) of 1.51, water specific gravity = 62.43, and log specific gravity (SGL) = (dry weight of 1.0 cu. ft. of green log volume)/62.43, the following equation can be used: Note: The geometrical proof (by the author) of the following general formula is not shown in this book.

$$\mathbf{MC\% = 100 \times ((1.0 / SGL) - (1.0 / SGF))}$$

And, for a SGF = 1.51

$$\mathbf{SMC\% = 100 \times ((1.0 / SGL) - 0.6622)}$$

SGL	SMC %
.30	267.1
.40	183.8
.50	133.8
.60	100.4
.70	76.6
.80	58.8

And to calculate the (kiln factor), KF = SGL × SMC %, we get the following table:

SGL	SMC %	KF
.30	267.1	80.13
.40	183.8	73.52
.50	133.8	66.90
.60	100.4	60.24
.70	76.6	53.02
.80	58.8	47.04

The term KF is used to analyze the internal geometry of a drying space in any convection-type drying process. KF is a measure of predicting the drying energy density rate inside a dryer.

Fiber-Saturation Point (FSP)
The fiber-saturation point (FSP) is the level of moisture content at which the only water in the wood is in the cell walls. This is about 29–30% moisture content (on a dry-weight basis). Above the FSP, the ability of the water to move within the wood is different than for below the FSP. Fiber shrinkage does not occur unless the moisture content drops below the FSP. From FSP to 5% MC, the shrinkage vs moisture content curve is approximately linear.

Species of Wood
The species of wood is a principal variable in wood moisture dynamics. Some species have closed cells, and some have open cells. The paths for transporting water from the roots to the leaves vary between and within species. Some species such as the southern pines are very porous and easy to dry. Some species like white oak are difficult to dry without defects.

Moisture Tracking Systems
This is a process in which the statistical averages and standard deviations of fiber moisture content, specific gravity, and density is tracked by species, tree diameter, age, geographic harvesting location, time of year, and weather – at each of the following processing centers:

At the forest

Entering the log storage yard
Entering the sawing, peeling, chipping operation
Entering the green fiber storage area
Immediately before entering the dryer
Immediately after exiting the dryer and before entering the dry fiber storage area
Exiting the dry fiber storage area and entering the dry-end moisture measuring system
Entering the final product process (grading, trimming, gluing, etc.)
Changes in moisture content during product shipment
Changes in moisture content while in the final warehouse or sales outlet
And so on throughout the expected use and life of the product

Sawing Methods

The sawing method is an important variable in wood drying.

The board **thickness** is the first criteria to consider. Both drying time and stress development are strong functions of board thickness.

The **orientation** of the board within the log is the second criteria to consider. Because of the differences in flow dynamics between grain direction, how a board is sawed from a log has a significant effect on how it dries. Edge-sawn and face-sawn boards dry differently in the same environment depending on their radial angle of cut in the log.

The sawed board **width** is the third criteria. Both drying time and stress development are strong functions of the beta ratio (board width/board thickness).

Heart vs Sapwood

Due to differences in specific gravity and initial moisture content between heart and sapwood, the drying times under the same conditions are different.

Capillary Flow

Capillary flow is a type of flow in liquids driven by surface tension forces. As the inner diameter of a tube in which a liquid exists approaches zero, the pressure created by surface tension will approach a value depending on the liquid, its diffused contents, and temperature. Because much of the flow of water in wood above FSP is capillary flow, the dynamics of capillary flow affects drying rates. Temperature has a strong effect on capillary flow dynamics.

Vapor Pressure Flow

Water has a specific vapor pressure depending on its state (ice, liquid, and gas) and its temperature. The vapor pressure of ice is insignificant in commercial drying operations. The vapor pressure of liquid water in wood is determined by its temperature. When liquid water is heated to a vapor pressure of one atmosphere (212 degrees F) it boils forming a gas.

Diffusion Flow

Diffusion flow is explained by Fick's law of diffusion. In diffusion flow, the flow rate is directly proportional to the moisture content gradient across a differential element of product fiber.

Thermal Expansion
When materials are heated, they expand. The relationship between expansion and temperature is the thermal expansion coefficient. Water has a volumetric thermal expansion coefficient higher than wood. Because of this, there is a differential expansion affect that can cause water to be expelled from fresh green wood during the period when the wood is heating up in a dryer. The volume of water expelled during the initial heating can be calculated if the initial moisture content of the wood, the specific gravity of the wood, and the change in temperature are known.

Drying Temperature Classes
Drying temperatures in wood dryers are classified (by the author) as follows:

33–120 F (low-temperature drying)
121–180 F (conventional-temperature drying)
181–211 F (accelerated-temperature drying)
212–250 F (high-temperature drying)
Above 250 F (hyper-temperature drying)

Low Temperature Drying
Low-temperature drying is drying at or below 120 degrees F dry bulb. Low-temperature drying involves air-drying, heated drying, vacuum drying, and dehumidifier drying.

Conventional Temperature Drying
Conventional-temp drying is drying at or below 180 degrees F dry bulb. Most hardwoods, stock over 2″ thick and upper grades, are dried at conventional temperatures.

Accelerated Temperature Drying
Accelerated-temp drying is drying at or below 211 degrees F. These schedules are used for both softwoods and some hardwoods.

High Temperature Drying
In high-temp drying, the dry-bulb temperature controller set point is kept between 212 and 250 degrees F. In southern pine high-temp kilns, the dry-bulb temperature entering the packages can exceed 250 degrees F, depending on the control scheme used. At high temperatures, the water inside the lumber is heated to the point where it boils creating internal pressure that creates both liquid and vapor flow to the surfaces of the lumber. During this process, a water-to-vapor profile is created inside the lumber. The shape of this profile is dependent on both time and heating rates. As the lumber dries, the vapor wave approaches the center of the board and flattens out. The effect of the FSP changes the shape of the wave. As the lumber approaches low moisture contents, the wood acts like an insulator to the heat for boiling off the water, and the drying rate slows down. Eventually, the wood moisture content will approach 3–4% if left in the kiln long enough.

Hyper-Temperature Drying

In hyper-temp drying, the dry-bulb temperature controller set point is kept above 250 degrees F. The most common applications are sheet and particle drying such as veneer driers and flash tube driers. Because drying rates are so fast in hyper-temp drying, controlling the final moisture content within acceptable limits without starting fires can be challenging.

Heat and Mass Transport and Exchange

During drying, there is an exchange between energy and mass properties. Mass-transfer principles in drying are pressure-gradient driven. Both liquids and vapor flow in wood are driven by pressure gradients. It takes a certain amount of energy to heat up and evaporate a pound of water, and it also takes a certain amount of energy to heat up a pound of dry wood.

Heat Transfer Principles

There are three types of heat transfer.

1. Conduction is heat transfer through a solid, liquid, or gas.
2. Convection is heat transfer to or from a body surrounded by fluids moving over the body.
3. Radiation is heat transfer from one mass medium to another by electromagnetic waves.

Energy Principles and Balances

Energy cannot be created or destroyed. However, its form can be changed. When heat is applied to wet wood, both the liquid water and the wood fibers gain energy depending on the temperature they reach. This form of energy is referred to as sensible heat. If the water evaporates from the wood, it requires additional energy to do so. This is referred to as the latent heat of evaporation. If the water vapor, now outside of the wood, is heated up to temperatures above the wood temperature, this is often called "tramp" superheat.

Specific Heat of Materials

The specific heat of a material is the amount of heat energy needed to raise one mass unit of the material one degree in temperature rise. In units of BTU/pound/F, the specific heat of wood is about .6, the specific heat of air is .24, the specific heat of liquid water is 1.0, and the specific heat of water vapor is .45. Specific heat values are not constant. They change slightly depending on the temperature of the material.

Bound Energy Level

Below the fiber saturation point, the amount of energy required to evaporate water increases. This increase is due to the molecular attractive forces between water and wood fibers. See Table 11.2 in the USDA Dry Kiln Operator's Manual for data on bound energy.

Changes of State and Energy Levels of Water

Water exists in the following three states: 1. solid (ice), 2. liquid, and 3. vapor (gas). For water to change from one state to another, energy is required. The amount of energy in water at different temperatures is shown in the following text.

Condition (at 1 atmosphere)	Energy level (BTU/pound)
Ice at 32 degrees F	−144.0
Liquid water at 32 degrees F (reference point)	0.0
Liquid water at 70 degrees F	38.04
Liquid water at 212 degrees F	180.07
Saturated water vapor at 212 degrees F	1150.4
Water vapor at 220 degrees F	1154.4
Water vapor at 300 degrees F	1192.8

Example: If liquid water at 70 F is heated to 300 F water vapor, the energy required is:

$$\text{BTU} / \text{pound} = 1192.8 - 38.04 = 1154.76$$

Changes of Energy Levels of Wood

When bone-dry wood is heated, its energy level changes depending on the species and the percentage of heartwood and sapwood. The following example is for a specific heat of 0.6.

Condition	Energy level (BTU/pound)
Pine at 32 degrees F	0.0
Pine at 70 degrees F	22.8
Pine at 212 degrees F	108.0
Pine at 240 degrees F	124.8

Psychometrics

Psychometrics is the study of air–moisture relationships. It is a branch of engineering for HVAC system analysis and design. The science of wood drying (heating, venting, and energy usage) involves the use of mass and energy continuity principles for both modeling the drying process as well as for designing drying equipment. See the ASHRAE Handbook for an in-depth review of the engineering principles of psychometrics.

Standard Temperature Terms in Wood Drying

There are several standard temperature terms in wood drying technology. Some of these terms have been coined by people to imply new technologies when in fact they are just other ways of expressing engineering principles found in all industrial drying systems.

Tdb – dry-bulb temperature
Twb – wet-bulb temperature
Tb – the temperature of the surface of wood fiber during drying
TDAL – the dry-bulb temperature drops across a load of wood during drying
Delta-T – the same as TDAL
CRT – constant rising temperature drying – rarely used
CT – constant temperature drying – common in many softwoods drying such as the southern pines
ST – stepped temperature drying – the most common practice in published wood drying schedules
TRAC – the air stream dry-bulb temperature rises across a heating coil
Ts – steam temperature
Tl – liquid temperature
Ti – initial temperature
Tf – final temperature
Tamb – ambient temperature

Energy Losses and Energy Efficiency

What are losses? It depends on the process and the objective of the process. For wood dryers, the objective is the removal of water from the wood. Since most types of commercial drying operations use the evaporative process, the ambient energy required for evaporation could be argued to be the benchmark for rating the energy-efficiency of commercial wood dryers.

Since at 70 degrees F, the latent heat of evaporation of water is 1054.3 BTU/pound, one could argue that this value should be the **benchmark** for rating commercial water dryers. Any additional energy used by any drying system beyond this value would be classified as losses.

Benchmark Drying Energy Efficiency

Total drying energy (**TDE**) used by commercial wood drying systems consists of the following:

Benchmark Drying Energy (BDE)

1. For evaporating water at 70 degrees F (1054.3 BTU/pound of water removed)

Drying Energy Losses (DEL)

2. For heating the wood during drying
3. For thawing out ice in frozen wood
4. For heating the liquid water in the wood
5. For heating air for wet-bulb control (venting process)

6. For superheating the water vapor leaving the kiln
7. For humidification spraying (both steam and water)
8. Heat-energy losses through foundations, buildings, and doors
9. Electrical energy losses (fans, blowers, conveyors, etc.)
10. All additional energy losses due to heating and utility support systems

    This additional energy can be a significant percentage of the total energy used.

    Steam boiler losses
    Steam system losses
    Condensate system losses
    Hot-oil system losses
    Hot-glycol losses

Thus,

$$\mathbf{TDE = BDE + DEL}$$

Wood dryer system rated energy efficiency **(REF)** is defined (by the author) as:

$$\begin{aligned}\mathbf{REF} &= \mathbf{(BDE\,/\,TDE) \times 100\%} \\ &= \mathbf{(1054.3\,/\,TDE) \times 100\%}\end{aligned}$$

In the preceding model, "energy" refers to all types of commercial energy (electricity, gas, oil, and wood) required for the **entire** drying system per pound of water removed.

Gross Energy Impact

In addition to the preceding **TDE**, future environmental impact studies will include the term **gross energy** that addresses the total shared global impact of the total energy required for putting the entire drying system in place. This is the **additional energy** that was required to mine the materials, transport, manufacture equipment, transport, install, and construct every piece of equipment that ended up making a complete drying system. Today, most industrial drying energy studies do not include this energy term, and thus produce final public reports that can be misleading for those not qualified to understand these reports.

The corn-to-ethanol industry and many offsite biomass-to-energy industries are two examples of government-promoted "energy" programs that do far more harm to the environment than good.

11.6 Air Delivery Efficiency (ADE)

All convection dryers use fans to circulate a mixture of air and moisture through the dryer and sticker openings for the purpose of transmitting heat energy to the wood.

The efficiency of the air delivery process is the ratio of the fluid stream work **consumed** at the lumber to the fluid stream work **produced** by the fans.

Air Delivery Efficiency = (sticker fluid work/fan fluid work) × 100%

Where:

sticker fluid work = sticker CFM × sticker head

and

fan fluid work = fan CFM × fan head

and

sticker CFM = sticker velocity × sticker area

thus

ADE = ((sticker area × sticker velocity × sticker head)/(fan CFM × fan head)) × 100%

Where:

Sticker area is the total cross-section area of all the sticker openings (sq. ft.)
Sticker velocity is the average exit velocity leaving the sticker openings (ft./min.)
Sticker head is the static fluid head across all the packages (wc″)
Fan CFM is the actual CFM produced by the internal fans
Fan head is the static head at the fan wall (wc″)

ADE is an **indicator** of the degree of design engineering put into convection dryers to improve the overall electrical and fluid stream efficiency. To do any ADE study properly, be sure to use the correct units for the data in your calculations.

ADE is dependent on the following principal factors:

The efficiency of the fan system
The geometric configuration of the dryer (track vs side loader vs side winder)
The psychometric loading (total width of heated packages/air stream pass)
The package grid factor (total sticker area/total load area)

ADE is an engineering term I coined during the early 1970s for comparing specific dryer configurations against others. For instance, a single-track dryer has a lower ADE than a triple-track dryer. A double-track dryer has a lower ADE than a triple-track dryer.

ADE is also related to effective package width (EPW). A dryer design that has a large EPW will have a higher ADE than the same dryer with a low EPW. For this reason, especially in softwood track-type dryers, a design that includes multiple tracks with reheat will have a high ADE.

ADE is improved by utilizing proven air-handling principles in the design of the interior **primary** plenums located inside the dryer. Plenums are large voids where airflow is collected, stabilized, and distributed to other parts of the dryer.

Convection Dryer Plenums and Fluid Energy Losses

In convection lumber dryers, there are **four** distinct **primary** plenums, each of which serves different **functions**. The following applies to dryer designs with ceiling-mounted fans.

1. Fan plenum

This is the air space located between the fan wall and the two upper sidewalls of the dryer. This plenum is the space located above a horizontal line level with the fan deck. This plenum serves two functions. It receives the air flow leaving the fans and delivers it to the upper baffle plenum located next to the wall.

2. Upper baffle plenum

This is the air space located between the upper (drop and ceiling) baffles and the two upper sidewalls of the dryer. This plenum is the vertical space located between the elevation of the top of the packages and the fan deck. This plenum serves two functions. It receives air from the fan plenum and delivers it to the load-to-wall plenum located below it. It also is a location where steam coils may be located.

3. Load-to-wall plenum

This is the air space located between the packages and the sidewalls of the dryer. This plenum is the vertical space between the floor of the dryer and the top of the packages.

This plenum serves one function. It receives air from the upper baffle plenum and delivers it to the packages of lumber.

4. Load-to-load plenum

This is the air space located between packages of lumber. This space is the vertical area between the floor of the dryer and the top of the packages. In multiple track kilns, reheat coils may be in this plenum.

The geometric configuration of plenums will determine how much electrical fluid energy is **wasted** due to air flowing through the dryer.

Plenum energy losses = Total fan energy × ((fan-head − sticker head)/fan-head)
Assuming no air bypasses the lumber packages.

Example:

A dryer has ten 15 horsepower motors that are all fully loaded when the kiln is loaded with lumber at 70 degrees F. The head at the fan wall is 1 wc″ and the total sticker heads add up to .375 wc″. What is the total plenum energy loss in electrical horsepower?

Answer:

$$\text{Plenum horsepower} = (10 \times 15) \times ((1.0 - .375)/1.0)$$
$$= 93.75$$

In this example, 62.5% of the total fan electrical energy is used just to circulate the air inside the dryer (with no air bypassing the lumber packages). Only 37.5% of the fan horsepower is used to force air through the sticker openings. Thus, in this example, ADE = 37.5%.

In commercial dryers with bypass CFM, the actual ADE is much less than shown in the preceding example. Air delivery efficiencies of 5–20% are not unusual in side-loader kilns and 20–40% in track type kilns.

This is why proper modern baffling and plenum designs are so important in all convection dryers.

11.7 Moisture Movement in Wood Drying

There are two simultaneous, but different, types of moisture movements occurring during drying. One is the moisture dynamics inside the wood, and the other is the moisture dynamics inside the dryer. To design and operate successful drying systems, both processes must be understood.

The drying systems found in lumber, plywood, and furniture plants are:

Pre-dryers for removing free water only, and degrade control
Kilns for removing free and bound water in a closed insulated chamber, and degrade control
Equalization and conditioning chambers for final wood moisture and grade control
Storage facilities for kiln-dried inventory storage, tempering, and grade control
Radiation dryers for heating the surfaces of wood and water
High frequency dryers for heating the interior of wood and water
High-pressure presses for squeezing water out of wood
Vacuum dryers for lowering the boiling point of water during drying

Many of the preceding technologies overlap and share similar heat transfer, fluid mechanics, and energy principles.

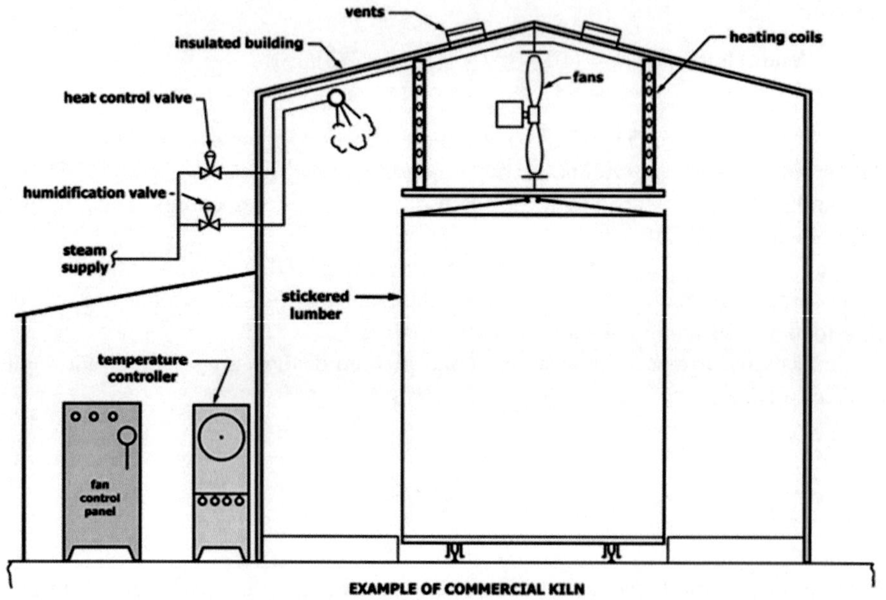

11.7.1 Moisture Movement in Lumber Pre-dryers, Kilns, and E&C Chambers

For the sake of expediency, only predominate commercial wood drying and conditioning systems are discussed.

Selection of Drying Schemes (Dry-Bulb and Wet-Bulb Temperatures and EMC Schemes)
There are five basic **schemes** for commercial lumber drying schedules.

Temperatures Controlled by Temperature Controllers
1. Constant temperature settings
2. Variable temperature settings based on drying time
3. Variable temperature settings based on wood moisture content
4. Temperatures determined by (heat input rates – heat demand rates)
5. Setting upper and lower limits on surface fiber moisture content

Depending on the dryer, the species and the thickness and width of the wood, the optimum drying schedule is chosen to both maximize dryer production and minimize degrade. Operator experience, the type of dryer, and the capability of the control systems dictate the best schedule and practices to use. For modern computer controls connected to management information systems (MIS), the opportunities for optimizing a drying process are numerous.

Batch Startup (Reaching the Initial Temperature Set-Points) with a Cold Dryer

During the startup of a cold batch dryer, the internal fans start rotating causing air to circulate around and through the packages of wood. Following the fans starting, the heat is turned on. The wood and its moisture then start reacting to the airflow and temperatures inside the dryer.

Following a brief period after the fans start, the dry-bulb and wet-bulb temperature sensors react to the air flowing over them. If heat is being put into the dryer, the dry-bulb temperature sensor and recorder will react to the heat. If no steam or water sprays are turned on, any water coming out of the wood will raise the humidity of the circulated air. If steam or water sprays are turned on, much of this water will condense and/or collect on the cold wood.

After several minutes of the air circulating through the lumber, the temperature recorder will show the wet-bulb temperature inside the dryer. Depending on many variables, including the design of the dryer, the dry-bulb and wet-bulb temperature recorders will start tracking what is happening inside the dryer. Eventually, if the heat system is on, after a period of minutes or hours, the dry-bulb and wet-bulb temperatures will reach the temperature controller's set points.

If wet heated wood is near saturation (high initial moisture content), liquid water may be forced out of the wood and cover the surfaces of the wood. Due to the ease by which water can travel longitudinally through most woods, significant amounts of water may be forced out of the ends of the boards. This is especially true for water near the ends of the boards.

Controlled Temperature Drying

During controlled temperature drying, the dry-bulb temperature controller automatically adjusts the heating system's input rate (BTU/hour). Likewise, the wet-bulb temperature controller automatically adjusts the vents and humidification sprays (if they are enabled) to maintain the wet-bulb temperature. If the wet-bulb temperature rises above the set point, the vents open. If the wet bulb drops below the set point, the sprays come on. For most dryers, the sprays are turned off during the first part of the drying cycle. However, if the species is a slow drying and extremely refractory wood, the sprays may be enabled.

During controlled temperature drying, either the dryer operator or automatic controllers adjust the temperature set points. The methods used are:

1. Constant dry-bulb and wet-bulb settings (used mostly for softwoods)
2. Stepped dry-bulb and wet-bulb settings based on dryer time
3. Stepped dry-bulb and wet-bulb settings based on wood moisture content
4. Constantly increasing dry-bulb and wet-bulb settings based on dryer time
5. Constantly increasing dry-bulb and wet-bulb settings based on wood moisture content

Controlled Heat Drying

Although thought to be rare, controlled-heat drying is common. This is a heat-balance process in which the heating capacity of the heat supply system dictates the rate at which the dryer removes the water from the lumber. If properly designed, this method can produce high quality lumber.

In a dryer I studied using this principle, the kiln did not have any temperature controllers. The kiln fans reversed every 4 hours. A small burner that operated at a fixed firing rate was connected to the kiln. The operator would shut the kiln off after a certain amount of time had elapsed and check the moisture content of the lumber. The operator learned after several kiln charges how to adjust the firing rate of the burner to prevent overheating the kiln at the end of the drying cycle. He used a 120-volt 7-day timer, connected to a fan starter circuit, to know how much drying time had occurred. At the start of each kiln charge, he reset the timer back to zero, and then started the burner back up. The drying system was both inexpensive and functional.

The controlled-heat principle can also be applied to large commercial drying systems operating on limited steaming capacity. One simply base-loads the steam generator, by the use of process-limiting devices, to the kilns using set-point controls, and then shed the remaining steam to floater kilns. If properly engineered, the steam generator will operate continuously at full rated capacity, producing the maximum possible drying production.

Perimeter vs. Interior (Package) Drying Rates

Most lumber dryers are batch package types. Additionally, most of the forest products industry has evolved toward using 4′- to 8′-wide packages. Because of this, the total travel distance for heated air to pass through the packages varies depending on the species and the operation. The hardwood industry favors narrow packages and side-loader (package) kilns. The softwood industry favors 8′ wide packages dried in track kilns, but not all softwoods are dried in track kilns.

Because of the psychometrics (heat and moisture exchange) involved in forcing hot air through packages of wet lumber, significant problems with wet lumber in the middle of packages can occur. Although dryers are designed to reverse their fans every 3–12 hours, the following factors will determine how much difference exists between the outside of the packages and the middle of the package during drying: The variables are listed in their order of affect.

The Permeability and Thermal Conductivity of the Wood

Depending on the permeability and thermal conductivity of the wood, the migration of moisture from inside the board to the surface will be restricted causing the temperature of the surface of the board to approach the dry-bulb temperature of the air being circulated through the package. If the temperature of the board's surface reaches the air's dry-bulb temperature, the heat transfer into the board stops, and the board stops evaporating water (the drying stops).

The Thickness of the Lumber

In high-temperature drying, the drying time is directly proportional to board thickness. In low-temp diffusion drying (below the FSP), the drying time is proportional to the square (second power) of the board thickness. In low-temp capillary-plus-diffusion drying above the FSP, the relationship between drying time and board thickness varies depending on the species, and the moisture content of the wood. The power function can vary from 1.5 to 2.0. Sample calculations for the effect of board thickness on drying time are given in the following text for narrow package drying:

1. A 10% increase in board thickness in high-temp drying.

    1.10 raised to the 1.0 power = 1.10
    The increase in drying time will be 10%.

2. A 100% increase in board thickness in high-temp drying

    2.0 raised to the 1.0 power = 2.0
    The increase in drying time will be 100%
    A 2″ thick board will take 2 times as long to dry as a 1″ thick board

3. A 10% increase in board thickness in low-temp diffusion drying below FSP

    1.10 raised to the 2.0 power = 1.21
    The increase in drying time will be 21%.

4. A 100% increase in board thickness in low-temp diffusion drying below FSP

    2.0 raised to the 2.0 power = 4.0
    The increase in drying time will be 300%
    A 2″ thick board will take 4 times as long to dry as a 1″ thick board

The Drying Schedule

The drying rate of wood in a specific dryer depends on two major factors.

1. The wet-bulb depression (dry-bulb temp – wet-bulb temp)
2. Whether low-temp or high-temp schedules are used

Using the same depression, switching from high- to low-temp drying in a dryer will increase drying time.

Total Effective Package Width (EPW)

Total effective package width is the total length of travel of air through packages without reheat. In softwood track kilns, where reheat capability is located between the tracks, a total effective package width of 8′ has proven successful. In many side-loader kilns drying oak, a total effective package width of 16′ has proven successful. The factors are drying schedule, sticker thickness, air velocity, board thickness, wood permeability, and wood thermal conductivity.

Sticker Thickness, Air Velocity, and CFM Modeling

Contrary to popular myths about air velocity, sticker thickness, and Reynolds number (a fluids dynamics parameter) it is the correct (minimum) amount of CFM through the packages that leads to successful drying operations. In commercial lumber dryers, water is evaporated from layers of wood stacked in packages. Heated air is forced through the package sticker openings and the heat energy in the air stream is transmitted into the layers of wood by convection. This convective heat transfer process is what drives the energy from the air stream to the lumber for evaporating the water from the lumber.

The mass flow rate of water leaving the lumber and entering the air stream can be a large portion of the mass flow rate of air entering a package. Because of this, conventional dry-surface forced-convection fluids dynamics and heat transfer models do NOT apply to this process. In effect what happens is that the water leaving the lumber upsets the conventional engineering film theory models for forced convection. This is why so many people in the wood products industry have been confused by discussions of the effect Reynolds and Nusselt numbers have on lumber drying rates and moisture content uniformity.

Saying this differently, in wide-package commercial lumber drying, it is the volume flow rate (CFM) of the heated air stream, not local heat transfer coefficients, that delivers the energy to and removes large amounts of water in lumber stacked in wide packages.

The relationships between sticker thickness, air velocity, and CFM are shown in the following text:

The CFM (cubic feet/minute) is the volume flow rate through the sticker opening.

CFM = air velocity × sticker opening area
= air velocity × ((sticker spacing – sticker width) × (sticker thickness))/144

The units are air velocity in feet/minute, and sticker spacing and dimensions in inches.

For a dry air stream entering a sticker opening, the relation between CFM, temperature depression, and available heating capacity is shown in the following text:

BTU/min. = CFM × air density × air specific heat × (dry bulb F – wet bulb F)

Summarizing the preceding into one equation to calculate the available BTU/hour heating capacity of a dry air stream entering a sticker opening:

BTU/hour = 60 × (sticker spacing – sticker width) × (sticker thickness) × air velocity × air density × air specific heat × (dry bulb F. – wet bulb F.)

Where all dimensions are in feet and air velocity is in ft./min.

For standard air at 70 degrees F:

BTU/hour = 1.08 × (sticker area sq. ft.) × air velocity ft./min. × (dry bulb F. – wet bulb F.)

Many studies to improve lumber drying by going to thin stacking sticks have been reported as advances in drying technology. These studies should be viewed with caution. Depending on the dryer, the drying schedule, the species, and the board thickness, one can have inadequate CFM (causing drying uniformity problems), or too much CFM (causing lumber degrading and high electrical costs).

Bypass CFM
In commercial wood dryers, three types of airflow are present.

1. The total CFM produced by the fans (fan CFM)
2. The CFM that flows through the packages (package CFM)
3. The CFM that **bypasses** the packages (bypass CFM)

And fan CFM = package CFM + bypass CFM

In most commercial lumber dryers, package CFM rarely gets as high as 50% of fan CFM. Only in track kilns in which full-course precision stacking and full-length sorting and stacking occurs does package CFM exceed 50%. The rest of the fan's CFM bypasses the packages.

In long pole drying track kilns, because of the large voids between the layers of poles, it is common for package CFM to be between 50% and 60% of fan CFM.

In sidewinder track kilns the fans are at ground level. A diverter baffle can be added to adjust the amount of bypass CFM to prevent excessive package CFM.

Bypass CFM can also be used to improve both dry-bulb temperature and depression control. In some kilns drying thick high-refractory species, bypass CFM may or may not be detrimental. In many cases, reducing package CFM by increasing bypass CFM prevents the heating system from over stressing the wood during time periods when stresses are prone to occur. There are numerous cases of people putting large fans in oak kilns resulting in increased degrade.

11.8 Measuring Air Velocity in Convection Lumber Dryers

Once a convection dryer has been installed, a final test of the delivered air velocity and thus CFM through sticker openings may be warranted in the contract. In many cases, the buyer may specify a guaranteed minimum air velocity for a certain size sticker. If such a clause has been included as part of the contract, several important variables should be included such that the true performance of the air delivery system meets the terms of the contract.

There are two situations the buyer and the contractor are presented with.

1. A new dryer installation
2. An upgrade of an existing dryer

In both cases, the principal factors that determine the delivered air velocity and CFM through the sticker openings are:

1. The total rated output CFM of the fans at the fan wall
2. The backpressure HEAD at the fan wall
3. The average tip clearance of the fan blades
4. The effectiveness of the baffles inside the dryer
5. The total board footage of lumber in the dryer during the tests
6. The actual thickness of the lumber
7. The actual thickness of the stickers
8. The relative roughness of the sawed surfaces of the lumber

Once the preceding variables have been established, the next set of conditions that the contractor and buyer should agree on are the temperature and density of the air stream during the final testing. Since ambient temperature and atmospheric pressure can vary widely depending on the geographic location of the dryer, any testing of air velocity from a fan system must consider both.

If the project is one in which a fan system in an existing dryer is being upgraded, many of the before and after principal variables may cancel out, and the only significant variables left are:

The total rated output CFM of the fans at the fan wall (before and after)
The backpressure HEAD at the fan wall (before and after)
The average tip clearance of the fan blades (before and after)

Adjustments to velocity data will be required when testing a fan system at different environmental conditions. For instance, a dryer tested at 70 degrees F at sea level will deliver more air mass flow rate than the same dryer operating at 90 degrees F at an elevation where the density of the air is much less, or the same dryer operating at 90 degrees F at sea level.

ACCURACY of AIR VELOCITY MEASURING INSTRUMENTS In addition, instruments used for measuring sticker air velocity must be calibrated against recognized standards for air velocity measuring instruments. In many cases, low-cost impulse meters are used for testing wood dryer air velocity, but not always. If a contract claims dispute or litigation is involved and testing is to be conducted by an independent third party, several instruments (each calibrated against a known standard) may be required.

Air turbulence and instrument accuracy, sensitivity, and response time can be significant issues if disputes arise about the accuracy of the velocity measurements taken at sticker openings. Most air streams leaving packages of lumber are turbulent causing considerable fluctuations in the velocities exiting sticker openings.

It is not uncommon to see the readings of a vane-type velocity meter vary by as much as 50% when positioned at a specific test location in a lumber dryer. If a vane meter is used, a protocol will be required for logging both the minimum and maximum instrument readings for each test location. In such a situation, the average of the two readings can be used for the effective velocity.

If an inclined manometer with a pitot tube is used, the selection of the pitot tube and the inclined manometer is crucial to accuracy. The following table shows the relationship between standard air velocity and dynamic pressure. It demonstrates the very low pressures encountered in measuring air velocities in wood dryers.

Dynamic Pressure for Standard Air – 70 degrees F @ 1.0 atmosphere

Velocity in feet/minute	*Inches of Water Column*
1750	0.1905″
1500	0.1399″
1250	0.0972″
1000	0.0622″
750	0.0350″
500	0.0155″
250	0.0039″
125	0.0010″
0	0.0000″

The pitot tube should be a design with both dynamic and static sensing ports. This type of pitot tube has two rubber connecting tubes that connect to two connections at the inclined manometer. Because the velocity pressure readings are so low in lumber kilns, the separate inclined manometer assembly should be secured to a firm mounting and zeroed out before taking measurements.

The advantage of using a pitot tube meter over a vane type is that the final instrument readings are far more stable and accurate. The only disadvantage is that the **position** and **orientation** of the pitot tube in the air stream exiting the sticker opening should be consistent from sticker opening to opening. Slight deviations in position or angle will cause variances in reading.

I suggest that whichever meter is used, a meter sensing bracket be used to assure the velocity in every sticker opening is measured the exact same way.

All lumber dryer air velocity readings should be taken ¼″ past the outer edges of the lumber and inside the packages. The velocity readings must be taken on the stacker-side of the package where every outer board is in vertical alignment with the boards located above and below. If the outer edges of the boards do not align themselves in a straight vertical line, the errors in velocity measurements will be significant.

Do not underestimate the importance of this issue. – Today, there are no industry standards for measuring air velocity in convection lumber kilns. Because of this, most civil court judges will throw out any claims of accuracy of sticker air velocity data.

Measuring Air Velocity Profiles in Lumber Dryers

When producing graphs of velocity profiles in a lumber dryer and total accuracy is not an issue, a small portable vane type meter can be used. Typically, velocity profiles are used to locate high and low velocity areas in dryers. In inferior dryer designs, especially with inadequate baffling, the average package exit velocity can vary over 50% from package to package causing significant drying problems. For these types of tests, the meter's calibration should still be checked before testing begins.

11.9 Wave Theory in Lumber Drying

If all sawed lumber were stacked in narrow packages of one board width, the dynamics of drying would be vastly different than what occurs in commercial drying operations.

If we could build a commercial dryer so that every board being dried was subjected to the same exact drying conditions as every other board, such a system would be vastly superior to the drying systems now in use. In such an ideal dryer, we would not have to worry about temperature drops across packages, fan reversals, heat distribution problems, or whether we were using too thin a stacking stick. In such a dryer, every board would dry at rates dictated by the accuracy of precision temperature controllers.

With such a dryer, life would be much simpler for kiln operators because the quality of the lumber exiting the dryer would be much higher than what occurs in commercial drying operations today. Such a dryer would allow for precise control of all the fiber's internal conditions that causes degrade. In such a system, the drying dynamics of every species, board thickness, and width could be both modeled and controlled for maximizing the yield from every log that enters the plant. Such an ideal dryer would eliminate the many deficiencies found in wood dryers today and allow wood scientists to fine tune computerized drying control models for each species of lumber entering the dryer.

However, such commercial drying systems do not yet exist. Instead, most lumber today is dried in dryers designed to hold large **wide** packages of lumber. The problem with this method of drying is that it introduces complex energy- and mass-balance transient dynamics into the drying process that can be far more technically challenging for dryer operators than the moisture dynamics in wood-fiber stress-control.

Almost any kiln operator can learn how to take moisture samples from a dryer and apply temperature schedules to minimize degrade. However, ask a typical lumber dryer operator to explain the complex heat transfer, and energy- and mass-balance engineering principles in large commercial lumber dryers and you will see something quite different.

In most commercial drying systems today, we now have a situation in which the tail is wagging the dog instead of the dog wagging the tail. Today, instead of minimizing degrade to get the maximum yield out of smaller and smaller lower-quality logs, we are maximizing dryer production to cover capital expenditures and operating overhead.

Up until about a century ago, most lumber drying was done in hand stacked piles of boards standing on one end. This method was used because it allowed air to circulate freely around each board. Although certainly crude by today's standards, such a practice allowed for the fastest drying possible, with Mother Nature controlling the drying rate. Some of these "drying systems" were successful and some were not. Much of the lumber that passed through these drying systems suffered extensive damage.

Eventually somebody got the idea of stacking lumber in horizontal piles. They discovered that stacking the lumber in horizontal piles with cross outs for air circulation was a better method than vertical piles. This was a significant advancement in lumber drying technology. This method would dry lumber in a "storage system" that required less acreage and provide restraint to minimize warp. This method of drying wood quickly caught on and led to the large air-drying yards still in use today.

Today, lumber is stacked in large wide packages for commercial drying. The evolution of wide-package technology took no less than a century of people trying this and that to see what worked and what did not. First, there was the eight-foot-wide package, then the four-foot-wide package, then the six-foot package, and then the infamous ten-foot-wide package. And to make things worse, everybody had a different theory about sticker thickness. Every sticker thickness from ½″ to 4″ was tried. Finally, people discovered that the number and type of fans used in the dryer had an effect on drying. And this went on and on for decades.

If you read through the many technical articles written on lumber drying since the 1950s, you will see many debates about package widths, thick and thin sticker thickness, laminar and turbulent air flow, Reynolds numbers, swirling fan designs, and degrade problems with kilns and air-drying yards.

After a century of hits and misses, the forest products industry is finally starting to understand the engineering principles involved in commercial lumber drying systems and even more importantly why it took so long to figure out something that is so obvious.

Lumber stacked in wide packages is vastly more difficult to dry successfully than lumber stacked in narrow packages.

In addition, I want to go on record at this point in this book with the following three statements about all types of commercial lumber drying operations:

Many of the problems seen in lumber drying during the last century were directly caused by a thin-sticker mentality promoted by people who did not understand the engineering principles in psychometrics, fluid dynamics, heat transfer, and thus lumber drying. Thin stacking sticks have historically been promoted to increase dryer holding capacity and flatten out velocity profiles when in fact the effects were extended drying times and significant increases in moisture uniformity problems.

If people would simply increase the thickness of their stacking sticks many of their lumber drying moisture control problems would vanish into thin air.

Furthermore, as long as money managers and hype dictate how lumber drying systems are purchased and operated, the wood products industry will continue to have major drying and degrade problems.

Let us assume for now that narrow package drying and thick stickers are lost causes, and penny-pinching mentalities and hype have taken complete control of lumber drying. Let us spend some time discussing the dynamics of wide package lumber drying systems. Let us talk about **wave theory**, and what it tells us.

Wave theory is an economic term I coined during the early 1970s for negative variables introduced into lumber drying that would not exist if all lumber were dried in narrow packages. Saying it differently, wave theory losses are those due totally to wide package drying.

It should be obvious to the reader that wave theory losses increase as the effective package width increases. The reader should also be careful to not misunderstand what wave theory predicts. Wave theory does not address the economic benefits of wide-package strategies, such as lower unit material handling and equipment costs. These benefits should be treated separately when analyzing a drying system.

We can think of this analysis like the following. Suppose we were going to design two separate drying systems. One was for drying 7/4 southern pine and the other was for drying 7/4 white oak. Let us examine the following six possible dryer design points to see what is happening:

Material to be dried	Maximum effective package width without reheat
7/4 pine	4′ vs. 8′ vs. 16′
7/4 oak	8′ vs. 16′ vs. 32′

The total capital and operating cost, losses, and benefits would be determined for each of the six drying systems. The next variable that would enter the picture is the problems with degrade and moisture control when wide effective package widths are used. Now double the sticker thickness and redo the preceding calculations for designing the drying system.

If you do this kind of analysis properly, you will be surprised at what you will discover about the effects effective package width and sticker thickness has on lumber drying economics.

11.10 Drying Curves in Wide-Package Lumber Dryers

The following three graphs demonstrate visually the complex transient heat- and mass-dynamics going on inside wide-package lumber drying. These are the wave theory curves that have puzzled so many people in the lumber drying industry and have caused so many myths about which drying methods are better than others.

Note that the curves in the following three graphs are not linear (straight lines), but instead are transients (constantly changing over time). These time-based transients cause considerable confusion for dryer operators whose job it is to both remove the water from the lumber and do it in such a manner that the lumber is not damaged during the drying process.

The first graph shows three parametric transient (decay) curves for the moisture content of lumber in a dryer operating at constant temperature settings. Curves A and B represent the projected moisture content of boards located at the outer edges of the package. Curve C represents the moisture content of lumber located in the middle of the package. At the start of the drying cycle, the difference between A and

B is defined to be DMC. For package lumber drying, DMC is determined by the effective package width (EPW), the sticker thickness (ST), and the air velocity (V) through the sticker opening, and K is a constant determined by the dry-bulb and wet-bulb temperatures of the air stream entering the package:

$$\mathrm{DMC} = \mathrm{K}\left(\mathrm{EPW} / \left(\mathrm{ST}^{*}\ \mathrm{V}\right)\right)$$

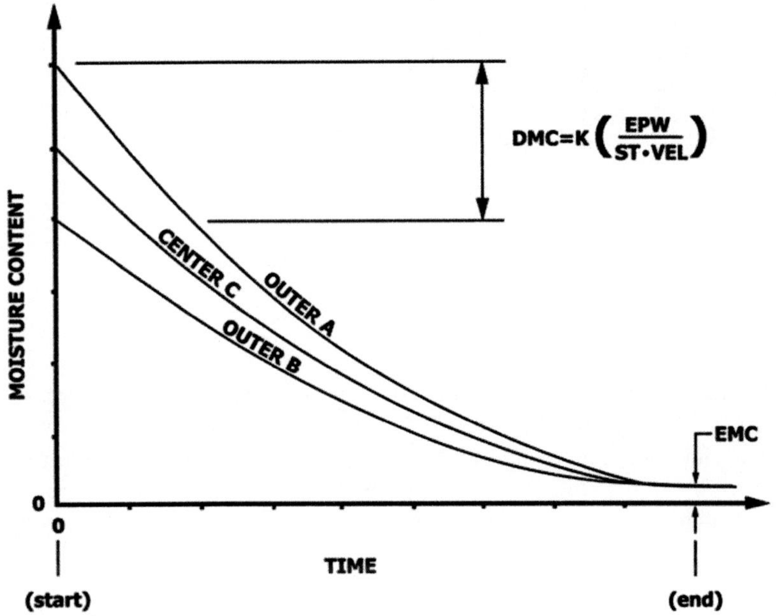

GRAPH #1

Now look at the next two graphs. These graphs demonstrate how the temperature of a heated air stream and the temperature drop across a package of wet lumber changes with time. Assume in this example, the fans do not reverse, and the entering temperatures are kept constant during the drying process.

In graph #2, curve #1 is the dry-bulb temperature profile moments after the drying starts. Curve #2 is 1 hour later. Curve #3 is several hours later. Curve #4 is when the exit air temperature depression is 40% of the entering temperature depression. Curve #5 is when the exit air depression is 66% of the entering temperature depression. Curve #6 is when the exit air temperature is equal to the entering temperature, meaning that the package's average moisture content has stabilized at the equilibrium moisture content of the air stream entering the package.

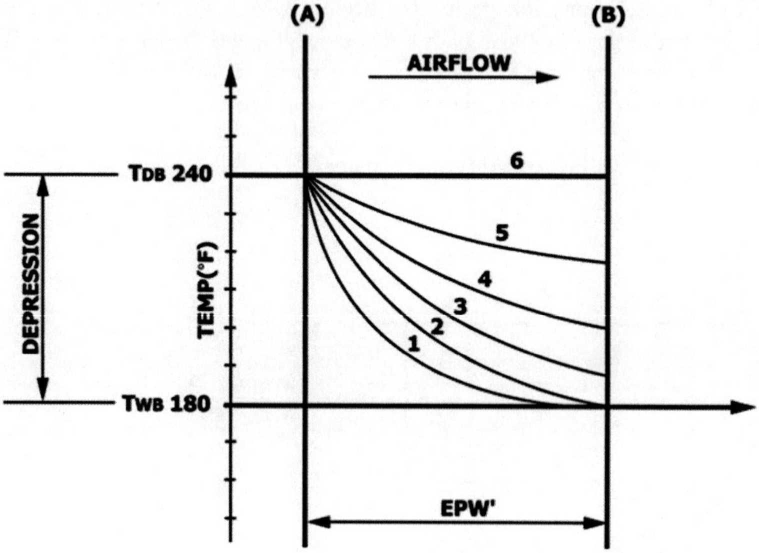

GRAPH #2
Now look at graph #3 that demonstrates how the dry-bulb temperature drop across a package of wet lumber is affected by effective package width and drying time. Note that the profile of the three curves is different and are shifted from the left to the right side of the graph. Note that for a narrow 4′ wide package, the air stream does not saturate before exiting the package but does for the 8′ and 12′ wide packages.

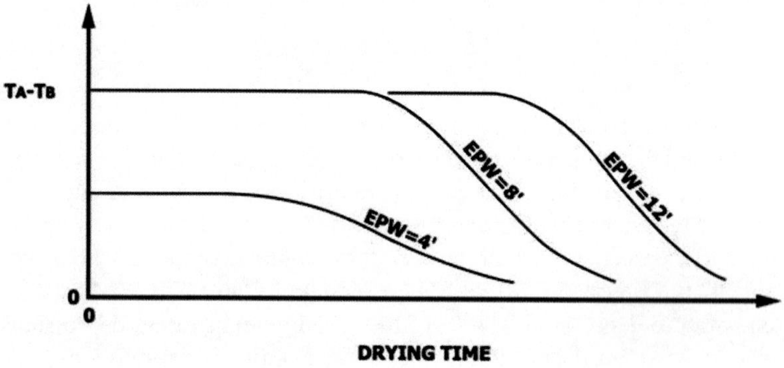

GRAPH #3
Many transient and temperature-dependent variables are involved in heat and mass transport dynamics in commercial wide-package lumber drying. These numerous and complex variables have caused considerable confusion and myths about all types of commercial lumber drying.

Research is needed to establish the **optimum CFM** for the different species and effective package widths found in the numerous and different types of commercial dryers. I use the word optimum because if too much CFM is put in a lumber dryer, the electrical energy needed to drive the fans can be very costly during the life of the dryer.

To do this research, a simple air-delivery model (that I will discuss later) could be used to predict the optimum CFM. Once reliable data is collected for different species dried with specific temperature schedules, it would be a simple matter to produce a table of optimum CFM for different species, board thickness, EPW, and drying schedules.

Because the same model could also be used to automatically slow down fans depending on the average moisture of the lumber in the dryer, it would be a simple task to develop a table of CFM curves for each species and drying schedule. When we consider all the wood dryers around the world, this would add up to a large amount of saved energy and reduced pollution from electric utility plants.

11.11 Minimum CFM for Wide-Package Lumber Drying

Let us now imagine we have two situations in lumber drying. One situation is the drying of a fast-drying species such as high-temperature drying of 7/4″ southern pine, and the other is the drying of a slow-drying refractory species such as 4/4″ white oak.

The question should be asked, what is the minimum required CFM for successful drying in both cases?

Since we know that the mass loading of water on the air stream determines how much air CFM is needed, we need to produce a set of curves that demonstrates the relationship between air velocity, sticker thickness, and effective package width for the different types of species and thickness to be dried. We also know that for fast drying operations (such as high-temp drying of softwoods), the water mass loading on the air stream will be the highest. Once we have these curves, we could predict the minimum required CFM for highly impermeable species and thicker boards.

But, before I get to these curves, we need to review an air-delivery saturation model that I developed in 1974.

Standard-Air Model 74 for Required Air Circulation Capacity in Wood Dryers

Back in 1974 when I was designing models for wood drying, I developed the following **standard-air** model for calculating the "minimum-allowable" and "targeted-design" levels of CFM in wood dryers. The model predicts the **minimum** required CFM through each sticker opening for successful drying of any species and board thickness. Once the CFM/sticker opening has been established, it is simply a matter of multiplying this value times the number of sticker openings facing a plenum space in a fully loaded dryer to determine the total package CFM. Once this value is known, the fan CFM can be calculated from package CFM + the bypass CFM at standard air conditions (70 degrees F and 14.7 psia).

Because there are thousands of variations of lumber drying air delivery systems around the world, model 74 was based on variables that I refer to as **benchmark parameters**:

1. Standard-air properties (density and specific heat for air at 70 degrees F and 14.7 psia)
2. 8′ wide packages
3. Square stickers placed on 2′ centers

These three benchmarks were chosen for several reasons.

To explain air handling dynamics in lumber drying to a large audience (those who design, manufacture, and operate wood dryers), a simple **benchmark** model is needed to start the dryer design process.

The design and testing of all air handling equipment is based on standard air conditions.

Most of the water removed in lumber drying occurs in 8′ wide packages.

Most stacking sticks are spaced on 2′ centers.

The use of square stacking sticks simplifies the calculations for modeling airflow through the sticker openings. The effect of using rectangular stickers on the model's outcome is insignificant as long as the sticker aspect ratio (width/thickness) does not exceed 2.0.

The reader should understand that model 74 is a saturation **benchmark** model designed for air dynamics at 70 degrees F and one atmosphere. If either the temperature or pressure of the air stream changes, the output CFM of the model will have to be adjusted accordingly. I will discuss these issues later.

Another feature of the model is the ability to produce **fan speed curves** for different drying schedules, species, and board thickness. The model addresses these variables and thus has the ability to automatically change the speed of fans depending on drying conditions. If an automatic fan control scheme is being used in a computerized drying operation, it is a simple matter of inputting data into the model and it will produce an output signal that can be used to speed up or slow down dryer fans to reduce electrical usage. This will be discussed later.

Procedure for Calculating Benchmark Fan CFM with Standard Air Model 74

Step 1 – Select the species of lumber to be dried
Step 2 – Select the thickness of lumber to be dried
Step 3 – Select a drying time (do not include E&C time)
Step 4 – Determine the pounds of water to be removed/actual cubic feet of lumber
Step 5 – Calculate the pounds of water to be removed/sticker opening
Step 6 – Calculate the average water evaporation rate/hour/sticker opening
Step 7 – Calculate the average heat energy rate for evaporating the water/sticker opening
Step 8 – Select a heat capacity peak ratio
Step 9 – Calculate the peak heat energy rate/sticker opening

Step 10 – Calculate the benchmark CFM/sticker opening
Step 11 – Calculate the total benchmark package CFM for a fully loaded dryer
Step 12 – Calculate the total benchmark fan CFM
Step 13 – Calculate the benchmark air velocity in the sticker openings

The preceding procedure involves 13 steps, 12 of which are used for arriving at the fan system CFM. Step 13 is included to calculate sticker air velocity. The following demonstrates the calculations:

Step 1 – Select the species of lumber to be dried.

Example: red oak, white pine, southern pine, ash, etc.

Step 2 – Select the exact thickness (inches) of the lumber to be dried.

Example: 1.06″, 1.58″, 1.71″, etc. Do not use nominal thickness (4/4, 8/4/ etc.)

Step 3 – Select a drying time (do not include E&C time).

Refer to published materials for drying rates of specific species and thickness using a specific temperature drying schedule in a test kiln. Typically, most published drying curves are determined for lumber stacked in narrow packages of less than four feet. Since commercial dryers using wide packages cannot replicate the results of experimental dryers using narrow packages, a longer "design" drying time will be required.

Step 4 – Determine the pounds of water to be removed/actual cubic feet of lumber.

Determine this value from actual tests where the dryer is to be installed or from published texts on specific gravity and moisture content. The wood specific gravity and the initial and final average moisture contents will have to be known.

Step 5 – Calculate the pounds of water to be removed/sticker opening.

Using the results from step 4, calculate the exact amount of water that each sticker opening will serve. One 2′ wide sticker opening serves one 2′ wide by 8′ long course of lumber.

Step 6 – Calculate the **average** water evaporation rate/hour/sticker opening.

Using the results from Step 5, divide this value by the drying time (hours).

Step 7 – Calculate the **average** heat energy rate for evaporating the water/sticker opening. Using the results of Step 6, multiply this value by 1054.3 BTU.

Step 8 – Select a heat capacity **peak ratio**.

This value is determined by the species, board thickness, and type of drying schedule. Refer to the following table to select a peak ratio.

Drying schedule	Peak ratio
Constant temperature schedule	2.80
Stepped temperature schedule	2.00
Constant-rising schedule	1.75
Heat-limited schedules	1.50

Step 9 – Calculate the peak heat energy rate/sticker opening – (**BTU/minute**).

Multiply the result of Step 7 times the selected peak ratio and divide the result by 60.

Step 10 – Calculate the benchmark CFM/sticker opening:

$CFM_{benchmark}$ = (peak heat energy rate/sticker opening)/.075/.24/DT
Where
Peak heat energy rate/sticker opening is the result of Step 9
.075 is standard air density (pounds/cubic foot)
.24 is standard air specific heat (BTU/pound/degree F)
DT = the temperature depression (dry bulb F – wet bulb F) used during the **earliest** stage of drying

Step 11 – Calculate the total benchmark package CFM for a fully loaded dryer.

Multiply the result of Step 10 times the total number of sticker openings facing one plenum in a fully loaded dryer.

Step 12 – Calculate the total **benchmark fan** CFM for the dryer.

Since:
Total benchmark fan CFM = total benchmark package CFM + total benchmark bypass CFM
and
Baffling efficiency (BE) = total package CFM/total fan CFM
thus
Total benchmark fan CFM = total benchmark package CFM/baffling efficiency (BE)

Baffling efficiency (BE) is a fluids dynamics factor strongly dependent on dryer and baffling system design, sticker thickness, stacking, and dryer loading practices from plant to plant. The following table can be used for estimating BE for 1″ thick stickers in 8′ wide loads of 7/4″ thick lumber: **Note: The numbers shown in the following text can each be improved with modern baffling systems to dramatically improve dryer fan energy efficiency and dryer operations. Baffling efficiency has a dramatic impact on fan energy efficiency. If BE could be increased to 1.00, the required amount of fan electrical usage/pound of water removed in convection dryers would be less than one-fourth of what exists in all convection lumber drying today.**

Batch Applications	**BE**
Modern full-course track kilns	
Over 80′ in length	.60
60–80′ in length	.55
40–60′ in length	.48
Track kilns with one cross-out	
Over 80′ in length	.54

60–80′ in length	.49
40–60′ in length	.43
Track kilns with two cross-outs	
Over 80′ in length	.48
60–80′ in length	.44
40–60′ in length	.38
Track kilns with three cross-outs	
Over 80′ in length	.42
60–80′ in length	.38
40–60′ in length	.34
Side-loader kilns	**.28 to .33**

Finally, using a thicker sticker will increase BE and reducing sticker thickness will reduce BE. The effect is dependent on the package grid factor (sticker thickness/board thickness).

Accurate field data is required for each type of dryer design to establish reliable values of BE. This is something every dryer manufacturer should do before marketing their dryers. Every dryer manufacturer should also publish tables of BE for specific dryer designs, sticker thickness, and board thickness.

Step 13 – Calculate the benchmark air velocity in the sticker opening.
Benchmark air velocity =benchmark sticker CFM/sticker opening area (sq. ft.)

Sample Calculation for Benchmark Fan CFM
Case #1

Southern pine, EPW = 8′, board thickness = 1.75″, drying time = 19 hours, specific gravity = .54, initial MC1 = 110%, final MC2 = 18%, temp depression = 60 F, sticker thickness = 1.0″, 84′ dryer length, 65 full courses high

Step 1 – Select the species of lumber to be dried

southern pine

Step 2 – Select the thickness of lumber to be dried

BT = 1.75″

Step 3 – Select a drying time (do not include E&C time)

drying hours = 19

Step 4 – Determine the pounds of water to be removed/actual cubic feet of lumber

$$\begin{aligned}\text{\# water/actual cubic feet} &= SG \times 62.43 \times ((MC1 - MC2)/100)\\ &= .54 \times 62.43 \times ((110 - 18)/100) = 31.02\end{aligned}$$

Step 5 – Calculate the pounds of water to be removed/sticker opening (in feet)

$$\begin{aligned}\text{water removed} &= 31.02 \times EPW \times \text{sticker spacing} \times \text{course thickness}\\ &= 31.02 \times 8' \times 2' \times (1.75/12) = 72.38\end{aligned}$$

Step 6 – Calculate the average water evaporation rate/hour/sticker opening

average evaporation rate = water removed/sticker opening/drying time
= 72.38/19 = 3.81

Step 7 – Calculate the average heat energy rate for evaporating the water/sticker opening

average heat energy rate = 1054.8 × average evaporation rate
= 1054.8 × 3.81 = 4018.8 BTU/hour

Step 8 – Select a heat capacity peak ratio

peak ratio = 2.8

Step 9 – Calculate the peak heat energy rate/sticker opening

Peak heat capacity rate = peak rate × average heat energy rate
= 2.80 × 4018.8 = 11,252.64 BTU/hour

Step 10 – Calculate the benchmark CFM/sticker opening

CFM min = peak heat capacity rate/(density × specific heat × temp. depression × 60)
= 11,252.64/(.075 × .24 × 60 F × 60) = 173.65

Step 11 – Calculate the total benchmark package CFM

The total number of sticker openings facing a plenum is 65 × (84/2) = 2730
Total design package CFM = 2730 × 173.65 = 474,064

Step 12 – Calculate the total required fan CFM for the dryer

Total required fan CFM = total benchmark package CFM/baffling efficiency (BE)
=474,064/.60 = 790,107

Step 13 – Calculate the benchmark air velocity in the sticker opening

V = benchmark sticker CFM/sticker opening area in sq. ft.
= 173.65/((23/12) × (1/12)) = 1087.2 ft./min.

NOTE: To ensure that the air delivery system is never overloaded by the dryer's heating system, I recommend you size the heating system such that it cannot produce more heat energy rate than 70% of the heat energy rate shown in Step 9. To calculate the dryer's heating system maximum allowable heat energy rate, multiply the output of Step 9 by the number of sticker openings times 0.7.

Dryer maximum BTU/hour = (70%/100%) × 11,252.64 × 2730 = 21,503,795

Sample Calculation for Benchmark Fan CFM
Case #2

Red oak, EPW = 8′, board thickness = 1.00″, drying time = 576 hours, specific gravity = .55, initial MC1 = 100%, final MC2 = 9%, temp depression = 2 F, sticker thickness = 1.0″, 84′ long kiln, 65 full courses high

Step 1 – Select the species of lumber to be dried

Red oak

Step 2 – Select the thickness of lumber to be dried

BT = 1.00″

Step 3 – Select a drying time (do not include E&C time)

drying hours = 576

Step 4 – Determine the pounds of water to be removed/actual cubic feet of lumber

water/actual cubic feet = SG × 62.43 × ((MC1 − MC2)/100)
= .55 × 62.43 × ((100 − 9)/100) = 31.25

Step 5 – Calculate the pounds of water to be removed/sticker opening

water removed = 31.25 × EPW × sticker spacing × course thickness
= 31.25 × 8′ × 2′ × (1.00/12) = 41.67

Step 6 – Calculate the average water evaporation rate/hour/sticker opening

average evaporation rate = water removed/sticker opening/drying time
= 41.67/576 = .072

Step 7 – Calculate the average heat energy rate for evaporating the water/sticker opening

average heat energy rate = 1054.8 × average evaporation rate
= 1054.8 × .072 = 75.95 BTU/hour

Step 8 – Select a heat capacity peak ratio

peak ratio = 2.0

Step 9 – Calculate the peak heat energy rate/sticker opening

Peak heat capacity rate = peak rate × average heat energy rate
= 2.0 × 75.95 = 151.9 BTU/hour

Step 10 – Calculate the benchmark CFM/sticker opening

CFMbenchmark = peak heat capacity rate/(density × specific heat × temp. dep. × 60)
= 151.9/(.075 × .24 × 2 F × 60) = 70.32

Step 11 – Calculate the total benchmark package CFM for a fully loaded dryer

The total number of sticker openings facing a plenum is 65 × (84/2) = 2730

Total benchmark package CFM = 2730 × 70.32 = 191,974

Step 12 – Calculate the total required fan CFM for the dryer

Total required fan CFM = total benchmark package CFM/baffling efficiency (BE)

= 191,974/.60 = 319,957

Step 13 – Calculate the benchmark air velocity in the sticker opening

V = benchmark sticker CFM/sticker opening area
= 70.32/((23/12) × (1/12)) = 440 ft./min.

NOTE: To ensure that the air delivery system is never overloaded by the dryer's heating system, I recommend you size the heating system such that it cannot produce more heat energy rate than 70% of the heat energy rate shown in Step 9. To calculate the dryer's heating system maximum allowable heat energy rate, multiply the output of Step 9 by the number of sticker openings times 0.7.

Dryer maximum BTU/hour = (70%/100%) × 151.9 × 2730 = 290,280

Case #3

Southern pine, EPW = 8′, board thickness = 1.75″, sticker thickness = 1″, specific gravity = .54, initial MC1 = 110%, final MC2 = 18%, same dryer as case #1.

Let us now look at benchmark air velocity for different drying times and temperature depressions.

Drying time (hours)	Temp depression (F)	Benchmark air velocity (ft/min)
19	60	1087
12	60	1721
19	80	815
12	80	1290

The tabulated values shown in the preceding text demonstrate the use of benchmark Model 74 for predicting drying times for different combinations of air velocity and temperature depression.

Checking Your Calculations for Model 74

There is a simple way to check your final calculations after you have gone through model 74 for a specific species and drying schedule. This involves calculating the required size of the heating system against known industry practices for wood dryers.

This is done by calculating the BTU/HR/nominal board feet of lumber in the dryer.

Example: In case #1 presented earlier,

Nominal board feet = 8′ × 84′ × 2 × 65 = 87,360
Heat system capacity = .7 × step 9 × # sticker openings

$$= .7 \times 11{,}252 \times 2730$$
$$= 21{,}503{,}795 \text{ BTU/hour}$$

Thus, the BTU/hour/nominal board feet = 21,503,795/87,360 = 246

Wood dryer manufacturers have historically maintained rules of thumb for sizing heating systems for dryers. These can be compared to model 74 before a heating system is designed.

Benchmark Application Curves For CFM, Sticker Thickness, and Effective Package Width

Now that we have reviewed model 74, we now know how to calculate the benchmark sticker CFM for quality lumber drying for 8′ wide packages with 1.0″ thick stickers. But what do we do if the EPW is not 8′? What do we do if the EPW is less than or greater than 8′?

To answer this question, I produced a set of CFM curves (see the next page) that demonstrates how to change sticker thickness for different effective package widths. These curves apply to **all** types of lumber drying **after** model 74 has been used to **calculate** the benchmark CFM/sticker opening.

The 10 CFM curves were developed from 10 **classes** of air velocity (defined by the author), each with a 1″ thick by 2″ wide sticker. Each curve is for a **specific** CFM class. The curves demonstrate visually how **minimum required** air velocity in sticker openings is affected by both sticker thickness and effective package width.

Class	Air Velocity	Sticker	CFM
I	100	1.0″	15.97
II	200	1.0″	31.94
III	400	1.0″	63.89
IV	600	1.0″	95.83
V	800	1.0″	127.78
VI	1000	1.0″	159.72
VII	1200	1.0″	191.67
VIII	1400	1.0″	223.61
IX	1600	1.0″	255.55
X	1800	1.0″	287.44

Looking back at the results of case 1 and case 2, we can refer to the curves and predict the required air velocity for different sticker thickness and effective package widths. The following are examples for case #1 and case #2:

Case	Species	Thickness	Drying time	EPW	Air velocity	Sticker thickness
1	So. pine	1.75″	19 hours	8′	1087	1.0″ (from model 74)
1	So. pine	1.75″	19 hours	16′	1087	2.0″
1	So. pine	1.75″	19 hours	24′	1087	3.0″
2	Red oak	1.00″	576 hours	8′	440	1.0″ (from model 74)
2	Red oak	1.00″	576 hours	16′	440	2.0″
2	Red oak	1.00″	576 hours	24′	440	3.0″

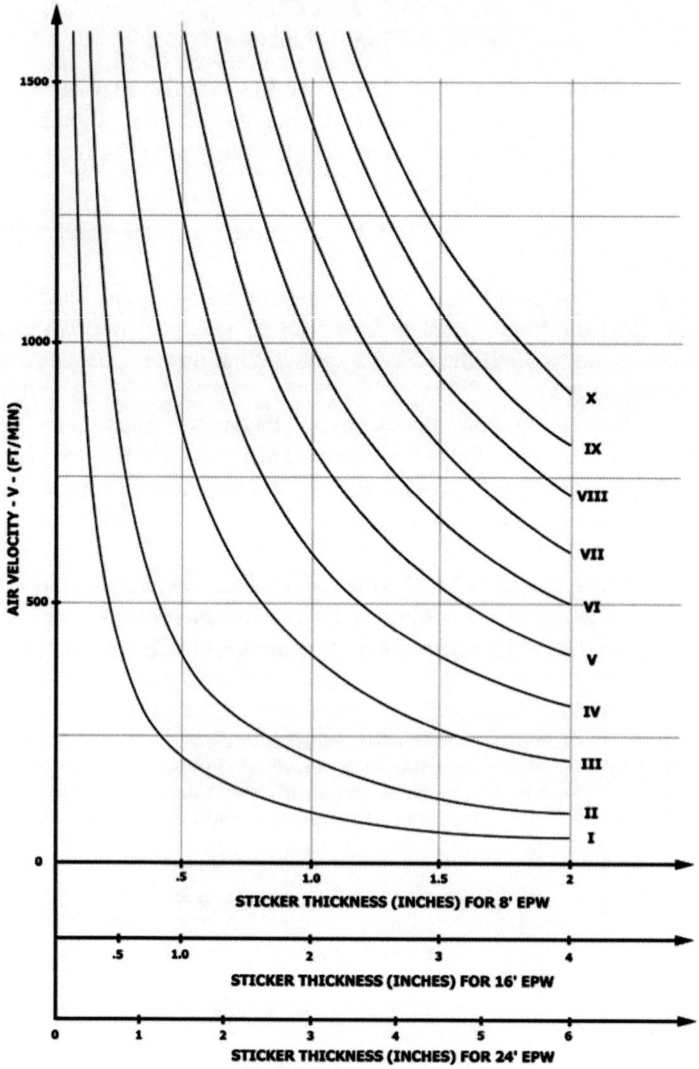

In addition to the preceding benchmark CFM application curves, the table in the following text was prepared to show air velocity for different sticker thickness in each CFM class.

Instructions for Using the Following Air Velocity Table

1 Calculate benchmark CFM for a specific drying application
2 Select the closest or next highest CFM class (vertical column)
3 Move vertically up or down the column to the applicable sticker thickness
4 Read the required air velocity for the drying application

Benchmark Air Velocity Table for Different Sticker Thickness and CFM Class

	CFM CLASS									
	I	*II*	*III*	*IV*	*V*	*VI*	*VII*	*VIII*	*IX*	*X*
CFM	15.97	31.94	63.89	95.83	127.7	159.72	191.67	223.61	255.55	287.44
Sticker	***Benchmark Air Velocity (ft./min)***									
.20″	500	1000	2000							
.30″	333	667	1333	2000						
.40″	250	500	1000	1500	2000					
.50″	200	400	800	1200	1600	2000				
.60″	167	333	667	1000	1333	1667	2000			
.75″	133	267	533	800	1067	1333	1600	1867		
.80″	125	250	500	750	1000	1250	1500	1750	2000	
.90″	111	222	444	667	889	1111	1333	1556	1778	2000
1.00″	**100**	**200**	**400**	**600**	**800**	**1000**	**1200**	**1400**	**1600**	**1800**
1.10″	91	182	364	545	728	909	1091	1273	1455	1636
1.25″	80	160	320	480	640	800	960	1120	1280	1440
1.50″	67	133	267	400	533	667	800	933	1067	1200
1.75″	57	114	229	343	457	571	686	800	914	1028
2.00″	50	100	200	300	400	500	600	700	800	900

Temperature and Pressure Corrections for Benchmark CFM

When air or water vapor heats up or cools down, the fluid medium's thermal and fluid properties change. When the atmospheric pressure changes, these properties also change. These changes will impact the drying capacity of a fluid stream passing through a sticker opening.

The drying capacity of an air/moisture fluid stream passing through a sticker opening is directly affected by the multiple of fluid density and specific heat. This term is the heat-capacity factor.

$$\text{Heat}-\text{capacity factor}=\left(p\times \text{Cp}\right)$$

where:
fluid density (p)
fluid specific heat (Cp)

The drying capacity of an air/moisture fluid stream passing through a sticker opening is directly affected by the atmospheric pressure at the geographic location of the dryer.

BTU/hour heating capacity is proportional to atmospheric pressure.
where:
Atmospheric pressure (P_{atm})

Fluid stream density is directly proportional to atmospheric pressure. As the atmospheric pressure increases, the fluid density increases. As the atmospheric pressure drops, the fluid density drops.

Pressure (psia)	Air density (#/cu ft)
14.7	.075
13.7	.070
12.7	.065
11.7	.060
10.7	.055

Fluid stream density is inversely proportional to fluid temperature. As the temperature increases, the fluid density decreases.

Fluid	Temperature (F)	Density *p* #/cuft
Air	70	.075
Air	212	.059
Steam	212	.037

Although the effect is small, changes in fluid stream temperature effects the specific heat of fluids.

fluid	temperature (F)	specific heat (Cp)
Air	70	.240
Air	212	.241
Steam	212	.445

Heat-capacity factors are shown in the following text for air and moisture fluid streams passing through a sticker opening. Temperature and atmospheric pressure affect both air and steam as shown.

	Air temperature F		Steam temperature F
Pressure (psia)	70	212	212
14.7	0.0180	0.0142	0.0165
13.7	0.0168	0.0132	0.0154
12.7	0.0156	0.0123	0.0142
11.7	0.0143	0.0113	0.0131
10.7	0.0131	0.0103	0.0120

To calculate the **combined** heat capacity factor for an air stream mixture, you need to know the dry-bulb stream temperature and the percentages of air and steam in the mixture. Refer to ASHRAE books and psychometric tables for accurate data and additional methods of analysis.

Integral Phase Analysis of Minimum Required CFM During Drying
Since most lumber drying occurs at temperatures above 70 F and less than one standard atmosphere, the heat capacity factors during actual drying conditions are always lower than the benchmark value of 0.018.

Because of this, the dryer designer should increase the benchmark CFM predicted by model 74 to a new **design** CFM to ensure quality drying. The design CFM is the CFM at which the fan system should be capable of producing during the **phase** of the drying schedule when the **highest** CFM demand exists.

Typically, for hardwoods, this phase occurs during the earliest period of drying. In high-temperature softwood drying, the peak demand for CFM occurs during the first half of the drying cycle.

For every type of wood drying application, the dryer designer should perform a series of model 74 calculations for the minimum required CFM for each phase of drying.

Once the benchmark CFM has been calculated from model 74, the actual design CFM for the dryer's fan system has to be increased according to the ratio of benchmark-to-design heat capacity factors.

Design CFM = Benchmark CFM × (0.0180/actual heat-capacity factor)

This part of the dryer design involves fan curves, dryer system curves, fan laws, stability analysis, and variable-speed controls. Only an engineer familiar with wood dryers, fan designs, fan laws, and stability analysis should be used for this part of the project.

Do not let people who do not understand these engineering principles design the dryer fans or their controls. They will do it wrong, and you will pay dearly for their mistakes for decades to come. It could cost you many hundreds of thousands of dollars during the first decade the dryer is in operation.

I recommend each design engineer develop their own computer software for model 74 calculations. Using this method, data can be input quickly. Then sit back and let the computer do all the work.

11.12 Model 74 Fan Speed Control Curves

Model 74 includes variables that determine the **minimum** required CFM for successful lumber drying. The model's output CFM value is based on both static and dynamic variables. Static variables are those that do not change significantly from kiln charge to kiln charge.

Wood specific gravity
Sticker thickness
Peak ratio
Transient flow factor
Baffling efficiency
Dryer loading efficiency
Temperature depression in constant-setting schedules

Dynamic variables are those that change significantly during each charge.

Temperature depression in hardwood drying
Drying rates in constant-setting schedules
Lumber moisture content

For these variables, both sensors and statistical routines can be used to collect data in drying operations. Once the data is determined to be reliable, it is then input into the model for specifying a sticker CFM. The computer will then vary the speed of the fans to reach the specified CFM for those conditions. Using this approach, the speed of the fans can be changed **automatically** during drying.

Example: Temperature Depression Changes While Drying Red Oak
In case 2 shown earlier for 4/4 red oak, a depression of 2 degrees F was used during the earliest stage of drying. The model calculated a recommended air velocity of 440 ft./min. for a depression of 2 degrees F. As the lumber dries and the depression is increased, it is a simple matter of inputting the new depression value into the model and thus reduce the fan speed. Upper and lower fan speed limit curves will have to be employed in the control logic to prevent overshoot.

Example: Drying Rates in Constant Setting Schedules
Since drying rates fall off exponentially in fixed temperature schedules, collecting data on energy input rates per board foot is a simple matter for a computer control system. Once sufficient statistical data has been collected on a specific dryer with a specific type of lumber, it is a simple matter to build a projected decay curve for moisture content and drying rates. From these curves, a second fan speed curve could be produced to adjust fan speeds automatically. The system could also look at temperature drops across loads for trimming. Upper and lower limit curves will also have to be used.

In addition to employing model 74, dryer plenums can be fitted with differential-pressure transmitters to monitor static pressure drops across packages. Since the CFM through a sticker opening is determined by the static pressure between plenums, a subroutine can be employed to both check and calibrate the model. If this self-checking method is used, the calibration event should occur at the start of each kiln charge when a known dry-bulb temperature is present. The fans should be running in the same direction during each calibration event to minimize transmitter-induced errors.

Once the model's self-test has been completed, the computer produces a report for the dryer operator alerting him to the integrity of the tests. If the integrity meets the minimum requirements, the fan speed controls will be enabled. If the self-test fails, the automatic fan speed controls will be left out during the drying of that charge of lumber.

In the future, expect to see the preceding and similar engineering technologies employed in all types of lumber drying. Models like 74 and many others will be used in computer systems to both improve dryer production and reduce degrade. Engineers and computers will change the way all lumber is dried around the globe.

11.13 Moisture Content Control with Stress Control

When wood dries below its fiber saturation point, shrinkage occurs. Depending on the species, the board thickness, the schedule used, and the level of moisture gradients created by drying, significant permanent damage to the lumber can occur. Numerous drying schedules have been developed during the last century to minimize drying degrade. The following are types of degrade seen in wood drying:

Type 1 – Radial and tangential defects

A. Tension defects (checks) – Checks are caused by fiber separation due to excessive tension created by excessive drying rates and too low EMC. These involve surface checks, internal checks (honeycomb), and end checks at the ends of logs or boards. End checks can be reduced or eliminated by coating the ends of the wood with moisture barriers. These coatings do not work in high-temp. drying.
B. Compression defects – Caused by cell wall failures during drying.

Type 2 – Longitudinal defects

A. Warp – Differential longitudinal shrinkage causing dimensional changes in the edge direction.
B. Bow – Differential longitudinal shrinkage causing dimensional changes in the face direction.
C. Twist – Differential circular shrinkage causing dimensional changes in rotation.

Type 3 – Tree-growth-induced defects

A. Shakes – Large checks caused by residual tree-growth-induced stress.
B. Loose knots – Separation of knots during drying.

Depending on the species and board thickness, the proper drying schedule should be chosen. The kiln operator's experience, dedication to detail, and the use of proven drying schedules are paramount to minimizing degrade. The failure to follow a highly disciplined approach can lead to significant lumber degrade and economic losses. Equalization (moisture control) and conditioning (stress relief) may be needed at the end of the drying cycle. Drying schedules and practices for minimizing grade losses can be found in the following books published by the Forest

Products Society, 2801 Marshall Court, Madison, WI 53705-2295. www.forest-prod.org.

Dry Kiln Operator's Manual
Dry Kiln Schedules for Commercial Woods: Temperate and Tropical

11.14 Moisture Movement in Lumber Storage Facilities

Storage facilities are needed to protect dried lumber from harsh degrading environmental conditions. Dry storage facilities are required for both the protection and tempering of kiln-dried lumber. The following is an overview of these systems.

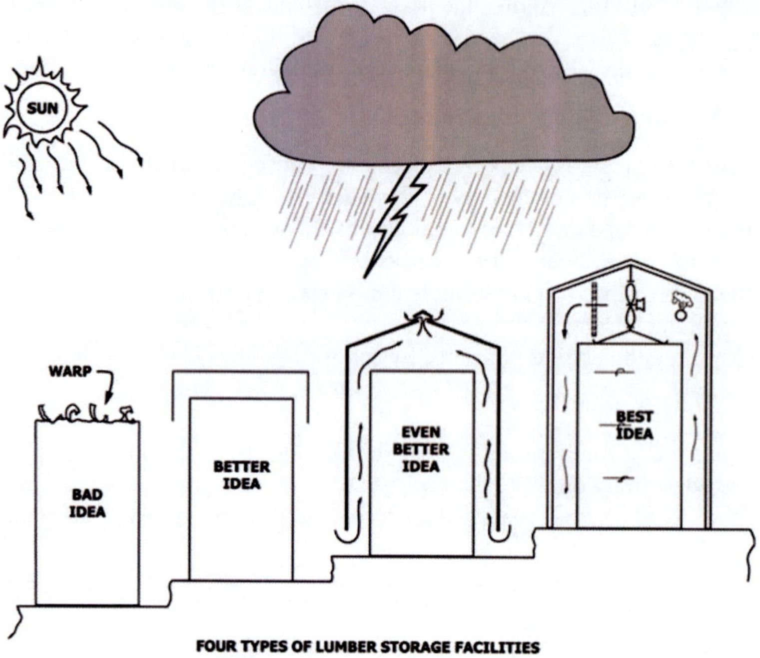

FOUR TYPES OF LUMBER STORAGE FACILITIES

A. **Open-Storage Space Involves the Following**

1. **Unprotected Storage**

   This practice should be avoided. Dried lumber has a high affinity for water and can suffer extensive degrade if left unprotected in an open yard. If moisture level grading practices are in place, considerable expenses will be involved in re-drying lumber exposed to rain or snow. If exposed to direct sunlight, the top courses of lumber packages can incur significant warp and checks.

2. **Shed Storage for Rain, Snow, and Sunlight Protection**
   This practice only protects the lumber from rain, snow, and sunlight. If the lumber is placed next to the edges of the shed, it can adsorb considerable moisture during blowing rainstorms. The edges of sheds should be fitted with skirts to protect the lumber.

B. **Closed-Storage Space (Buildings)**

1. **Temperature and Humidity Not Controlled**
   This is just a step above A.2. The lumber will still be exposed to the equilibrium moisture content (EMC) of the outside weather. This can lead to excessive loss or regain of moisture.
2. **Temperature and Humidity Controlled (in an Insulated Building)**
   This is the best practice for kiln-dried lumber storage. In many cases, small heaters, humidifiers, or dehumidifiers can be installed to control the EMC in the building. This is accepted practice for furniture manufacturing plants.

Capacity Requirements
The importance of adequate dry storage capacity cannot be overstated. There should always be ample storage capacity to meet the maximum expected weekly production level of the plant. Softwood plants maintain at least 1 week of inventory between the kilns and the dry finishing area.

A rule of thumb for sizing dry storage capacity is multiply what the accountants tell you by 1.5 and you may have enough capacity. Later on you will discover that you should have doubled the original figure.

11.15 Pre-drying Conditions and Equipment Needed for Successful Lumber Drying

Log Storage and Rotation
Too much log storage or improper storage techniques can lead to significant problems in lumber drying. If the logs are stored for long periods of time, they can suffer end checking and significant loss of moisture. Water sprinklers can be used to prevent drying and stain. The storage system should be designed to rotate the log inventory as quickly as possible. Rotation is required to prevent old logs being mixed with new logs entering the sawmill.

Sawing of Green Logs
Board thickness, reduction of board thickness variations, and board-sawing solutions (within the logs) will dictate which drying system should be used. If sawed boards with a wide variance in board thickness enter the lumber dryers, the final moisture content (after drying) will also have a variance due to just this one factor. The ratio of board width to board thickness (beta ratio) will also dictate which schedule to use.

Sorting of Green Sawed Products for Successful Lumber Drying

Sawed boards must be sorted by products for successful lumber drying.

A green product is a unique combination of species, grade, thickness, width, and length. Example – A southern pine 1 × 4 × 16′ board is one product.

In a southern pine mill that produces 2″ boards that vary in width for 4″ to 12″ and lengths that vary from 8′ to 24 × , there are a total of 1 × 5 × 9 = 45 different products.

Green Lumber Stacking Ahead of Lumber Kilns

There are four separate practices used in stacking green lumber for drying, each one of which has its pros and cons.

1. **Package course control** – A course is one layer of lumber going across a stacked package.
   (a) Solid course stacking (most common method) – Boards are jammed against each other.
   (b) Air space stacking (for slow drying stock) – Allows air to pass between board edges.
2. **Package height control** – Package height is the total top to bottom size of the package.
   (a) Full course packages (**best method** for improving plant profits and energy efficiency)
   (b) Multiple-packages with cross-outs (causes significant problems in kiln drying) – The number of separate packages stacked vertically inside a side loader kiln can vary from 3 to 4 to 5, or onto trams ahead of track kilns can vary from 2 to 3 to 4 to 5. If this type of package stacking and kiln loading system is used, the kiln's performance and energy efficiency will be degraded significantly because of the large amount of heated air flow that bypasses the tight stickered packages of lumber. Thus, when using the multi-package method, the green stacked package height is kept small (3′ to 6′) to allow fork trucks to be able to lift, transport, and load the kilns for drying the packages. – Furthermore, even though large fork trucks or straddle carriers can be built to handle tall full course green packages of lumber, the cost of these large heavy mobile machines and the many safety and handling issues with this package handling method is both difficult and dangerous. Additionally, large fork trucks and straddle carriers and the required level surfaces they operate on are expensive and costly to maintain especially in extreme weather conditions.

Up until 2012, when counterflow jogging batch softwood kilns first appeared in the USA, full course stacking was common in many US lumber manufacturing operations. Then once the counterflow kilns appeared, every mill using these kilns were forced to go to the multi-package type of handling system. The reason being, this was the only way to both load and unload the packages off the trams at each end of these kilns. One unfortunate result of this was the high cost of the additional fan electrical energy required to keep sticker air velocity at a level required to achieve the required drying production rates.

Air space stacking is common practice in the drying of hardwoods and large timbers.

All modern softwood plants should use full course stacking and handling equipment to reduce degrade, reduce handling costs, and maximize kiln production and energy efficiency. Older softwood operations, treating operations, and most hardwood operations use multiple-packages with cross-outs.

The use of multiple-packages with cross-outs in high-temp kilns will cause considerable moisture variances and degrade losses. If multiple cross outs are used, the use of equalization cycles following kiln drying is warranted. Warning: If you use E&C cycles in high-temperature dryers, expect to see massive corrosion of mild steel building components and heating equipment.

Stacking Stick Thickness

Stacking stick thickness has a strong effect on the following:

1. Drying production capacity
2. Drying time
3. Drying uniformity

Drying production analysis (due to stick thickness) is determined from a series of equations involving heat transfer mechanics, fluid mechanics, fan system design, and package grid analysis. Drying production analysis predicts three optimum sticker thickness sizes:

1. The optimum thickness for dryer production
2. The optimum thickness for degrade control
3. The optimum economic thickness (determined from items 1 and 2)

The following two curves demonstrate holding capacity and dryer production rate in a softwood track kiln. – Field tests are required to plot these curves accurately for a specific kiln design operation.

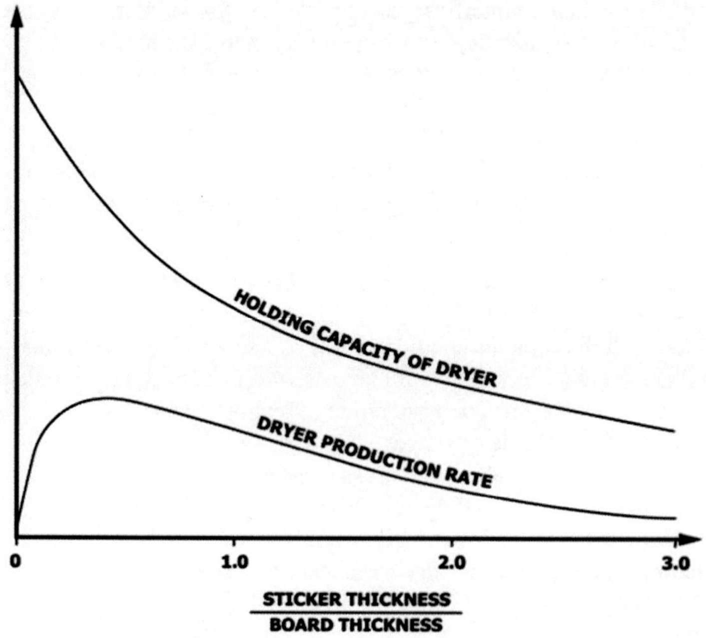

Stacking Efficiency

Stacking efficiency is the ratio of actual to maximum possible board footage of lumber that can be put in a lumber kiln. Imagine a track kiln being loaded in such a manner that there are no open spaces between the ends of the packages, no boards left out of a package, no air spaces between the courses (due to cross outs), and the doors at the end of the track kiln can barely be closed. Also imagine every sticker and every board in the kiln being exactly the size they were supposed to be. In such a situation, the kiln is fully loaded and as near to stacking and loading perfection as possible.

In such a case, we could say that the stacking efficiency "at the kiln" is 100%. Let us suppose that under these conditions, the nominal board footage in the kiln is 150,000. Now imagine the same kiln being loaded in such a manner that only 135,500 board feet is in the kiln. In this situation, the stacking efficiency at the kiln is 90.3%:

$$\text{STACKING EFFICIENCY} = (135{,}500 / 150{,}000) \times 100\%$$

$$\text{STACKING EFFICIENCY} = 90.3\%$$

For modern pine operations, using track kilns over 68′ long, stacking efficiencies above 95% are not uncommon. For most side-loader (package) kilns, stacking efficiencies of 40–60% are common due to poor stacking and loading practices.

High kiln stacking efficiency is mandatory for increasing lumber dying energy efficiency.

Stacker-to-Kiln Inventory Control

Once lumber is stacked in packages with stickers, significant air-drying can occur before the packages enter the dryer. If exposed to the elements (wind, sun, rain, and temperature extremes), the lumber will react accordingly. Significant amounts of drying and degrade can occur if the exposure time is excessive. For this reason, totally enclosed pre-dryers may be installed to remove moisture and protect the wood from degrade.

This is a common practice in hardwoods. For some types of softwoods, both air-drying and pre-drying may be used. In some pine operations, air-drying is still in use, but is becoming rare. For high-temp drying, mixing pre-dried lumber with fresh lumber from the sawmill causes moisture and degrade problems during drying.

The use of equalization and conditioning cycles after drying is justified if moisture variances are too great. This is a common practice with Australian pines.

11.16 Loading the Dryer

11.16.1 Loading a Pre-dryer

Due to the large holding capacity of pre-dryer systems, inventory rotation and moisture management is a significant issue. A records system should be developed in which the history of each package is tracked within and through the pre-dryer. The reason is (usually) very little HVAC or CFD analysis goes into these drying systems, levels of inventory can change dramatically, and their drying performance is usually erratic.

First, a pre-dryer is defined to be any "system of drying" located upstream of a commercial dryer. This includes the following classes and types of pre-drying systems:

Class I – Pre-drying Without Fans, Heating, or Dehumidification Systems

Five Basic Types

Air-drying open yards – common in hardwood operations
Air-drying sheds with only roofs – side-loader types
Air-drying sheds with only roofs – track types
Air-drying sheds with roofs and side skirts – side-loader types
Air-drying sheds with roofs, and side skirts – track types

Class II – Pre-drying with Fans, Heating, or Dehumidification Systems

Eight Basic Types

Air-drying sheds with roofs and fans – side-loader types
Air-drying sheds with roofs and fans – track type

Air-drying sheds with roofs, side skirts, and fans – side-loader type
Air-drying sheds with roofs, side skirts, and fans – track type
Pre-drying sheds with roofs, side skirts, heat, and fans – side-loader type
Pre-drying sheds with roofs, side skirts, heat, and fans – track type
Pre-drying buildings with fans, and dehumidification systems – side-loader type
Pre-drying buildings with fans, heat, and dehumidification systems – side-loader type

Air circulation inside air-drying sheds is usually crossflow, and possibly dependent on prevailing winds.

Air circulation inside pre-drying buildings can be downflow, crossflow, or a combination of the two. Some systems include a building end-to-end return-air system.

If you survey the many pre-dryers in just the USA, you will find everything imaginable from old metal warehouses with no building insulation to modern computerized inventory control systems.

And, because virtually every pre-dryer on planet Earth is different, you are on your own when tackling this problem. Studying this book front to back is the best way to tackle your pre-dryer problems. You will also discover that most people around the planet do not have a clue to what is happening inside their pre-dryers. Note how lost the fellow shown below seems to be as he stares at the stacks of lumber inside his pre-dryer.

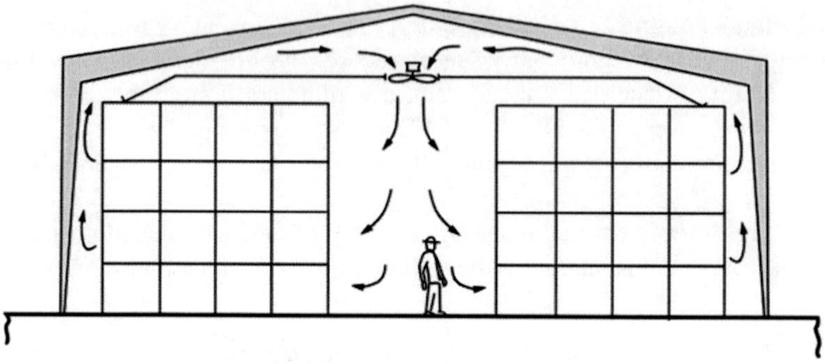

11.16.2 Loading a Lumber Kiln

The first objective in loading a lumber kiln is separation by board thickness and width. Sufficient inventory should be kept ahead of the kiln to allow separation. The second objective is to achieve 100% stacking efficiency. Although this level of stacking efficiency is difficult to do consistently in commercial drying operations, it is a tool by which management can set goals for stacking and drying operations. Management should implement stacking-efficiency quality-control charts to reduce drying losses.

The failure to load kilns properly will cause the following:

1. Increased incidence of moisture control problems
2. Increased degrade
3. Loss of drying capacity
4. Increased energy usage per board foot
5. Increased risk of fires in direct-fired kilns

11.17 Restraint Drying

The application of restraining forces to wood will reduce exterior dimensional changes during drying. To be successful, the restraining forces must be maintained throughout the entire periods of drying, equalization, conditioning, and cooling. Wood creep dynamics play a key role during restraint drying. Top forces on the package can be applied by concrete pads, steel plates, or pressurized mechanical devices that adjust for lumber shrinkage.

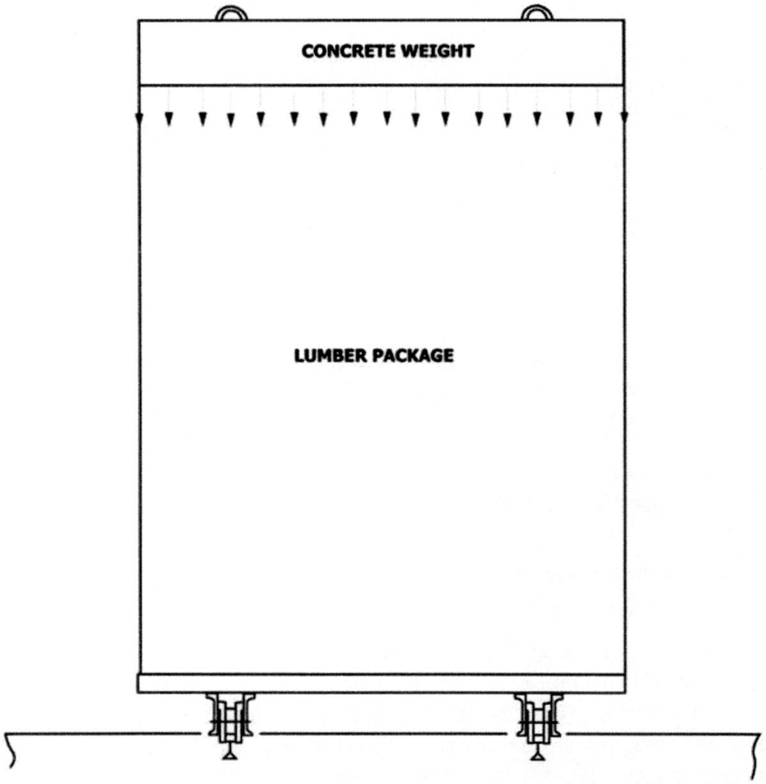

The most common methods used in restraint drying are as follows:

1. **Precision stacking of sorted lumber in tall packages.**

This method uses the weight of the lumber above a course to restrain the course during drying. Using this method, the restraining forces diminish depending on the vertical location in the package. The bottom course has the greatest and the top course has no restraint.

2. **The use of heavy weights resting on top of the lumber packages.**

This method uses weights loaded on the top layer to provide additional restraining forces on the package. The required amount of weight per square foot of lumber course area varies depending on the lumber species, board thickness, and final target moisture content.

The benefits of restraint drying will be reduced by:

1. The presence of wide statistical variations in either board thickness or sticker thickness
2. The failure to follow precision sorting and stacking practices

The economics of restraint drying is determined by the following:

1. The total costs of loading and unloading the weights
2. The loss of kiln holding capacity due to the presence of the weights
3. The degrade reduction due solely to the weights on the packages

The potential savings in degrade for a specific species and board thickness in a specific kiln can be determined by comparing the amount of degrade in the top three courses to the amount of degrade in the middle three courses. Both degrade and moisture content data should be measured. If degrade is affected by the final moisture content, adjustments must be made for the effect of moisture content.

In commercial drying systems, the benefit of top loading only impacts the top 3–5 courses of lumber. The species, thickness, and final moisture content of the product impacts the economics of top loading.

11.18 Lumber Dryer Buildings

11.18.1 Introduction

Lumber dryer buildings are complex structures that require considerable design and construction expertise only available from people with years of experience in the industry. Any attempt to cut corners in this area of lumber drying will lead to failures and losses that in many cases cannot be corrected without major capital outlays and lost drying capacity.

Only highly experienced reputable people with proven track records should be used for designing, manufacturing, and installing these structures. Taking shortcuts on dryer building design and construction will lead to significant financial losses.

The following are recommended guidelines for the design, manufacturing, construction, and maintenance of all types of dryer buildings.

11.18.2 Safety and Code Requirements

1. Follow the local applicable building codes

A. **Structural Requirements of Local Building Codes**

Wind Loads

This is a major factor in the design of dryer buildings. The following Fujita scale demonstrates expected wind speeds.

Fujita scale	**mph**
F2 – significant	113–157
F1 – moderate	73–112
F0 – gale	40–72

Snow Loads

Snow loads can reach 100 pounds per square foot of roof area in northern climates.

Earthquakes

In some locations, building codes include design criteria for earthquakes.

B. **Allowances for Corrosion**

Doors, door hardware, door fronts, lintel beams, and column bases are especially prone to corrosion due to persistent problems with condensation of moisture in these areas. This is a significant problem in wood dryers for corrosive species, and if humidity sprays are used. In some cases, the use of mild steel structures should be ruled out due to the inherent problems with corrosion.

Aluminum is for many types of drying a superior choice of building materials.

C. **Allowances for Thermal Expansion and Contraction**

Cyclical thermal expansion and contraction of structural members in wood dryers require a sufficient number of expansion and slip joints be designed into the structure.

D. **Design Review and Approval**

The final structural design should be reviewed by a licensed structural engineer to ensure code compliance for the location where the dryer will be built.

2. **Comply with the Latest OSHA Safety Requirements for Lumber Kilns**

Minimum Safety Clearance Requirements for Personnel

There should be no less than 18″ clearance between moving packages in track kilns and stationary parts of the kiln. Protruding package stickers, steam and condensate lines, steam traps, door columns, baffles, and baffle posts should be designed such that no less than 18″ of clearance exists while packages are being moved through the dryer.

Personnel Escape Doors

There should be adequate personnel escape doors fitted to areas accessed by kiln operators and maintenance personnel. All escape doors should be fitted with latches such that the door can be opened from either side of the door. Each floor level dryer plenum area should have no less than one escape door. If the dryer has baffled temperature heating zones, each zone must have an escape door. Fan decks should be fitted with escape doors. Prepare formal safety instructions for dryer operator and maintenance personnel so that there is no confusion about which areas of a dryer are required to have personnel escape doors including how many people must be present before any person can enter a dryer.

Main Kiln Door Safety Guards

Safety guards should be installed above main doors to prevent doors from falling onto personnel.

Safety Guardrails

If permanent stairways or ladders are installed for access to the dryer roof, the roof
should be fitted with OSHA safety rails.

3. Fire and Explosion Protection

The National Fire Protection Association (NFPA) is the nationally recognized organization for fire protection standards. In many jurisdictions, the NFPA codes have been adopted as law. The following items should be used in conjunction with the NFPA standards. In some of the items discussed in the following text, the NFPA standards and industry practices overlap.

A. NFPA Codes

1. Sprinkler Systems

   NFPA sprinkler systems should be installed in dryers prone to having fires. This is especially true in both high-temp and direct-fired kilns. If the kiln is direct fired with wood waste, the probability of fires is extremely high.

Only licensed fire protection contractors should be used for the design and installation of sprinkler systems.

2. Fire Hydrants and Hoses

   Each kiln should have NFPA-approved fire hydrants and hoses in close proximity to the kiln.

3. Furnaces and Burners

   Each furnace must be designed, manufactured, installed, maintained, and inspected per NFPA and all other applicable local safety standards. Records of inspections should be maintained and kept in a secure fire-proof vault.

B. Industry Standards and Practices

1. High-temp safety limits with alarms should be installed on each kiln. The safety limit should be factory-set to trip at maximum kiln temperature plus 50 degrees F. Review the specifications for these limits with your property insurance carrier.
2. High-temp safety limits and controllers should be installed on heat supply ducts. The duct controller should be set at 400 degrees F. The high-temp safety limit should be set at 450 degrees F. The return air duct should have a safety limit set at maximum kiln temperature plus 50 degrees F.

Warning: If dry-bulb temperatures above 280 F enters packages of lumber, the risk of fires increases dramatically. It is highly recommended that dry-bulb controller set points for the temperature entering lumber never be set above 240 F. If exit-air temperature control schemes are used, the dry-bulb controller set point should never be above 220 F. Although exit-air control schemes will reduce drying times, such practices can start fires and require very accurate temperature measurement and controls. Any upset in the temperature loop can cause excessively high dryer temperatures that could cause a fire. Consult with your property insurance carrier before using exit-air control schemes in a lumber dryer. If an upper limit is to be set for entering air temperatures when using exit-air control schemes, a second control loop will be required. Setting this upper controller at 250 F will allow the exit-air scheme to operate while minimizing the risk of fires. Have your property insurance carrier approve each of these setpoints.

3. Center heat risers should not come within 24″ of lumber packages.
4. Heat duct surfaces should be insulated if located less than 24″ from lumber packages.
5. Steam lines should not be located within 12″ of wooden structures. Long-term exposure of wood to hot surfaces can cause spontaneous combustion of wood.
6. Personnel training:

   Written procedures should be prepared for fighting fires in dryers.

   A company fire brigade should exist for fighting dryer fires.

   Coordination with public fire departments should be established.

7. Hazards of flashback explosions when opening dryer doors:

Flashback explosions can occur when dryer doors are opened in the presence of fires.

These explosions are caused by a sudden rush of oxygen into an oxygen-starved fire. The result is an explosion that can blow the dryer doors open and/or dislodge them from their supports. Serious injuries to people can occur. During dryer fires, large dryer doors should only be opened by using heavy equipment.

8. Remote load baffle operators:

   In kilns prone to fires, remote load baffle operators should be installed to raise the ceiling baffles off the lumber packages so the packages can be removed without damaging the baffles.

9. Pushing loads out of a track kiln during a fire:

   Written procedures should be prepared for pushing lumber packages out of a track kiln during fires. The procedures should include instructions on how to raise the ceiling baffles, open the doors, and push the packages out of the kiln. Review these procedures with your local fire department.

11.18.3 Dimensional Requirements of Dryer Buildings

The correct selection of building size is crucial to successful dryer operations. The following practices should be used during the planning stage for new dryers and dryer upgrades.

11.18.3.1 Holding Capacity

The holding capacity of wood dryers is affected by many variables, each of which should be understood to get the most holding and drying capacity out of a dryer. Each of the variables listed in the following text impact the economic performance of a dryer in ways that may not be readily apparent. For purposes of clarity, a review of common terms used in the industry is given.

Actual board footage – This is a volume of wood that is 1″ × 12″ × 12″, or 1/12 of a cubic foot.

Nominal board footage – This is a term that is "by name only." Nominal board footage is different from actual board footage. In softwoods, a finished 1.5″ × 3.5″ × 12″ long board is sold as .6666 nominal board feet, when in fact it is .4375 actual board feet. In hardwoods, lumber is sold based on the thickness in 1/8″. Hardwoods are sold as 4/4, 5/4, 6/4, 7/8, etc., when in fact in the rough condition, they are often of a thickness up to 15% over their marketed thickness. As far as a wood dryer is concerned (a fixed volumetric chamber), the use of nominal board footage holding capacity can be misleading. Wood dryers dry actual board footages of wood. Both the heating systems and fan systems must be designed to remove a certain amount of water from a given package space and both are determined by the actual board footage, not the nominal board footage.

MBF = 1000 board feet.

Board Thickness

The actual board thickness of lumber entering a dryer varies depending on sawing techniques and quality control practices. In modern softwood operations, the green board thickness can be controlled to within several thousands of an inch. In hardwoods, the variance can be as much as 15% of the target thickness. Dryer performance is diminished by excessive board thickness or wide variations in board thickness within a load of lumber.

Sticker Thickness

Target sticker thickness can vary from .5″ to 3.5″ depending on the drying operation. In some operations, laminated 5/8″ stickers are used with very little variance in thickness. In some drying operations, 1.5″ thick dressed lumber is used for stickers. In softwood lumber operations target sticker thickness can vary from .687″ to 1.5″ depending on the design of the fan system, the baffle systems, the species, the board thickness, and the type of drying schedule used. Typically, select hardwoods, steel, or aluminum are used for durability.

In some operations, strips of plywood are used for stickers. In high-temp kilns, considerable loss of strength can occur following several passes through the kiln, causing the stickers to become brittle. This causes them to eventually break or split, especially near knots. It is common to find wood sticker thickness to have a variance of 20% of their target thickness, causing moisture control problems during drying. Variation in sticker thickness or inadequate sticker thickness has a significant degrading impact on lumber dryers.

The maintenance of uniform stickers of sufficient thickness is paramount to quality drying. In many drying operations, the erroneous decision is often made to use thin stickers to gain additional holding capacity in the dryer, and/or more uniform air velocity. Unless the dryer is designed for the thin stickers, significant problems with drying time and drying quality can be created by using thin stickers.

Nominal Package Width and Actual Package Width

Package widths can vary from 4′ to 12′ depending on the lumber stacking and handling equipment. The **actual** package width is a **multiple** of the actual board width.

Example: An 8′ wide package has 23 boards across the package. Each board is 4″ wide. The actual package width is 23 times .3333′ = 7.66′. In this example, the nominal package width is 8′, and the actual package width is 7.66′.

Package Height

Package height can vary from 3′ to 17′ depending on the lumber stacking and handling equipment. The dimension is measured from the bottom of the bottom board to the top of the top board.

Load Height

Load height is the distance from the floor of the dryer to the top of the top package.

Package Cross-out Thickness

This is 3″ to 4″. Most operations use 3.5″ × 3.5″ squares. Some operations use square steel tubing. Some operations use square aluminum tubing. The thickness of the cross-out should be at least ½″ thicker than the thickness of the fork truck blades (at the shank) to prevent the forks from binding between packages.

Safety Warning Do NOT grease or oil the forks on fork trucks to allow the use of cross-outs equal to or thinner than the shanks of the fork truck. This practice is dangerous. This could cause a package of lumber to suddenly slip off the forks while the truck is moving.

Package Length

Packages of stacked lumber are longer than the nominal length of the lumber in the packages. If 16′ long 2 × 4s are stacked into a 15′ tall package, the ends of the boards will not all be even, nor will they be perfectly square with the bottom layer of the package. Several things will occur at one time. First, the 16′ boards are probably trimmed to 16′–2″. Next, the boards are not all even ended by the stacker to form smooth package ends. In addition, the stacker's platform will not be square with the platform's lift cables or hydraulic cylinder. Next, the tracks going through the kiln will not all be level.

The impact of these factors is that the 16′ package can take up to 16′–8″ of kiln space once it is loaded into the kiln. The loss of kiln capacity alone due to poor stacking and loading of 16′ lumber can be significant. The cumulative impact of these manifests itself at the kiln doors. If a track kiln is designed to hold five 16′ packages on one track, the nominal package length is 5 × 16′ = 80′. The actual required kiln length will be 5 × 16.666 = 83.33′. Of this value, 2″ per package length is lost in trimming. The remaining net loss is 2.5′ of kiln holding capacity. Over one year's time, these factors will cause a considerable loss of kiln capacity and unnecessary costs.

Stacking Efficiency

Stacking efficiency refers to the actual versus maximum possible loading capacity of a dryer. In modern softwood operations using track kilns, stacking efficiency varies between 94% and 97%. In most package (side loader) kilns, stacking efficiency varies from 40% to 60%.

11.18.3.2 Pre-dryers

Required Holding Capacity

The drying capacity of pre-dryers should be based on the maximum expected initial moisture content of the lumber and the most conservative drying rate. These two conditions occur during the coldest months of the year. The holding capacity (board footage) of pre-dryers are determined by board thickness, sticker thickness, package width, total package height, ceiling height, building width, building length, and stacking efficiency factors.

Plenum Width

Plenums serve two purposes in dryers. One is to provide sufficient space by which circulated air is allowed to collect prior to entering the packages of lumber. The other purpose is to provide a sufficient space for maintenance personnel. Plenum width should never be less than 4′ in any kind of commercial dryer. The following chart should be used for sizing the minimum allowable width of plenums in all types of lumber dryers.

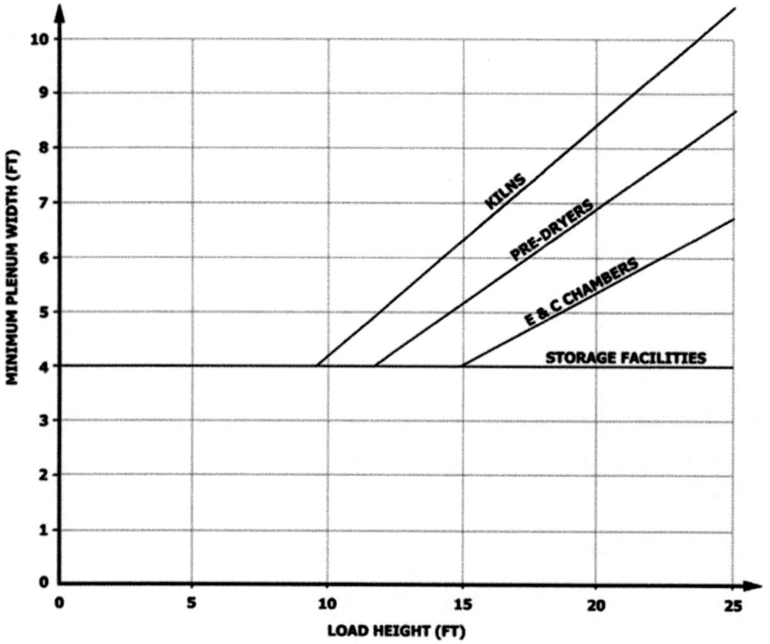

Maximum Effective Package Width (**EPW**)

Effective package widths more than 16′ should be avoided unless the sticker thickness complies with model 74 discussed earlier.

Depending on the species, the thickness of the lumber, and the thickness of the sticker, effective package width can vary significantly. Only by objective testing of model 74 in a specific dryer can the maximum effective package width be determined reliably.

Package Handling Criteria

The pre-dryer should have sufficient workspace for handling the largest package of lumber.

The door height and width should accommodate the largest expected package.

The aisles for fork trucks should accommodate the largest package including the turning of fork trucks within the aisles.

The height of the fan deck and roof support structures should be sufficient to accommodate the handling of packages, including the highest point on the mast of the fork truck.

11.18.3.3 Track Kilns

Holding Capacity

The holding capacity of track kilns is determined by board thickness, sticker thickness, package width, total package height, the number of tracks, kiln length, and stacking efficiency factors.

Kiln Width

Kiln width is defined as the total exterior width of the building. Kiln width is calculated by adding the following dimensions:

Total plenum widths
Total load widths
Total width of the reheat equipment between tracks
Total OSHA clearances
Total wall thickness (both walls)

Example: A high-temp double-track direct-fired kiln with 8′ package widths, 3″ thick wall panels, and a 15′ door height. The center heat riser is 2′ thick.

Total wall thickness = 2 × .25′ = .5′
Total plenum widths = 2 × 7′ = 14′
Total load widths = 2 × 8′ = 16′
Total reheat space = 2′ + 2′ + 2′ = 6′
Total outer width of the building = 36.5′

Kiln Length

The definition of kiln length (for track kilns) is the measured door-to-door length (of the kiln) from the inside of one door to the inside of the opposite end door. This value should be determined by the optimum best fit calculated from the statistical distribution of the actual package lengths that will be loaded into the kiln. For each batch-type lumber mill, there is one optimum kiln length determined by the mix of lumber lengths to be kiln dried.

Quite often the optimum kiln length does not match the standard lengths offered by dryer manufacturers. Some kiln manufacturing companies have standard design lengths such as 42′, 54′, 68′, 84′, 104′, and 120′. Some manufacturers offer different standard lengths. Typically, track kiln design lengths are dictated by the width of the prefabricated wall panels (usually about 4′) and the standard door front designs. Failure to custom design a track kiln to the optimum length at a specific lumber mill will result in persistent lost holding capacity, lost drying capacity, excessive bypass CFM around the lumber packages, wasted electrical fan horsepower, and increased moisture control problems.

Determining the Optimum Track Kiln Length

For each lumber manufacturing operation, there is a range of product lengths that a particular mill produces. Although this range does not change, the statistical mix does. From this distribution, a weighted average package length can be determined. Once this value has been determined, the length of the track kiln should reflect a multiple (integer) of the value.

In the following table, example variations in actual package lengths are shown for 8′ and 16′ packages. Such a table should be constructed for the specific average package length that the kiln will be required to hold.

Example Track Kiln Load Length in Feet

PL	**# PL**									
	1	2	3	4	5	6	7	8	9	10
8.000	8.000	16.00	24.00	32.00	40.00	48.00	56.00	64.00	72.00	80.00
8.125	8.125	16.25	24.37	32.50	40.62	48.75	56.87	65.00	73.25	81.25
8.250	8.250	16.50	24.75	33.00	41.25	49.50	57.75	66.00	74.25	82.50
8.375	8.375	16.75	25.12	33.50	41.87	50.25	58.62	67.00	75.87	83.75
8.500	8.500	17.00	25.5	34.00	42.50	51.00	59.50	68.00	76.50	85.00
16.000	16.00	32.00	48.00	64.00	80.00	96.00	112.00	128.00	144.00	160.00
16.125	16.12	32.25	48.37	64.50	80.62	96.75	112.87	129.00	145.12	161.25
16.250	16.25	32.50	48.75	65.00	81.25	97.50	113.37	130.00	146.25	162.50
16.375	16.37	32.75	49.12	65.50	81.87	98.25	114.62	131.00	147.37	163.75
16.500	16.50	37.00	49.50	66.00	82.50	99.00	115.50	132.00	148.50	165.00

To determine PL (the average package length), I suggest that four packages be loaded tight on a track and the overall length measured. Then divide the total measurement by four to get the average.

Once the true (actual) load length is determined, a determination of which type of door baffle will have to be made. From this, add sufficient additional length to the kiln for the door baffles.

If this analysis is done properly, the kiln length will "match" the plant's stacking practices. One will also discover from such an analysis that long track kilns are easier to load properly than short kilns. This is one of the many reasons why long single-track kilns perform better than short double-track kilns. However, single-track kilns cost more and use more electricity and energy per board foot dried.

Door Height

The definition of door height in a track kiln is the vertical distance from the top of the dryer floor to the bottom of the lintel beam at the door front. Door height involves the following measurements:

1. Vertical height of the tracks above the dryer floor
2. Vertical height of the kiln truck
3. Vertical height of the cross out on the kiln truck
4. Vertical height of the lumber packages
5. Vertical height of the cross outs between packages
6. Standard 6″ safety clearance between the top course of lumber and the bottom of the lintel beam

Example: A modern track kiln is designed for full-course packages containing 75 layers of 1.69″ thick lumber stacked with 1″ thick stickers. What is the door height?

1. Vertical height of the tracks = 3.25″
2. Vertical height of the kiln truck = 5.5″
3. Vertical height of the cross out on the kiln truck = 4″
4. Vertical height of the lumber package = 200.75″

   75 × 1.69″ = 126.75″ – the boards
   74 × 1.0″ = 74″ – the stickers

5. Vertical height of the cross outs between packages = 0″
6. Safety clearance between the top of the packages and the bottom of the lintel beam = 6″

$$\text{Door height} = 3.25 + 5.5 + 4'' + 200.75 + 0'' + 6'' = 219.5, \text{or } 18^{\prime} - 3.5.$$

It is also common to see kiln designs in which the tracks going through the kiln are sloped from the green end of the kiln to the dry end. Tracks are sloped to reduce the force required for moving the heavy packages of lumber on the tracks. In these designs, the standard 6″ safety clearance (item 6) at the dry end of the dryer is considerably more than 6″. The reason is that lumber shrinks during drying and thus has a lower load height than when green. The combined result of sloped rails and dried lumber is that any warped boards on the top course will have ample clearance to not hang up on the lintel beam at the dry end of the kiln.

In many southern pine KD19 dryers, the actual clearance between the top of the package and the green end lintel beam can be as little as 3″. At this high of final moisture content, the amount of warp in the top courses is significantly less than KD15 dryers and thus far less chance for boards to hang up on dry-end lintel beams. If such a small clearance is used, the statistical control of both sticker thickness and board thickness must be carefully controlled.

In addition, the upper and lower travel limit switches on the green lumber stacker lift platform must be maintained accordingly.

Load Height

The term load height is door height minus clearance height.

Economics of Door Height

Door height plays a significant role in the economics of drying operations. In older hardwood and softwood operations it was common to see door heights in track kilns as low as 10′. In modern softwood operations, most stacking is full course height in which the packages are completely handled by transfer cars. In recent track designs, door heights vary from 14′ to 16.5′.

The advantages that modern tall full course designs have are less capital cost per board foot dried, less unit handling costs, less degrade, less total energy used per board foot, less labor handling costs per board foot, and far less moisture problems.

Eventually, most softwood lumber mills will adopt tall full course stacking using track kilns with door heights of 16′ or higher, and most all of the now counter-flow lumber kilns will have been upgraded or replaced by parallel flow kilns.

11.18.3.4 Side-Loader (Package) Kilns

Holding Capacity

The holding capacity of package kilns is determined by board thickness, sticker thickness, package width, package height, total package height, the number of cross outs used, kiln width, kiln depth, and stacking efficiency factors.

Kiln Width

Kiln width is defined as the total **exterior** width of the building. Kiln width is calculated by adding the following dimensions:

Total load length
Total load clearances (wall panel to packages)
Total loading clearances between packages
Total wall thickness (both walls)

Because side loader package kilns are loaded by fork trucks, a significant amount of the kiln's available load space is wasted. In these kilns, the required clearances between package ends for fork trucks to handle the packages can vary from 12″ to 18″.

Example: Package kiln for three 16.5′ packages, 1′ wall clearance, 1′ load clearance, and 12″ wall thickness:

Total package length = 3 × 16.5′ = 49.5′
Total load clearance = 2 × 1′ = 2′
Total wall clearance = 2 × 1′ = 2′
Total wall thickness = 2 × 1′ = 2′
Kiln width = 55.5′

In the preceding example, the amount of wasted space in the width of the kiln is 4.0′.

Determining Optimum Load Space Width

The width of side-loader kilns varies from 20′ to 60′. Because of the inherent small available load space for package lengths, adequate planning must be done prior to the construction of the kiln. The interior width of the kiln has to "fit" the mix of package lengths that will be loaded into the kiln as well as the required clearances for loading the packages into the load space. The fork truck drivers should be involved in the final decision about the total interior space inside the kiln for the different packages that will be loaded into the kiln.

Kiln Depth

The definition of side-loader kiln depth is the measured door-to-rear wall of the kiln, measured from the inside of the front door to the outside of the rear wall. Kiln depth is the sum of the following:

Total plenum widths
Total load widths
Space between packages
Rear wall thickness

Example: A package kiln is to be built for drying 1″ oak, with two 8′ wide packages, 12″ thick rear wall, and a 20′ door height. What is the depth of the dryer?

Total plenum width = 2 × 7′	= 14′
Total load width = 2 × 8′	= 16′
Space between packages = 1 × 1′	= 1′
Rear wall thickness = 1 × 1′	= 1′
Kiln depth	= 32′

Determining Maximum Load Depth

Package kilns are narrow batch type dryers intended for lumber with slow drying release rates. Because of their low cost, package kilns are cost effective means of drying large amounts of slow drying lumber. Therefore, they are common in hardwood and timber drying operations. Unlike track kilns that can deliver high air velocities through narrow packages, package kilns cannot deliver high air velocities through a package of lumber, and they also do not have reheat capability between packages. Thus, they are not suited for drying 1″ and 2″ softwoods, except for small treating operations in which conventional temperature schedules are used.

For hardwood operations, the maximum load depth (the total feet of air travel through packages) must be determined by a combination of experience, and dryer engineering. Model 74 was developed for this reason. The board thickness, sticker thickness, and air velocity through the packages are the key variables that determine how a certain species and thickness of lumber will perform in a kiln. It is common to see 8/4 oak dried successfully in package kilns in which 24′ of total effective package depth is present. It is also common to see large softwood timbers dried successfully in package kilns in which the total effective package depth is 16′. Kiln operator experience and **conservative** package kiln designs are the key factors for successful drying.

Kiln Height

The height of package kilns is dictated by the load height. In many package kilns, the load height can exceed 20′. Although fork trucks can reach these heights, the practice of such designs creates handling problems for fork truck drivers. If the load height increases, the risk of loads tipping over increases. The risk of heavy packages damaging the kiln's walls and structures also increases.

Door Height

The door height of package kilns is dictated by the upper travel limit of the fork truck's mast. Consideration must be made about the maximum height of the mast with a package on the forks. Failure to provide sufficient height at the bottom of the lintel beam at the door front can lead to a serious accident in which the fork truck can be turned over resulting in property damage and injury or death to people near the truck. The rule of thumb is to measure the highest possible height the mast and a package can reach and add 12″ to this dimension. This will be the designed height of the bottom of the lintel beam above the concrete pad at the door. In addition to the required lintel beam height, the fan floor and the front ceiling baffles are subject to damage if they are located too low. Sufficient clearances must be provided to protect them from both the fork truck's mast and the highest package.

Center Door Columns

In some package kilns exceeding 40′ in width, kiln manufacturers often install center door columns to support the weight of the wide upper door front and roof structure. The problem with center door columns is that fork trucks will encounter them on a regular basis leading to the possibility of an accident. The presence of the column also increases the difficulty of loading the kiln in the area behind the column.

Center door columns should never be designed into a package kiln. The kiln's entire door-front should be supported by one truss designed to span the width of the kiln.

Door structures are also prone to heavy condensation of corrosive water and acids from the drying operation. Sufficient additional safety factors for corrosion should be designed into the trusses and side columns. Designs should also consider heavy snow loads that can accumulate on the roof during periods when the dryer is inactive.

11.18.4 Equalization, Conditioning, and Steaming Chambers

11.18.4.1 Introduction to E&C Chambers

E&C Structures are often referred to as steaming chambers. Due to their nature, they are not really dryers, but instead are designed specifically for equalizing and conditioning lumber that has left a lumber kiln. These systems can also be used for steaming color treatments of wood and sterilization to kill pests.

E&C structures are subject to extensive corrosion caused by water and steam humidity sprays. If not built entirely of corrosion-resistant materials, one can expect major problems with corrosion, extensive repair costs, and lost production. Aluminum is the best choice for metal structures, fan walls, and baffles. All fasteners should be of stainless steel. Fans should be made of aluminum. Line-shaft or cross-type fan shafts should be stainless steel. Oil-type sleeve bearings should be used for line-shaft fans. Ball and roller bearings will suffer frequent failures if used in E&C chambers. Fan motors should be located on the outside of the chamber.

The walls and roofs should be designed to prevent condensation on the inside of the walls and roof. All E&C chambers must include an impermeable interior moisture barrier to prevent moisture from entering the walls and roofs. If prefabricated panels are used, the exterior of the panel should have weep holes to let water drain out of the panels.

The geographic location of these units plays a role in their design. In colder climates, the floor slab can be fitted with embedded hot water/glycol heating pipes to keep moisture from condensing on the slab.

Typically, interior temperatures are rarely over 160 degrees F dry-bulb with a 155 wet-bulb.

Accurate temperature sensors are needed for precise temperature control.

11.18.4.2 Application of E&C Chambers

E&C Chambers Are Used for Five Reasons

1. To protect capital-intensive wood dryers from lost drying capacity due to E&C cycles
2. To protect capital-intensive wood dryers from corrosion from E&C cycles

3. To improve final moisture content and degrade control
4. For steaming wood prior to plastic bending/forming
5. For color and/or pest treatments

Holding Capacity

The same calculations for package and track kilns should apply to E&C chambers. Since the air velocities in these units are usually one-third of the dryers they serve, large plenums are not mandatory.

Side-Loader Type E&C Chambers

Typically used for re-drying and E&C in softwood operations.

Track Type E&C Chambers

This design sits on the same tracks just downstream of a track kiln. In batch operations, sufficient space exists between the kiln's doors and the doors of the chamber to transfer the load of lumber from the kiln to the chamber. The tracks inside and adjacent to the chamber should be of a corrosion resistant design to minimize corrosion. The tracks should be located above the floor of the chamber for ease of replacement.

The following page shows a track-type E&C chamber. Note the line-shaft fan system and the two separate steam spray lines. The roof vents are included for venting the chamber during maintenance and repairs.

E&C CHAMBER

11.19 Dry Storage Facilities

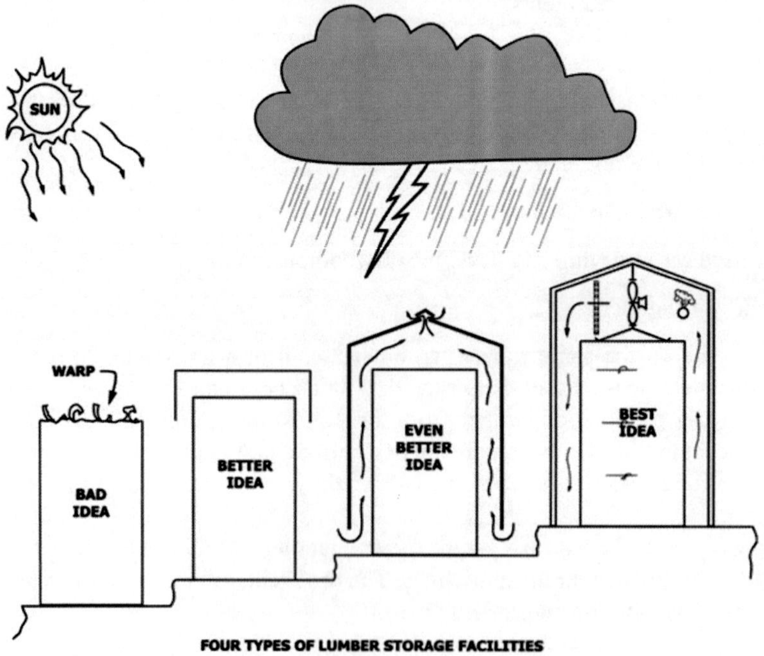

FOUR TYPES OF LUMBER STORAGE FACILITIES

The design of dry storage facilities is dictated by the species and annual capacity of the plant.

Required Holding Capacity

The holding capacity of dry storage facilities for most softwood plants are based on the maximum expected weekly plant production level. The product mix will affect the required capacity. The holding capacity is determined by board thickness, sticker thickness, package width, total package height, ceiling height, building width, building length, and stacking efficiency factors.

One rule of thumb for sizing the holding capacity of dry storage facilities is to figure what is needed for a typical plant operation and add 25–50% to it. The day always comes when you will wish you had installed the additional holding capacity during the design and construction phase of the plant.

Plenum Width

Plenums serve two purposes in all types of dryers. One is to provide sufficient space by which air is allowed to circulate around the packages of lumber. The other purpose is to provide a sufficient space for maintenance personnel. Plenum widths should never be less than 4′. The same plenum width criteria for pre-dryers should be used if the dry storage facility is equipped with temperature and humidity controls and air circulating fans.

Maximum Effective Package Width

Due to the minimal amount of moisture dynamics inside dry storage facilities, there is no upper limit on effective package width.

Package Handling Criteria

The dry storage facility should have sufficient workspace for handling the largest package of lumber.

The door height and width should accommodate the largest package.

The aisles for fork trucks should accommodate the largest package including the turning of fork trucks within the aisles.

The height of the fan deck and/or roof support structures should be sufficient to accommodate the handling of packages including the highest point on the mast of the fork truck.

If the dry storage facility includes temperature and humidity control, the design should be like lumber pre-dryers, except with smaller fans and heating systems.

Main doors may be fitted with energy conservation curtains.

The interior of the facility and door areas should be well lighted.

11.20 Calculations for Holding Capacities of All Types of Lumber Dryers

Preliminary Meeting

Before a dryer is purchased, both the buyer and the engineer should meet to review the stacking, sorting, and handling practices for the plant where the dryer is to be installed.

Determine the Actual Cubic Feet of Lumber in the Dryer

The difference between actual and nominal board footage should be discussed for the actual packages that the dryer will be required to hold. The objective of these discussions is to arrive at the total actual cubic feet of lumber that the dryer will be required to hold.

Drying Rate and Moisture Content Requirements of the Dryer

Depending on the species, board thickness, sticker thickness, total feet of air travel, air velocity, and the drying schedule, the size and quantities of dryers will have to be determined. Both the initial and final moisture content of the lumber being dried will have to be known. Do not speculate about the initial moisture content. Collect reliable moisture content information.

Determine the Dryer's Load Space
The required load space inside the dryer should be reviewed. This will vary depending on whether the dryer is a track or side-loader design. Exact measurements of package height, widths, and lengths, cross outs and kiln trucks will be required.

Determine the Stacking Efficiency
In modern softwood operations, stacking efficiencies of 94% to 97% are common in track kilns. In side-loader designs, stacking efficiencies of 40% to 60% are common. An analysis should be performed to establish the stacking efficiency at the plant where the dryer is to be installed.

Final Holding Capacity Analysis
After calculations have been agreed on for the required dryer load space, the standard dryer designs offered by the dryer contractor should be compared to the requirements of the plant where the dryer is to be installed. Adjustments in dryer dimensions should be made accordingly.

It is at this point where the common practice of cramming (putting too much electrical horsepower into fan systems) should be avoided. Do not overlook this final analysis of the drying system because the future costs of electricity will increase dramatically, and dryer OEMs focus only on getting orders, not long-term electrical efficiency.

11.21 Classes of Large Wood Dryer Building and Enclosure Designs

There are four main **classes** of large dryer building and enclosure designs each of which is determined by how the main structural elements and exterior thermal and moisture barriers are structurally configured.

Note: Large softwood continuous dryer structures may exceed 40′ width by 40′ height by 300′ length.

Class I Building – The building's thermal and moisture barrier is located **outside** of the main structural components designed for handling local wind and snow load code requirements.

In this building system, the entire main structure is erected, then fitted with internal fans, and heat distribution components, and then **exterior** insulated panels or field-erected insulated assemblies are attached (by loose straps or clips) to exterior stringers and purlins of the main structure. In this design, the exterior insulated panels are not considered primary structural elements. This is the **current predominate method** used for large wood dryer buildings. The problems with this building design are the highly corrosive environment inside dryers and the numerous thermal expansion and contraction cycles that must be dealt with during the life of the dryer. The advantages of this design are the ease by which the exterior roof panels can be installed and maintained (from above), and the cosmetic appeal of having the entire dryer structure enclosed inside either all aluminum, stainless-steel, or painted insulated panels.

Class II Building – In this building system, the thermal and moisture barrier is located **inside** of the primary structural components designed for handling local wind and snow load code requirements.

In this building system, the entire main structure is erected, then fitted with internal fans, and heat distribution components, and then **interior** insulated panels or field-erected insulated assemblies are attached (by loose straps or clips) to stringers and purlins of the main building structure. Currently, this (improved) method of construction is **rare** for large dryer buildings. The benefits of this dryer building design are:

One – None of the primary building structural elements are exposed to the highly corrosive environment inside the dryer, and none of the main building structural elements experience any thermal expansion and contraction issues that must be dealt with during the life of the dryer. Thus, this building system can last for many decades with minimal maintenance, and it can also be removed and relocated after many decades of use.

Two – Any qualified local building contractor can design and build the main building structure. There is no need to have a dryer OEM build this major part of a new dryer, thus saving the owner a significant amount of money.

Three – This design allows the entire dryer building and its support equipment be enclosed inside an outer weather-protection enclosure mounted to the exterior of the main structure. This outer enclosure can also be custom designed for cosmetic appeal if the owner chooses such and can also be installed after the dryer operation is put into use. This outer enclosure (chamber) can also be slightly pressurized by low-head fans to prevent dryer wall leaks from entering and causing damage to the outer protective enclosure chamber or any dryer equipment located inside the outer chamber.

Class III Building – In this **(integrated)** design, the main structural building elements include both interior and external elements. In this design, the external insulated panels serve as thermal and moisture barriers and structural elements by **rigid connections** to the interior elements. This design requires the designer provide calculations and drawings demonstrating the final building complies with local wind and snow code requirements. Even though these designs do exist, they are rare.

Class IV Building – The thermal and moisture barrier is integrated with the main structural components by using a **field erected sandwich** of either aluminum or stainless-steel sheets formed over interior insulation boards. Although these systems do exist, they are rare in large lumber dryers because of cumulative thermal expansion issues but used in some types of small side-loader hardwood dryers. Additionally, in installations requiring building code compliance, the designer must provide structural calculations and drawings for applicable local wind and snow load requirements.

11.22 Life Expectancy of Dryer Buildings

The life of dryer buildings depends on the type building design, the drying operation, and the operating philosophy of the owner. In many drying operations, dryers are often viewed as burdensome high-cost operating centers because of their high utilization rates – 24/7 for over 360 days/year.

In most drying operations, dryers are expected to operate around the clock without significant downtime. It is common to see modern softwood dryers in operation 90–95% of the available hours in a year. In many hardwood operations, the utilization rates can vary from 75% to 95% of the year. Because of this and the wood-products industry's historical philosophy to not install excess drying capacity at plants, the buildings for dryers are traditionally expected to last forever.

All wood dryer buildings do eventually wear out because of heat, moisture, weather, wood acids, and abuse. In many cases, dryers are operated at temperatures far above what they were designed for to get more drying capacity out of the system.

In many cases, little to no attention is given to the building, resulting in structural collapses. Another significant factor is the erosion and settlement of concrete foundations. When these occur, the decision must be made to either try to repair the damaged structure in place or replace it. Often, the upper structure is patched until the patches cover most of the structure. In some cases, patches are installed over older patches to avoid the cost of replacing the structure. In many cases, no attention is given to dryer building repairs of any kind.

Typically, high-temp prefabricated double-track pine kilns can last 20 years. Some have lasted over 40 years, but these are rare. If common wall designs are used, expect problems with roof leaks after the first year. If a free-standing single-track prefabricated design is used, expect to see them last over 30 years in southern pine operations. In hardwood operations, moisture, extreme cold, and corrosion is the biggest causes of building failures. If all aluminum structures are used, expect 30 or more years of life. If inferior insulation is installed in exterior panels or not supported properly, expect insulation settlement to start after 2 years.

Corrosion in Dryers
Condensation on Metals

Any part of a metal kiln structure is subject to corrosion. Any part of a structure that is subject to condensation or exposure to liquid water droplets is subject to aggressive corrosion. It is common to see hardwood kilns with aggressive corrosion in areas where condensation is continuous and see other parts of the structure relatively free of corrosion.

Although mild steel is susceptible to corrosion, other metals can also be attacked. Zinc is useless in lumber dryers because of attack by wood acids. Galvanized metals will quickly corrode making them useless or unsafe. Aluminum can be attacked by water that is highly alkaline. Low-quality stainless steel can be damaged by acidic or alkaline water. If the level of condensation or collection rate of water droplets on the surface of the metal is high, expect high corrosion rates.

Wood Acids

There are numerous acids that occur naturally in wood. Tannic acid is very high in the oaks. Some species of wood are highly corrosive when dried and some are not. Hemlock is very corrosive. Most pines are not.

Water and Steam Spray Systems

The chemistry of the water supply or steam supply to a dryer will impact how much corrosion occurs in the dryer. If humidification sprays are used, small droplets of water will mix with the circulated air stream and fill the entire interior space of the dryer. This is an especially difficult problem with high-pressure water nozzles marketed to avoid the cost of steam production and spraying systems.

If low-pressure steam is used for humidification, it must be de-superheated by water injection, and this also leads to water droplets entering the dryer. Once water droplets from any source enter the dryer's air stream, depending on the initial PH of the water, they can cause corrosion.

Lack of Effectiveness of Protective Coatings

Protective coatings in lumber dryers are temporary fixes. Eventually, due to temperature extremes, oxidation, and attack by chemicals, they harden and develop cracks. Once the cracks occur, air-born moisture droplets carry the corrosive chemicals to the surface of the metal. In low-temperature dryers, bitumastic and oxide coatings offer excellent protection. However, even these have a limited life. In high-temperature dryers, protective coatings that protect against E&C corrosion are virtually nonexistent. Unless a high-temp dryer is completely built of aluminum and stainless steel, long-term protection against E&C corrosion is virtually impossible.

Effect of Temperature on Corrosion Rates

Absolute temperature (Tabs) is a measure of the kinetic energy exchange between molecules. Kinetic energy is a function of molecular velocity and velocity is the variable that drives chemical reaction rates. At temperatures below 50 degrees F, water-born chemical reactions found in lumber dryers slow down to very low rates. As dryer temperature increases above 50 degrees F, the rate of corrosion increases dramatically.

Steel Components of Dryer Buildings

Mild Steel Components

In most lumber dryers, mild steel is the metal of choice for frames, decking, and baffles. Due to its low cost, high strength properties and workability, it is the predominate metal used for wood dryer construction.

Design Allowances for Corrosion of Mild Steel

The applicable building code will dictate the maximum allowable stress for the different components of the dryer. The codes will specify limits on tensile stress, compressive stress, bending stress, shear stress, and deflection in a corrosion-free environment. Because of the risk of corrosion, additional safety factors (reduction in load stresses) must be applied.

The first classifications are:

Type I – Dryers without E&C cycles

Mildly corrosive wood species

Highly corrosive wood species

Type II – Dryers with E&C cycles

Mildly corrosive wood species
Highly corrosive wood species

The second classifications are.

Type A – Components free of condensation and wetting
Type B – Components subject to condensation and wetting

The third classifications are:

Type 1. – Indirect-fired (partial list)

Steam heated
Hot-oil heated
Hot-water heated

Type 2. – Direct-fired

Natural gas
Fuel oil
Wood waste

Using these classifications (by the author), the recommended additional safety factors (by the author) for different types of dryers can be scaled, going from the least corrosion safety factor to the highest.

Corrosion classification	Required additional corrosion safety factor
Least corrosive dryer type	0%
Mean corrosive dryer type	100%
Most corrosive dryer type	200%

In the three classifications presented, the dryer with no additional safety factor above the minimum code requirements is a natural gas, direct-fired dryer for a mildly corrosive wood species in which no E&C cycle is used. The dryer with the most required safety factor of 200% is a dryer for a highly corrosive wood species (such as oak) in which an E&C cycle is used. In this dryer, the structural components subject to condensation and wetting require a maximum stress level of 1/3 of the maximum allowable stress level dictated by the building code. The safety factor only applies to those components in the dryer subject to condensation and wetting, and not the entire dryer structure.

American Institute of Steel Construction (AISC)

The AISC was developed decades ago for establishing minimum design standards for steel used in construction. The AISC manual should be used in conjunction with the applicable local design codes.

Fasteners, Straps, Clips, and Hinges

In all dryer designs, numerous fasteners, straps, clips, and hinges are used to attach the different structural components of the dryer together. Due to the relatively low cost of these items, these items should be liberally oversized to prevent failures.

Dryer Door Fronts

Door fronts pose difficult problems for dryer designers. They are especially prone to corrosion because of the heavy condensation that occurs on cold steel members located near or in the door front.

In addition, dryer doors are constantly plagued with air leaks. Depending on the position of the roof vents during their operation and the direction of the airflow inside the dryer, either hot moist corrosive air or cold dry air can pass through the leaks at the doors. This creates significant problems with condensation of acids on steel located near or at the door fronts. Lintel beams, lintel beam trusses, and door front columns can experience severe corrosion in hardwood kilns.

The use of protective coatings will slow down the corrosion but will not prevent it. The use of steam tracer lines to warm lintel beams and door front columns have been used by kiln designers, but even these often fail due to the lack of maintenance of the condensate traps on the tracer lines. Another design is to use a heated water/glycol solution circulated through pipes located inside the lintel beam and the door columns.

Column Base Pads

The base pad of dryer columns is another point where condensation is a problem. All dryer support columns should be mounted on an elevated concrete pad to prevent condensation. The minimum elevation above the adjacent floor of the kiln should be no less than 12″ in softwood dryers and 18″ in hardwood dryers. In kilns where condensation is occurring, the steel should be examined monthly for corrosion, and repairs made accordingly. Concrete anchor bolts should be of stainless steel in dryers subject to corrosion.

Heating System Components

Fin pipe and steam piping can experience heavy corrosion in kilns where E&C cycles are used. If the fin material is subjected to heavy corrosion, the ability to transfer heat to the air stream will be reduced resulting in coil replacement costs and dryer downtime. The ease of replacing corroded fin pipe should be discussed with the dryer contractor prior to the construction of a dryer. Although some have suggested coating fin pipes with a mixture of diesel oil and graphite to slow the corrosion down, this is a temporary fix and will not prevent the final replacement of the fin pipe.

In direct-fired dryers, using E&C cycles, expect to see extensive corrosion of heat supply ducts. Because of the high temperatures encountered in heat supply ducts, very little can be done to stop the corrosion. Direct-fired dryer designs with E&C cycles should be avoided. If these designs are used, insist that the dryer contractor provide a design that allows for easy replacement of ducts.

Preparation of Steel

All steel used in corrosive wood dryers should be adequately prepared for the application of protective coatings. Loose mill scale, welding flux, oils, and dirt should be removed prior to the application of protective coatings. Sandblasting should be done on dryer components subjected to heavy corrosion. In dryers not subject to corrosion, the cost of expensive prep and coatings is a waste of time and money. In such dryers, an initial coat of red oxide paint is sufficient.

Protective Coatings for Steel

In the dryer industry, commercial coatings is an industry within itself. Considerable expertise exists to find the best coating for any wood dryer applications. Paint and dryer vendors should be consulted with for the best coating at the least applied cost. Often, wood dryer designers and contractors do not possess the necessary expertise in this area. Some of the many commercial coatings are listed in the following text.

Asphalt-Based Compounds

These are oil refinery heavies. The wood dryer industry started using these coatings following WWI. They provide low-cost surface protection to steel when properly applied. They can be obtained from dryer and paint vendors. Some of these coatings are filled with bright aluminum pigments to improve appearance and durability. Asphalt-based compounds can be applied by using dip tanks, rollers, or spray.

Coal Tars

Coal tar derivatives are used extensively in all types of roofing construction. They can be used to protect dryer structures exposed to heavy condensation. Dryer temperatures above 180 degrees F will shorten their life.

Carboline Coatings

These are thick coatings that can be applied to areas subject to excessive corrosion.

Carboline coatings provide a thick, flexible, heat-resistant barrier to moisture and air.

Metal Oxide Epoxies

Red oxide epoxy can be used in dryers not subject to excessive corrosion.

Aluminum oxide epoxy can also be used in dryers not subject to excessive corrosion.

Phenolics

Plasite and heresite are used on metals subject to heavy acid corrosion. They are expensive but offer superior protection. They are used on dehumidifier condenser coils.

Chromates

Zinc chromate is used for high-temperature supply ducts. The metal should be sandblasted before painting.

Coatings for Embedded Items in Masonry

If mild steel is to be embedded in concrete or masonry walls, it should first be coated with a protective coating of heavy asphalt-based material. Since most embedded steel items are relatively small, they should be dipped into vats containing the coating material and allowed to dry before use.

Stainless Steel Components

Because of the high cost of stainless steel, its most common use in dryers is for burner flame shields and fasteners (bolts, nuts, and washers) subject to rapid corrosion. All foundation anchor bolts, clips, and embedded column bolts should be stainless if the dried species is corrosive. If the dryer is to be drying a large amount of lumber, stainless steel can be used in select locations such as interior wall panel sheets, or locations where aluminum may not be the best choice.

Aluminum Components of Dryer Buildings

Aluminum is an excellent metal for wood dryers. If hardwoods are dried and/or E&C cycles are used, aluminum is the choice of metals for longevity. All-aluminum dryers can last for decades with little to no maintenance.

However, aluminum is expensive and there is always the possibility of dryer manufacturers taking shortcuts to protect their profits. To prevent this from happening, the final design should be reviewed by a licensed structural engineer familiar with aluminum structures.

Corrosion of aluminum in wood dryers is usually caused by exposure to alkalines and chlorines. Both can come from the water or steam supply to the dryer. If E&C cycles are used, the use of highly alkaline water for humidification will cause both face corrosion and pitting of aluminum, especially close to where the spray nozzles are located. Aluminum corrosion can also be caused by close contact with concrete, lime, brick mortar, concrete blocks, or alkaline soils. These contain alkaline that will corrode aluminum. If water drains from any source containing lime, such as masonry walls, onto aluminum, expect corrosion to occur. Excessive levels of chlorine in water can also corrode aluminum.

For highly corrosive dryers, clad aluminum alloys should be used to offer additional protection. The exterior cladding acts as both a physical barrier and ion protector. The selection of alloys for structural items should be based on yield strength and cost. The selection of alloys for wall, roof, and door panel skins should be based on corrosion resistance and cost.

Wet Bulb Wells

Wet-bulb wells are a potential source of constant water spillage and corrosion problems. Most wet-bulb wells are made of brass. Some are made of stainless steel.

Accidental overflow from wells, due to plugged drain lines and excessive water feed rates, are a constant nuisance. If not kept clean and operating properly, spillage will occur and the concrete and metals in the vicinity of the well will suffer both corrosion and erosion. Catch pans should be located under wells to catch spillage. The pan can be used to both detect spillage and protect adjacent structures.

Tracks, Kiln Trucks, and Trams

New kiln trucks should be dipped in a bitumastic kiln coating. Tracks and their clips are a potential problem if E&C cycles are used in the dryer. Due to their constant contact with the cold floor of the dryer, tracks can suffer extensive corrosion. Tracks inside kilns that dry corrosive species should be located on top of the floor of the dryer for ease of replacement. Stainless anchor bolts, clips, and nuts should be used for easy replacement of the tracks.

Concrete Foundations

Dryer foundations often fail before the building resulting in expensive repairs, and/or replacement of the entire dryer. Foundation failures can involve a combination of corrosion and erosion resulting in costly structural failures and settlement.

Prior to designing a dryer foundation, geotechnical soils testing should be performed by a licensed engineer and soils testing lab to determine the bearing strength, potential rise, and settlement characteristics of the soils under the foundation. In some cases, supporting piers may be required to keep the top of the foundation level during the life of the dryer.

The drainage requirements of dryer foundations should be established to handle rainwater runoff and dryer process discharges. The plasticity index (PI) of the soils under the foundation should be established to determine the maximum potential heave and settlement effected by variances in soil moisture content.

Follow the American Concrete Institute (ACI) guidelines for concrete mix used in all dryer foundations. Because of aggressive acids from drying, embedded rebar can corrode. Anchor bolts and floor wall clips used in corrosive dryers should be stainless.

Load-bearing column-pedestals should be used to isolate the dryer columns from the cold floor of the dryer. Failure to do so could result in corrosion of column bases. The finished surfaces of concrete inside wood dryers should be covered with a commercial sealer to protect the concrete and prevent moisture penetration. These sealers will have to be inspected frequently and repaired.

The top surface of outer wall runners (longitudinal concrete beams) should be protected with stainless-steel shields that also act as a lower sealer/connection to outer wall panels. This shield should be both fitted and sealed around anchor bolts/studs for wall columns.

Exterior walls of foundations should be protected from standing water by sloping the top surfaces of soils or slabs away from the kiln. The exterior walls of foundations should also be insulated against cold weather and adjacent soils per ACI guidelines.

Thermal Expansion in Dryer Buildings

All dryers go through numerous hot and cold cycles. They are subject to large thermal strain and must be designed with sufficient expansion joints.

It is common practice in the prefabricated dryer industry to not include expansion joints in the upper half of steel-framed dryers under 100′ in length. The theory is that since the free-standing metal frame is supported by tall columns, differential expansion problems only occur at the base of the building where the columns attach to the concrete foundation.

This can lead to buckling of wall stringers located close to the foundation or binding of attachment clips for exterior panels. To prevent these from occurring, the lowest set of wall stringers should be attached with fasteners mounted in holes of sufficient diameter to allow for slippage. The bolts should be fitted with self-locking nuts tightened to a level that will allow slippage of the stringers. Using this method, the following spacing of lower joints should be used:

Steel Frames

Dryer temperature change (F)	Spacing of joints (ft.)
100	30
200	15
300	10

Aluminum Frames

Dryer temperature change (F)	Spacing of joints (ft.)
100	15
200	8

Prefabricated Panels

Dryer temperature change (F)	Spacing of joints (ft.)
100	12
200	6
300	4

Prefabricated insulated wall panels and aluminum extrusions should **float** on the main frame of the dryer's structure by using clips and straps. Allowances should be made for movement close to foundations.

Masonry Construction

Avoid building any industrial dryer out of masonry blocks because they are a defective design subject to foundation movements and thermal stresses.

When small dryers are built of concrete blocks, the following expansion joint spacing should be used to reduce the occurrence of cracks in block walls:

Dryer temperature change (F)	Spacing of joints (ft.)
100	20
200	10

Because of the expansion problems, every mortar course of the wall should have k-bar embedded for reinforcement. If this is not done, the blocks will develop cracks.

Concrete Floor Slabs

Slabs are slow to change temperature due to their mass and high thermal conductivity to underlying soils. Heat-adsorption transient computer models should be used to predict maximum slab temperature changes. This process is different between batch, jogging-batch, and continuous dryers.

Building Corners

Large stresses develop at the corners of dryer buildings because of cumulative strains caused by thermal expansion. Corner expansion joints should be included to prevent failures.

Common-Wall Kiln Failures

This is a practice that is inherently defective and problematic. Common wall lumber kilns are a totally defective design used only for the purpose of reducing front-end capital.

Because common-wall kilns do not always operate at the same temperature at the same time, it is a certainty that large stresses will develop at the common wall joint causing tears and leaks.

Wall and roof panels located at common walls should include slip-joints to allow for expansion differences. However, these will eventually fail and leak.

Do not let any person convince you to install common-wall kilns unless a detailed analysis of the slip joints has been done and approved by the buyer's engineer.

11.23 Temperature Rating of Dryer Buildings

The four temperature ratings for dryer construction are: 140, 165, 250, and 350 F.

Low temp 0–140 degree F designs apply to dryers of all-wood construction. These dryers should be fitted with high-temperature limit controllers in their roofs to prevent the heating system from exceeding the rating of the building. Wood building designs are common for dehumidifiers and low-temperature drying. Special attention should be given to the types of fasteners used for nailing the building components together. Zinc-plated fasteners and nails should not be used because of the corrosive problems with zinc in wood dryers.

Conventional temp 0–165 degree F designs apply to dryers of masonry construction. These dryers should be fitted with high-temperature limit controllers in their roofs to prevent their heating system from exceeding the rating of the masonry building. These designs can be used for low-temperature drying not exceeding 165 degrees F. Sufficient expansion joints and wall sealers must be used to protect the walls from moisture and cracks. The life of a masonry kiln building is dependent on the maximum temperature the building is subjected to. Reducing the maximum dryer temperature will significantly increase the life of the building. Avoid these kilns if at all possible.

High temp 0–250 degree F designs apply to high-temperature lumber drying. Prefabricated metal designs are the only design that will withstand these temperatures. The rating and support of the exterior panel insulation boards is a key factor to longevity. Common walls should never be used due to the problems with differential expansion between buildings. Expansion joints should be used to prevent damage to the numerous building components. These dryers should be fitted with high-temperature limit controllers to prevent their heating system from exceeding the rating of internal fan motors and for the prevention of fires. High-temperature safety limits, set at 275 degrees F, should be installed in the sidewalls.

Hyper temp 0–350 degree F designs are for veneer drying. Only prefabricated designs with high-temp panel insulation boards are capable of these high temperatures. Sufficient expansion joints are mandatory to prevent damage to the numerous building components. I suggest that only dryer manufacturers with a proven track record in the industry be used for these dryers.

11.24 Insulation Requirements of Dryer Buildings

Energy reduction and the prevention of condensation on the interior of walls and roofs are significant issues in all types of dryers. The dryer application dictates how much minimum insulation R-value is required.

If fossil fuels are used as a heat source, an economic analysis of energy cost versus insulation cost should be performed to determine the optimum effective R-value of the building. If wood waste is the source of energy, the optimum R-value will be less than for fossil fuels. If the dryer is in a plywood plant or paper mill, the optimum R-value will be higher and affected by the economics of the plant's steam cycle.

The total annual heat loss of dryers is an integral function of the temperature drop across the walls and roofs and drying time.

All pre-dryers and kilns should have an effective R-value such that the total annual cost of the energy losses through the walls and roofs do not become a significant economic cost. Analysis of the total cost of all the energy sources for the dryer must be considered. The following minimum effective R-values should apply for **energy conservation** only.

Wood-waste-fueled dryers should have a minimum R-value of 8.
Fossil fueled dryers should have a minimum R-value of 20.
Electrically heated dryers, such as dehumidifiers, should have a minimum R-value of 30.

Equalization and conditioning chambers are relatively short time cycles in which the cost of the heat loss through the walls and roofs are not significant. Instead, the prevention of condensation on the walls and roofs is the issue. The R-value should be sufficient to prevent condensation on the inside of the walls and roofs during the coldest days of the year. In cold climates an R-value of 30 may be required.

Foundations of dryers are also sources for energy loss. To minimize foundation heat losses, the following practices should be followed:

1. Use only designs that drain water and dryer discharges away from the dryer.
2. Do not locate dryer foundations in low areas subject to standing water. If the water table is high, you can expect foundation problems. French drains and sump pumps may be used to keep the water table both stable and sufficiently low enough that the bearing strength of the soil is not reduced.
3. Do not locate dryer foundations in areas subject to runoff from storms.
4. Insulate the exterior walls of foundations adjacent to soils.
5. Do not install conventional insulation boards under slabs. They will cause settlement. If a foundation design is to be used to reduce heat loss through the foundation, I recommend a licensed structural engineer be used. None-load bearing slabs can be poured on a thick slab of lightweight concrete to reduce heat loss, but even these require considerable engineering to prevent failures.

11.25 Condensation Inside Walls, Roofs, and on Floors

Condensation inside the walls and roofs of dryers will eventually lead to the failure and settlement of insulation, corrosion of metals, failure of masonry walls, freeze damage, and excessive heat losses. The mechanism of condensation is caused by the dry-bulb temperature at the condensation point dropping below the dew point temperature.

The causes of condensation inside dryer walls and roofs are:

1. Too much insulation causing cold spots inside the walls and roofs
2. High wet-bulb temperatures inside the dryer
3. Low ambient air temperatures
4. Moisture entering the walls and roofs through leaks caused by:

   A. Failure of interior moisture barriers
   B. Pin hole leaks in the interior skins of prefabricated panels

Condensation Inside Dryer Roofs

This is a common problem that may not be visually obvious to either dryer operators or maintenance personnel because the effects are often hidden. Because of rain, snow, and low temperatures, the problem can grow until the roof is saturated with water. Generally, for most dryers exposed to the elements, condensation inside roofs is difficult to stop without the use of a ventilated cover over the roof. Some dryer vendors provide these covers, and some do not. Both prefabricated and built-up roofs are subject to internal condensation if not sloped properly and protected by a ventilated cover.

Condensation Inside Dryer Walls

The condensation problems in roofs also exist in walls. Because walls are vertical, water that collects inside the wall has the force of gravity pulling it downward. Thus, the wall should be designed with drains for releasing water that collects inside the wall. Prefabricated wall panels should have external drain holes at their bottom. Masonry walls should have external weep holes in the bottom course. If condensation inside prefab panels continues for long periods of time, insulation boards inside the panel may lose their binders and be subject to attack by water-born chemicals. Either can cause the insulation to lose its strength and swell or settle to the bottom of panels.

Many dryer vendors use vertical panels (for esthetics and cost reduction) and some use horizontal panels. Horizontal panels with an external vertical weather cover are the most durable of all designs. Properly designed, this type of wall construction will last for decades with little to no maintenance.

Condensation on Dryer Floors

This occurs at some point in all dryers except dehumidification designs. Commercial coatings should be used frequently to protect the concrete from water and chemical attack. Embedded heating coils can be placed in the floor slabs during construction to keep the temperature of the top of the slab warm thus reducing the amount of condensation. Do not put insulation boards under concrete slabs in wood dryers. This can cause both foundation heaving and settlement.

11.26 Dryer Roof Design

The design of dryer roofs involves the following issues:

Prevention of slip and falls

Personnel safety can be compromised if the roof is sloped to such an extent that a safe footing is not possible on a wet roof. Edge safety guards will be required on steep roofs to prevent accidents.

Long-term durability against weather conditions

Prevention of damage from wind uplift forces
Prevention of damage from hail stones
Prevention of leaks from rain and snow
Ability to withstand high snow loads

Energy Conservation

Prevention of water and air leaks
Minimum effective R-value
Prevention of water accumulation inside the roof membrane

Dryer Equipment Operation and Maintenance

Ability to withstand the maximum internal temperature of the dryer
Prevention of corrosion of roof support structures
Accommodation of roof vents and actuators
Accommodation of fan drives and motors
Accommodation of safety platforms and walkways for maintenance personnel

Types of Roof Construction

Prefabricated panels
Conventional built-up roofs for low-temp dryers

Roof Slope

The roofs of dryers can be either: flat, gabled, or curved. The following is a review of their relative merits:

Flat roofs are the least expensive to build and are common in low-temp kilns. Although the term flat is used, for water drainage reasons, no dryer roof should ever be totally flat. All roofs should be sloped slightly to handle rain runoff. The type of dryer will dictate which direction the roof slopes.

Gabled roofs are common in prefabricated designs. The fan wall is located under the peak of the roof. If the slope is excessive, the performance of the kiln fans can be significantly reduced. Steep roofs can also cause vortex shedding leading to non-uniform air velocities through the packages of lumber. This will cause one end of the kiln to have higher sticker air velocities than the opposing end.

Curved roofs are used to enhance the flow of air inside the dryer by reducing the static backpressure on the kiln fans. Curved roofs are the most expensive, but also the best design for reducing the electrical consumption of high-air-velocity kiln designs.

11.27 Dryer Door Design

Properly designed dryer doors are required for safe successful dryer operations. The following issues should be addressed during their design.

Sliding Versus Hinged Designs

Dryer doors are available in five basic designs. During the last century, the designs used on dryers in the USA are:

1. Horizontal-sliding movement, suspended from overhead rails
2. Vertical lift sliding, suspended from steel cables, chains, or mechanical rams
3. Roll-up doors, like garage or warehouse doors
4. Vertical-hinge designs – hinges located at door columns
5. Horizontal-hinged designs – hinges located at the door lintel beam

Of the five designs, the #1 horizontal-sliding suspended door is the most predominate design. The second most predominate design is the #4 vertical-hinge design. The #4 vertical-hinge design is common on track and small package kilns. For door heights above 16′, the design most often used is the #1 horizontal sliding door. For door heights under 16′, either sliding or hinged designs can be used. If the dryer is a large package type, the #1 sliding door is the most predominate design, although some designers have used both the #4 and the #5 hinged designs. The #2 vertical-lift and the #3 roll-up designs have been used, but their use is rare in the USA but common in Europe.

Safety Issues with Dryer Doors

Because of their large size and weight, dryer doors can cause injuries or death if not designed, operated, and maintained properly. If any part of a door component fails suddenly, the door may fall. Sudden gusts of winds can cause large horizontal sliding doors to be lifted off their support rails and latches. Such an event can happen suddenly and without warning. If not latched, hinged doors can suddenly swing open or closed during wind gusts.

During dryer fires, flash back explosions can occur if a dryer door is opened. These explosions have blown heavy horizontal sliding doors up to 30′ away from the dryer. People can be injured or killed by dryer doors during flashback explosions.

Sliding doors can fall if their lift mechanism, door carrier, or latches are worn out or out of adjustment. The #1 horizontal-sliding doors should be fitted with OSHA-approved safety guards to prevent the door from tipping over.

The #5 horizontal-hinged designs have been used on large package kilns, with a cable winch lift system. In these designs, the dryer door is hinged at the lintel beam and swings out and up using a cable winch system located at the rear of the dryer. Due to the high probability for a major accident caused by wind gusts, structural or equipment failures, I do not recommend this door design for any dryer.

Some dryer designers have used the #2 vertical-lift design on track kilns. These doors must be fitted with automatic safety latches to protect people from failures of the lifting equipment. I do not recommend this door design for any dryer due to the high probability of a serious accident caused from a failure to inspect and maintain the equipment. However, although these door designs are common in many multi-zone jogging-batch softwood dryers, they require frequent scheduled safety inspections to prevent accidents.

Health Issues with Dryer Doors

All heated dryers, especially wood dryers, emit toxic emissions, including cancer-causing compounds. These emissions should be directed away from people working around the dryers. Dryer doors should also seal tight. They should not leak vapors from the dryer or allow cold air to enter the dryer through door leaks. All dryer exhaust venting should occur at or above the dryer's roof level. The higher the discharge point, the safer it is for people working around dryers.

Drying Problems Due to Door Leaks

Door leaks in dryers cause inconsistent drying, cold spots, added energy usage, and accelerated corrosion around the doors. Any concrete located around the door will suffer accelerated corrosion and erosion due to door leaks.

Door Gaskets and Seals

Dryer doors should be fitted with gaskets rated for the maximum temperature inside the dryer. High-temp dryers require silicone rubber or stainless-steel mesh rope. Low-temp kilns can use natural rubber gaskets, but these fail often. Because the designed geometry of the doors involves the thickness of the door seals and gaskets, many doors and lifting devices will not function properly if the door gaskets and seals are not maintained and/or replaced when gasket failures occur. This is especially a problem with the horizontal sliding door designs. Some dryer doors use a flexible rubber sheet located at the bottom of the door for a seal. Such seal designs should allow for vertical adjustments, flexibility for the presence of foreign debris on the ground, and easy replacement of the seal.

Door Leaks at Kiln Tracks

There are three designs used in track kilns to minimize air leaks at the location where tracks pass under the kiln's main doors. The most common design is to slot the bottom base angle of the doorframe. Another method is to slot the steel tracks under the door to allow the lower edge of the bottom door angle to drop into the slot. Another method, patented by Frank Cook, now deceased, is the lift-out rail section located under the door. I prefer the slotted base angle method. A seal can be added to the tracks to minimize leakage where the base angle fits over the tracks.

Door Rain Guards

Dryer doors should be fitted with rain guards located above the top of the doors to prevent rain from draining over the top of the door and into the dryer causing cold spots and corrosion.

Door Front Foundation Design

There are three main issues in the design of door front bases and their foundations.

1. The ability to provide a tight seal to minimize air leaks.
2. The ability to provide drainage of water out of and away from the dryer.
3. The prevention of settlement of concrete at the door front.

Package Kiln Door Front Foundation Design

1. Use a flexible gasket at the bottom of the doors to minimize air leaks
2. Slope the concrete slab away from the door front to drain water away from the kiln
3. Use a reinforced slab of sufficient thickness to prevent settlement

Track Kiln Door Front Foundation Design

1. Use a base angle attached to the concrete with anchors to provide a durable surface for the bottom door gasket to seal against.
2. Install a drain trough at the door front to drain water away from the kiln.
3. Use a deep reinforced concrete beam under the door front and the adjacent tracks to prevent settlement. This beam should have the drain trough formed into it.

11.28 Catastrophic Dryer Implosion Prevention

Wood dryers can implode if they are started with a cold charge of lumber with hot humid air accumulated in the top of the dryer. These implosions occur during the startup of the kiln fans.

Once the fans start moving the moist air, a vacuum is created when the hot mixture of air and moisture enters the packages of cold lumber. The hot air is quickly cooled down and most of the moisture condenses on the cold lumber. The result is a sudden partial vacuum inside the dryer that can cause doors or walls to be sucked inward. Usually the doors fail first, but there have been incidents involving long track kilns in which entire masonry walls collapsed.

To prevent dryer implosions, all dryers should be fitted with a fan startup control scheme to prevent implosions. Some of the schemes used are listed in the following text.

Dryers with Roof Vents
In these dryers, all the roof vents are opened 100% for a period long enough to allow the hot air and moisture to escape from the dryer. This waiting time will increase with the size of the kiln. Typically, after 4–5 minutes, the fans are started. Then, after another 30 seconds, the vents close.

Roof vents may also be designed to stay open anytime the kiln is shut off to prevent the accumulation of hot air in the top of the dryer.

Roof vent actuators should be fitted with a proof-of-vent-open safety-switch to ensure the system is operating properly before the fans start.

Dryers Without Roof Vents
Although rare, some wood dryers are designed without roof vents. In these dryers, the kiln fans can be started gradually to prevent implosion. However, there must be sufficient tramp building leakage to use this method. The starter contactors for the fan motors can be jogged for two short periods of time to gradually start the hot air circulating through the cold lumber. The first jog period should be about two seconds followed by a five-second rest followed by another two-second jog period followed by another five-second rest period. After several timing periods, the fan motors can then be started.

11.29 Replacement of Steam Coils, Fan Motors, and Fans

All dryer buildings should incorporate designs that allow easy replacement of steam coils, fan motors, and fans. Fan decks in track kilns should be fitted with personnel doors at the door front gable panels. Door front panels should be a design allowing easy removal for kiln repairs. Overhead rails should be installed in kilns to allow safe removal of heavy fan motors. Overhead steam coils should NOT be mounted above major structural beams. They should be located below them allowing for their easy replacement. OSHA cat walks should be installed at door fronts for access to fan decks.

Personnel Platforms for External Fan Drives
Dryers that have wall-mounted external fan motors should have a work platform installed for maintaining fan drives, bearings, couplings, and motors. The platform should be fitted with OSHA-approved handrails and caged access ladders or approved stairways. The platform should be fitted with an overhead rail for transporting electrical motors to a location at which they can be lowered by a chain hoist to the ground. The platform and its support columns should be freestanding. Because of the possibility of hidden internal corrosion, the platform should not rely on the internal structure of the kiln for support. The platform should be of sufficient width such that personnel access around fan motors and/or belt safety guards are possible.

11.30 Dryer Foundation Design, Construction, and Site Management

Structural and Site Considerations
Licensed structural engineers should be used to either design or review foundation designs for all dryers. Retain a geotechnical engineer to evaluate the soils under the site where the dryer is to be installed. Construction cost, life expectancy, wind codes, earthquake codes, temperature extremes, floods, equipment vibration, dead loads, live loads, etc., all must be considered.

The Types of Soils and Site Conditions Must Be Investigated
The terrain and drainage plan are the first consideration in the design of a dryer project. Topographic maps should be reviewed to determine stability and achieve proper drainage. Proper sloping and drains will be needed. Will subsurface drains or sump pumps be needed? Try to avoid using sump pumps because they are often not inspected and can fail.

Will perimeter drains be needed? What will be the worst-case weather conditions? Is the project located in a flood area? What is the risk for mud slides and creep flow down hills? Retainer walls located near the dryer should be reinforced concrete. Do not use treated wood retainer walls. They will fail. The maximum possible hydraulic pressure behind retainer walls must be considered.

What factors will exist under the dryer? See the Thornthwaite Moisture Distribution maps in the unified building code (UBC) for soil moisture and moisture gradients in soils. Will there be any perched water tables present after the project is completed?

Atterberg limits in soil testing include the following:

Liquid limit is the water content at which a soil is almost liquid.
Plastic limit is the lowest water content at which a soil is plastic.
Shrinkage limit is the smallest water content that can occur in a saturated soil sample.
Plasticity index = (liquid limit – plastic limit) is a measure of the soil's volumetric expansion due to moisture gains
Shrinkage index = (plastic limit – shrinkage limit)

At the doors of all dryers, both the dryer slab and the exterior slab should be supported by a 30″ deep reinforced turndown with slip joints for shear loading. Do not allow reinforcement steel to be less than 3″ of the surface of the concrete. Use 3500 psi concrete. Consult with a concrete expert in the area where the dryer will be located. If highly acidic woods are to be dried, all anchor bolts should be stainless steel. Interior concrete column support pads should be at least 12″ tall (above the slab) to prevent condensation at the base of the column. The perimeters of dryer slabs should be insulated in cold climates.

Foundations Subject to Excessive Vibration from Rotating Plant Equipment
Vibration can be rotational, vertical, and horizontal. Determine the static bearing capacity of the soil, without vibration present. Then use a safety factor dictated by the mass-spring amplitude coefficient. In some cases, vibration isolation may be the

only solution. If rotating plant equipment transmits vibration into the ground, the vibration can radiate for large distances causing damage to structures. This is a problem with many saws, chippers, and heavy material handling systems in lumber and plywood plants. Equipment anchor bolts added after the slab is poured and set should be high-strength epoxy filled. Do not use expansion-type anchors if vibration is present. They will fail.

11.31 Underground High-Temperature Concrete Ducts – 450 degrees F Max

Prior to laying the concrete pipes, lay 6″ of washed gravel in the trough. Wrap the joints with a stainless-steel wrap. Backfill the voids with mixed gravel to the lower half of the pipe. Cover this with compacted dirt. Even with this, some settlement is to be expected. Do not use any part of a heat supply concrete duct to support a kiln foundation. Use high-temperature rope inserted between the pipes for expansion relief.

11.32 Foundations for Fuel Silos (Hog Fuel, Shavings, etc.)

Use a heavy-reinforced monolithic mat slab for silos. The principle loading on top of the slab is the material density times the holding capacity of the bin. The perimeter will also have to be increased to handle the additional weight of the structure. Core samples must be taken of the underlying soils before designing the slab. Perched water tables can cause tilting and collapse. Earthquake codes also apply.

11.33 Flat Slabs for Lift-Truck Traffic

Typically, use an 8″ thick slab with #5 rebar on 12″ centers in both directions. The perimeter edges are subject to failure more so than the interior of the slab. Perimeters should include a 24″ turndown, reinforced with #5 rebar. In high traffic areas, expansion joints should include 1″ diameter steel bar slip joints, for shear, located on 12″ centers. All slab designs should be reviewed by a structural engineer following soils tests.

11.34 Quality Control in Foundation Construction

Lack of pretesting and poor quality-control practices are the most common causes of foundation failures. The best foundation designs are based on calculations involving soil stability and bearing strength and concrete strength. If soil conditions are not controlled failures will occur. If the concrete mix is not controlled, failures will occur.

Most Common Problems That Lead to Failures

Moisture in soils and concrete mix are major factors that must be controlled.

The type of concrete mix should be site specific. See the ACI guidelines.

Contractors rarely install rebar according to design specifications. Contractors often hire cheap field labor that cause failures.

11.35 After Construction Is Completed

Once the dryer is put in operation, the moisture content of the soils directly under driers can approach zero due to the high temperatures and low relative humidity. If the kiln sits dormant for several months, the foundation can heave if located above highly expansive soils. Do not let wet-bulb wells drip water onto the concrete inside the kiln. Review these issues thoroughly with the engineers doing the soil tests and the foundation design before finalizing on a final design. In some types of track kilns, foundation movement due to changes in soil moisture content can cause problems with tracks and tram cars.

11.36 Air Circulation Fan System Dynamics in Convection Lumber Dryers

Back in 1972, when I started to practice engineering in the lumber drying industry, it was not long before I discovered that most of the lumber kilns in the USA were seriously lacking in both air circulation and temperature control. Once I started seeing kilns in operation across the country, it became obvious to me that most lumber kilns were, to put it mildly, designed by people who simply did not understand lumber drying. Back during the 1970s, many of the kiln manufacturers were there mostly for the purpose of selling replacement parts and bailing themselves out of lumber dryer problems.

That was also a time when sales hype about lumber drying was rampant and the personal relationships between dryer salesmen and their customers was more important than the dryer's performance. It was a time when knowing the first name of the mill owner's children was more important than having a quality dryer design.

Sales price was also a big factor. If a kiln salesman could offer a dirt-cheap dry kiln, his chance of closing the sale was guaranteed.

Because so many people in the industry knew very little about lumber drying and the massive cost of degrade, kiln salesmen across the country became masters at blowing hot air and made a ton of money doing just that.

By 1975 I had designed the first 14-hour southern pine kilns for drying 2 × 4 lumber. Up until then most southern pine kilns were struggling to get drying times below 28 hours. This achievement was reached when Walker Wellford (R&D Director at the Moore-Memphis Company) told me about the high-efficiency AEROVENT reversible propeller fan.

To maximize the benefits of this fan, I designed the first high-temperature cross-shaft fan drive with external motors. This design allowed kiln temperatures to peak safely up to 300 degrees F without damaging internal fan motors. Eventually, almost all kiln manufacturers adopted their own versions of this high-performance fan system.

By 1976, many Moore-Memphis track kilns had package air velocities more than 1200 feet/minute with 1″ thick stacking sticks. Drying times of 15 hours was common for 7/4 southern pine lumber using entering air temperature controls. With exit air temperature control, drying times of 11–12 hours was possible with 1–1/4″ stickers. By 1978, 7/4 southern pines were being dried successfully in 5 hours in a Moore-Memphis experimental kiln.

Successful lumber drying requires exceptional expertise in air handling engineering. To achieve successful air handling in a dryer, every aspect of the entire air delivery system must be understood. Everything from the fan motors to the baffles at the floor of the dryer must be carefully studied and designed accordingly. Failure to pay attention to the smallest details can result in a design that is inadequate, inefficient, or over-designed.

Sufficient air circulation is needed in lumber dryers to deliver heat energy to the surfaces of the lumber. Not only must the amount of air circulation be adequate, but it must also be uniform throughout the length and height of every lumber package inside the dryer. If either the amount or the distribution of air circulation is inadequate, problems with dryer production and moisture control will occur. In lumber drying, there are no short cuts to air handling problems.

11.37 Air Heat Dynamics in Convection Lumber Drying

Let us take a brief look at the **heat dynamics** of air movement in lumber drying.

There are four principal terms that need to be understood.

Air velocity is a measure of the speed of air (expressed in feet/minute).

Volumetric flow is a measure of the volume flow rate (expressed in cubic feet/minute).

Mass flow rate is a measure of the air mass flow rate (expressed in pounds/minute).

Energy flow rate is a measure of the heat energy flow rate (expressed in BTU/ minute).

Air velocity is measured using pitot-tube manometers, hot-wire anemometers, or vane meters. Depending on the design of the lumber dryer, the air velocity through lumber packages can vary from 100 ft./min. (for extremely slow drying) to over 1500 ft./min. (for hyper-temp softwood dryers).

Volumetric airflow rate (cubic feet/minute) is calculated by multiplying the air velocity through a sticker opening in a lumber package times the cross-sectional area of the sticker opening. The term CFM means cubic feet/minute. Sticker CFM = sticker opening area (square feet) × air velocity (feet/minute).

The following table demonstrates how sticker CFM is determined by air velocity and sticker thickness in one 23″-wide sticker opening:

Air velocity (ft./min.)	Sticker thickness (in.)	CFM
100	.50	7.98
100	1.00	15.97
300	1.00	47.91
1500	.50	119.79
1000	1.00	159.72
1500	1.00	239.58
1500	1.50	359.37

From this table, we can see that the air flow rate through sticker openings in commercial wood dryers can vary from as low as 7.98 CFM to as high as 359.37 CFM. That is a 45:1 ratio from the highest to the lowest CFM.

Air mass flow rate is calculated by multiplying the CFM times the density of the air in the flow stream. Density units are pounds/cubic feet. For standard air at 70 degrees F, the density is .075 pounds/cubic feet. As air heats up, it expands, and its density drops. As air temperature drops, its density increases.

Air energy flow rate is calculated by multiplying the air mass flow rate times the specific heat of the air in the flow stream. For standard air, the specific heat is .24 BTU/pound/degree F. The following table demonstrates energy flow rates entering a sticker opening 23″ wide for different combinations of air velocity and sticker thickness. The entering air stream has been heated 1.0-degree F above the wood temperature.

Air velocity (ft./min.)	Sticker thickness (in.)	CFM	Mass rate (#/min.)	BTU/min
100	.50	7.98	.6	.14
100	1.00	15.97	1.2	.28
300	1.00	47.91	3.6	.86
1500	.50	119.79	9.0	2.16
1000	1.00	159.72	12.0	2.88
1500	1.00	239.58	18.0	4.32
1500	1.50	359.37	27.0	6.48

The following table demonstrates energy flow rates for different air velocities and sticker thicknesses in an opening 23″ wide for an entering stream heated 1, 10, 20, 40, and 60 degree F above the wood temperature.

Air velocity (ft./min.)	**Sticker thickness (in.)**	**Energy flow rate (BTU/min.)**				
		1F	10F	20F	40F	60F
100	.50	.14	1.4	2.8	5.6	8.4
100	1.00	.28	2.8	5.6	11.2	16.8
300	1.00	.86	8.6	17.2	34.4	51.6
1500	.50	2.16	21.6	43.2	86.4	129.6
1000	1.00	2.88	28.8	57.6	115.2	172.8
1500	1.00	4.32	43.2	86.4	172.8	259.2
1500	1.50	6.48	64.8	129.6	259.2	388.8

From this table, we can see that the BTU/min air delivery rate entering sticker openings varies from .14 to 388.8. That is a 2777:1 ratio from the highest to the lowest heating rate. In commercial wood dryers the lowest heating rate is probably .56 and the highest is probably 250. That is a 446:1 ratio from the highest to the lowest heating rate demonstrating the wide variance in drying rate conditions in commercial wood dryers.

This table visually demonstrates that there are literally millions of possible combinations of air velocity, sticker thickness, and heating rates in commercial dryers. The table also explains why there has been so much confusion in the lumber drying industry during the last century about the effects that sticker thickness, air velocity, and heating rates have on lumber drying quality and drying rates.

We can go even further with this kind of analysis. Let us suppose we prepared a table that showed average energy flow rate through a 23″ wide sticker opening for drying two different species and thicknesses of wood dried in commercial dryers. The table would look like the following for the first foot of air travel into the package when green lumber is saturated with water:

Energy Flow Rates in the First Foot of Air Travel

Species	**Board thickness (in.)**	**Drying rate (%MC/min.)**	**Energy flow rate (BTU/min.)**
w. oak	4/4	.003	.17 – 1st foot
so. Pine	7/4	.350	34.76 – 1st foot

Now calculate the energy flow rate for an 8′ wide package using these heat rates across the package.

Species	Board thickness (in.)	Energy flow rate (BTU/min.)	Energy flow rate (BTU/min.)
w. oak	4/4	.17 – 1st foot	1.36–8′ package
so. pine	7/4	34.76 – 1st foot	278.08–8′ package

Note that the required energy flow rate for 4/4 oak is well within the preceding table for air flow energy rates. However, note that for 7/4 southern pine, the required energy flow rate is just under the 388.8 figure for 1500 ft/min, a 1.5″ sticker, and 60 F depression. I will now explain the significance of this.

The Effect of Air Mass Flow Rate and Effective Package Width

The multiple of air velocity, sticker thickness, and air density gives air mass flow rate. If we multiply air mass flow rate times the specific heat of air times the depression (dry-bulb temperature minus wet-bulb temperature), we get the entering (available) energy flow rate for drying the lumber as the air passes through the package.

However, once the heated air stream enters the leading edge of the package, complex heat transfer and fluid dynamic processes determine how the available energy will be transferred into the lumber across the width of the package.

If the surfaces of the boards across the package are heavily wetted with water, the lumber will act like a pool of water and the heat transfer rates into the different boards across the package can be predicted relatively accurately.

During this process, the heat transfer rate into the boards and the evaporation rate will be determined by air velocity, sticker thickness, and the temperature depression of the air stream above each wetted board. Then, as the air stream progresses further across the package, several things occur at the same time. First, the relative humidity of the air stream becomes higher. Second, the dry-bulb temperature of the air stream drops. Next the density, specific heat, thermal and viscous properties of the air-moisture mixture will change significantly. Then, as the mass flow moves further into the package, because of the effect that all of these have on the convective heat transfer coefficient at the surface of the board, the (local) heat transfer rate at each board will change.

Finally, after a distance determined by all the preceding factors, the air stream approaches 100% relative humidity. When this happens, the air stream approaches saturation, and the result is that the drying rate of all the lumber downstream of this point is essentially zero.

Theoretically, the drying rate never fully reaches zero, but for all practical purposes, the drying stops. This is why after running the kiln fans in one direction they have to be reversed for operation in the opposite direction.

Following several fan reversals and some drying time, the surfaces of the boards across the package start to dry out. When this occurs, the temperature of the surfaces of the boards starts approaching the dry-bulb temperature of the mass stream above each board. When the difference between the dry-bulb temperature of the mass stream above a board and the board surface temperature is zero, the board stops drying. At this point in the drying cycle, the moisture content of the surface of each board will approach an equilibrium value dependent on the dry-bulb and wet-bulb temperature of the mass stream above that board.

Depending on the drying dynamics and moisture content of each board, the difference between the entering and exiting dry-bulb temperatures across the package will change. During air stream saturation that occurs in the early stages of drying, the temperature drop across the package will be constant.

This temperature drop is determined by air velocity, sticker thickness, effective package width, and the integrated package moisture content. In some softwood dryers designed for 8′ wide packages, the initial temperature drop across the package can approach 60 degrees F.

The temperature drop is also dependent on many air delivery factors, baffling practices, loading practices, reliability of the heating system, etc. Because of these changes between dryer loads, temperature drop is not a reliable industry-wide indicator for controlling final board moisture content accurately. In softwood operations, with very tight controls over sawing practices, board thickness, sticker thickness, stacking, and dryer loading, temperature drop across the package (often called TDAL or Delta-T by non-engineers) can be a statistical **indicator** of the average moisture content of the package. However, in the lumber manufacturing business, temperature drops can be difficult to rely on day after day.

11.38 Dryer Uniformity Factor (DUF)

In all wood dryers, there is a distinct relationship between the heat-energy circulation capacity (CC), the water evaporative load (EL), and the drying uniformity factor (DUF). The dryer uniformity factor DUF is a dryer performance coefficient that I developed during the early 1970s for establishing minimum design standards for both hardwood and softwood dryers. DUF is based on energy, mass, and heat transfer equations, including empirical data. DUF is a measure of how uniform a dryer can dry lumber.

$$\text{Circulation cap.}(\text{CC}) = \text{stream density} \times \text{sticker area}$$
$$\times \text{air velocity} \times \text{specific heat} \times \text{temperature depression}$$

DUF is the ratio of **circulated energy** capacity (CC) to the **evaporated energy** load (EL):

$$\text{DUF} = \text{CC} / \text{EL}$$

For each species, board thickness, package moisture content, and drying temperature schedule, there is a minimum DUF by which wood dryer designs should comply with. For all drying above the fiber saturation point, significant problems with moisture content variances between the middle and the outer edges of stacked packages will occur if the DUF drops too low. When drying below the fiber saturation point, the DUF drops linearly with the average package moisture content. However, even though the DUF approaches zero when the average package moisture content approaches zero, sufficient velocity should be maintained through the packages and overheating coils to keep the heat energy distributed throughout the dryer. It has been my experience that the DUF during the final stage of drying should never be less than ½ the design DUF for drying above fiber saturation point. Such a control scheme reduces the dryer fan's electrical energy usage during the last hours of drying by 87.5%.

The benefits of DUF modeling for a specific species and board thickness are:

1. The prevention of wasted electrical fan horsepower in all types of drying
2. The prevention of excessive moisture gradients in hardwoods
3. The reduction of dryer fan, bearing, and motor failures

The three sample curves shown in the following text demonstrate the DUF model. The curves also explain why kiln operators learned over 50 years ago they could save electricity by slowing dryer fans down near the end of drying cycles.

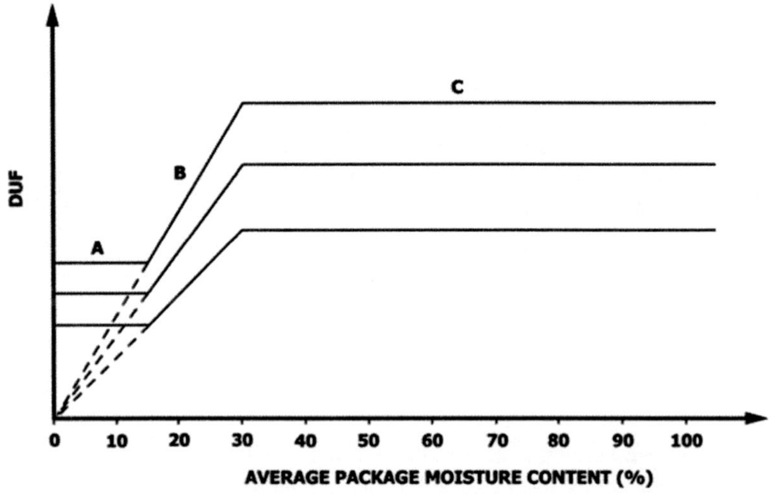

Because the moisture movement dynamics inside lumber varies significantly depending on the species, research is needed to establish minimum DUF curves for each species and board thickness. With the increasing use of computer technology, automatic control systems could be employed by which a dryer operator could load a specific dryer with one species of lumber and then enter: board thickness, board width, sticker thickness, effective package width, and initial moisture content. After the operator clicked on the start command, the computer would then automatically select the proper temperature schedule and fan speed. As the drying progressed, the fan speed would be changed automatically.

The effect of board thickness, width, and beta ratio on dryer production and final moisture content in all dryers, board size (thickness and width) has a strong effect on dryer production, degrade, and DUF. In addition, the beta ratio (board width divided by board thickness) also affects the drying rate and levels of degrade. The effects are different between low-temp and high-temp drying.

Increases in either board thickness or board width will increase drying time. Increases in dryer temperature and/or temperature depression will decrease drying time.

Because most wood is dried in commercial convection dryers, the total exposed surface area of each board to the circulated air stream effects the ability of the heated air stream to deliver heat energy to each board for evaporating the water out of the board. If the boards in a package are stacked tight together, the circulated air stream cannot reach the surfaces between the boards.

For this reason, dryer operators often air space thick lumber in packages to improve the flow of air around the side surfaces of the lumber. In all types of drying, especially thick lumber, air spacing reduces the standard deviation of the moisture content of the material exiting the dryer. The effect is not as dramatic for thin lumber such as 4/4.

11.39 Sticker Thickness Versus Air Velocity Testing Methods for Convection Dryers

Situations in lumber drying often arise in which the relationship between sticker thickness and air velocity needs to be established. The question is often asked, if the sticker thickness is changed, what will be the effect on the air velocities through the sticker openings, and thus the change in CFM, and thus the effect on drying?

Assume a dryer exists in which 1″ thick stickers are being used and air velocity testing reveals the average velocity through the sticker openings is 1000 ft./min. If the sticker thickness is increased to 1.25″, what will be the new average air velocity?

The answer to this question is based on two principal factors:

1. The shape of the fan performance curve (head vs. CFM)
   This is determined by the speed, type and number of fans, and their pitch settings.
2. The shape of the system curve (backpressure head vs. CFM) at the fan wall
   This is determined by the dryer design, stacking efficiency, bypass CFM, and grid factor.

For a specific load of lumber in a dryer, the air velocity through the sticker openings will be determined by the delivered differential head from one plenum of the dryer to the opposite plenum. If only the sticker thickness is changed, the differential head between the two plenums will also change. The effect is driven by many factors not readily apparent to those unfamiliar with dryer engineering and fluid dynamics.

If we assume that a fan design has been properly engineered into a dryer, operates at a stable portion of the fan curve, and does not hunt, then the following procedures can be used.

The Three-Test Method

One simple test method is to load the dryer three times with lumber using different **test** stickers and measure the change in air velocity profiles through the sticker openings. If this method is used, it is very important that the dryer be loaded with the exact same type of lumber and the exact same load space during each of the three tests. The locations for taking velocity readings must also be the same during the tests. I recommend vertical velocity profiles be taken at no less than four locations across the load area, with the fans running in the same direction during the tests. The dryer temperature during each test will also have to be recorded for temperature compensation calculations.

This test method should be used for sticker thickness test increments of ¼″. The first test uses the current sticker thickness, the second test uses a ¼″ thicker sticker, and the third test uses a ½″ thicker sticker. From these three tests, graphs can be produced that demonstrate sticker velocity and CFM vs sticker thickness for a specific dryer.

Since velocity profiles are driven by baffling, bypass CFM, and plenum fluid dynamics, do not be surprised to see the shape of the velocity profiles change significantly during these tests.

Fluids-Dynamics Method

If the three-test method is not used, a fluids dynamics model can be used to predict average air velocity changes due to sticker thickness changes of less than 30%. However, due to the large variations in dryer designs, and the number of assumptions that have to be made, this method is best used on dryers in operation where sufficient data can be collected to justify the exercise.

11.40 Optimum Sticker Thickness for Maximum Drying Production Rate

Dryers are often installed in manufacturing facilities in such a manner that additional dryer production is needed but additional dryer size is not possible. In many cases, the limiting factor is a lack of available real estate. In some cases, the existing dryer size, fan system, temperature schedule, and board thickness are fixed, and the only remaining variable is to reduce sticker thickness to try to increase dryer holding capacity and drying production rate.

The following two curves demonstrate a typical relationship between sticker thickness and dryer holding capacity and drying production rate for a kiln drying 7/4 pine lumber.

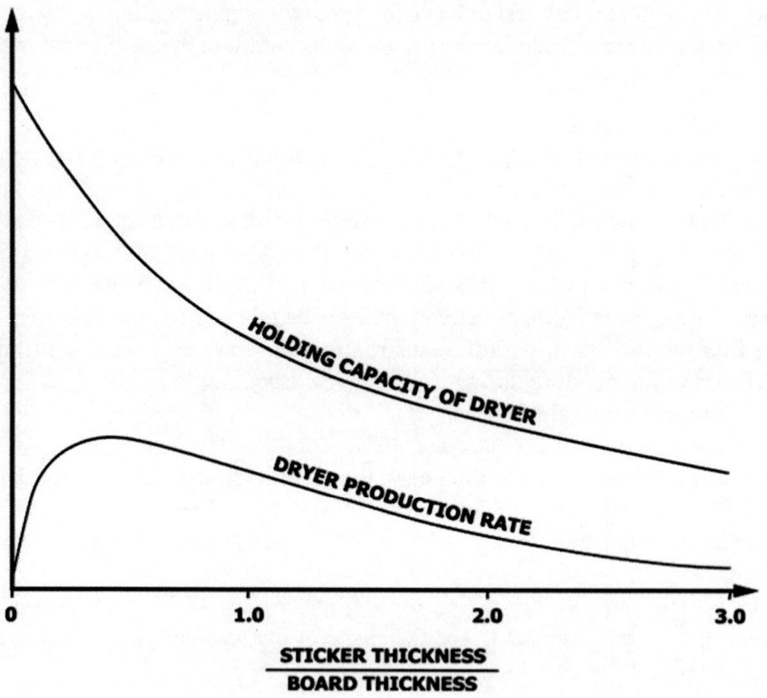

The heat-delivery dynamics of sticker thickness are determined by the following eight variables:

1. The targeted board thicknesses
2. The effective package width
3. The initial moisture content of the lumber entering the dryer
4. The specific gravity of the lumber
5. The temperature schedule used
6. The dryer's fan performance curve (CFM vs HEAD)
7. The stacking and loading efficiency of the packages in the dryer
8. The surface roughness of the boards (sawing techniques and maintenance)

If we assume the variables listed in the preceding text are constant, the simplest approach for determining the optimum sticker thickness for drying production rate (for a specific dryer located at a specific plant) is to experiment with different thickness stickers until an optimum size is found.

I recommend sticker thickness in increments of 1/8″ be used in such tests. Accurate final moisture content and grade data is required to determine the optimum production-rate sticker thickness for a specific board thickness. I suggest you start your testing with a sticker 1/2″ thicker than what you are currently using and then gradually experiment with thinner stickers. During these tests, you will discover that problems with both moisture content and degrade will increase with thinner sticks, **IF** you do the tests correctly.

11.41 Optimum Economic Sticker Thickness

Unfortunately, there is a tradeoff when selecting a sticker thickness for optimum production rate. Not only does sticker thickness affect dryer production, but it also has strong effects on final moisture content distribution and lumber degrade. This leads us to the next test for determining the optimum economic sticker thickness.

In this test, both the economics and heat-delivery dynamics of sticker thickness are determined by the following 11 variables:

1. The targeted board thicknesses
2. The effective package width
3. The initial moisture content of the lumber entering the dryer
4. The specific gravity of the lumber
5. The temperature schedule used
6. The dryer's fan performance curve (CFM vs HEAD)
7. The stacking and loading efficiency of the packages in the dryer
8. The surface roughness of the boards (sawing techniques and maintenance)
9. The total electrical and fuel usage of the entire drying system
10. The total labor and overhead costs per board foot dried
11. The total degrades due to drying alone

In this test, each of the preceding 11 variables must be known for each sticker thickness used in the tests. Once the data is collected correctly, you will discover that, in most commercial dryers, the optimum economic sticker thickness exceeds the optimum production rate sticker thickness by 20–50%.

If drying capacity is still a problem after you find the optimum economic sticker thickness, this is when a complete analysis of the heating and package air delivery systems will be needed. In many cases, upgrading the fans and baffle systems plus improving the stacking and dryer loading practices will solve both problems, but not always. For many softwood operations, it usually does, but for hardwoods, the only viable solution may be to add additional dryers.

11.42 Location of Fan Systems for Convection Lumber Dryers

Dryer fan systems are either located above the load of lumber, or at ground level. Up until the early 1907s, there was a large quantity of ground-level fan designs in the lumber industry. Today most dryers use the overhead design.

The ground-level fan design requires less front-end cost but has a reputation for drying uniformity problems and fires especially in direct-fired designs. Because the fans are at ground level, they are easier to maintain but are also a significant safety hazard. Ground-level dryer fans must be fitted with protective cages to protect people and falling boards from getting too close to the fans. The fans, fan housings, blade containers, and drives must be liberally over-designed to prevent equipment damage or injury from fan failures.

The overhead fan design is more costly to build and maintain but avoids the problems with ground-level fan designs.

Safety/Risk Analysis

Discuss the fan safety issue with your dryer operators, your safety manager, your attorney, and your insurance carriers before installing or operating a ground-level fan system. I suggest you not use them unless very stringent safety and maintenance practices are enforced at your plant. Even with stiff safety practices in place, ground-level dryer fans may cause a serious or fatal accident.

Track-Type Dryers

In track kilns built during the last century, fans have been installed in the ceilings above the loads of lumber and have also been installed at ground level between the tracks in multiple-track dryer designs.

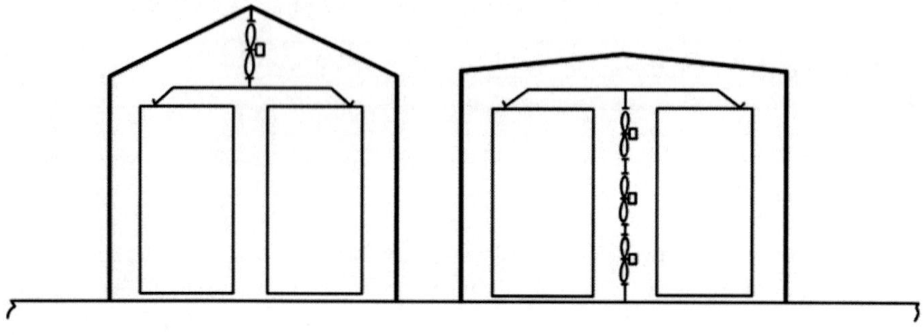

Side-Loader (Package) Dryers

Like track kilns, the fans in package kilns have also been installed both in the kiln's ceiling and at ground level. The ceiling design became the most predominate design during the last 75 years. There are some dryer manufacturers who offer back-to-back designs in which the main doors are installed on opposite sides of the dryers and the fans are installed between the lumber load areas.

Back-to-back designs have been used for the last 75 years. They have the same pros and cons as discussed earlier. High winds can create safety problems for large doors used on back-to-back designs. For this reason, I do not recommend they be used unless the dryer is in an area surrounded by other large buildings near the dryer.

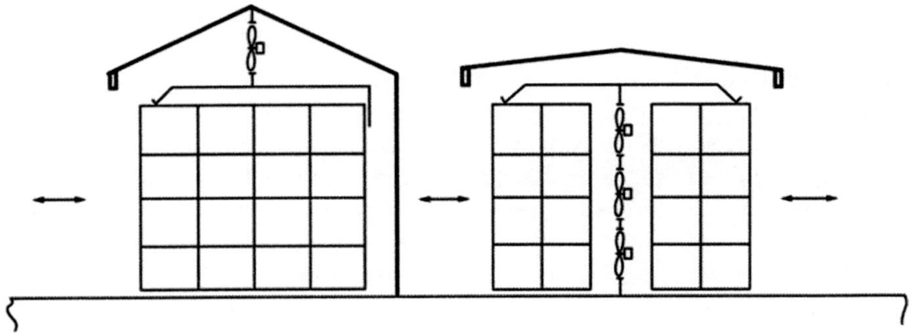

Sidewinder Dryers

A sidewinder design can use either a track or side-loading scheme. Sidewinder designs are used in small drying operations requiring low construction cost. Their advantages are:

1 All equipment is located at ground level
2 They only use one or two fans total
3 They only use two banks of steam coils if steam heated
4 They only have one heat supply duct if direct fired

Sidewinder dryers can be either of a track or package design. However, there is a limit to how long they can be. Properly designed, they are the most maintenance free dryers of all types.

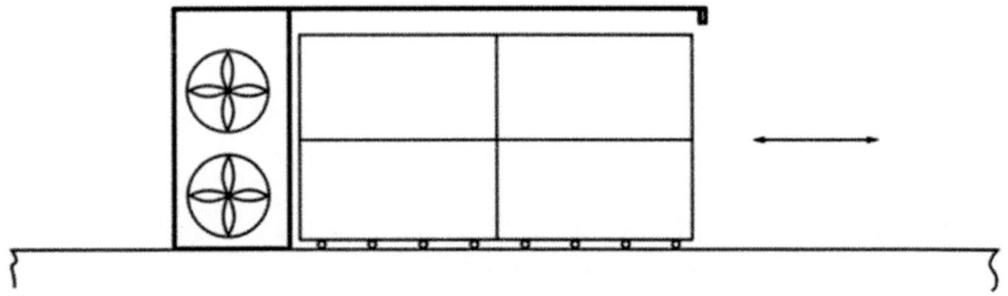

SIDEWINDER TRACK DESIGN

11.43 Fan Safety Rules

Fans are extremely dangerous machines that can cause serious injury or death. This is especially true when maintenance is being performed on them. Because wood dryers have numerous fans, the risk for failures and accidents are high. The following fan safety rules are recommended:

Rule #1 – Prepare written safety rules for dryer operators to follow. Keep a current copy of these rules posted in the dryer's control room.

Rule #2 – Insist that dryer operators report dangerous activities by people to plant management.

Rule #3 – Comply with the National Electrical Code when both designing and maintaining dryer fans. Safety lockouts must be used when working on fans. Each fan motor should have an electrical disconnect located next to it to prevent accidents.

Rule #4 – Never work alone. Always have no less than two people present when working on dryer fans.

Rule #5 – Never get on the fan deck or near fans while the fans are running. Never enter a fan plenum area (with ground-level fans) while the fans are running.

Rule #6 – Ground-level fans should be fitted with personnel safety guards and blade containment shields.

Rule #7 – Minimize fan failures to reduce the probability of an accident

Choose the right fan for the application
Choose the right fan drive for the application
Precision balance each fan assembly prior to its use
Torque all fasteners to the manufacturer's specs
Check minimum tip clearance to prevent failures
Test jog replaced fans in both directions before starting the fan system

Rule #8 – Provide safe access to fan decks using gable doors, fan deck doors, and fixed ladders inside the dryer. Access to fans, shafts, bearings, and motors should not require maintenance personnel to place themselves in precarious positions in which accidents can happen.

11.44 Designing the Fan System for the Dryer

There are four major issues in the design of dryer fan systems. These are listed in the following text in their order of importance.

1. Safety is the number one issue in designing fan systems for all types of wood dryers.
2. Design integrity is required to prevent failures and the loss of dryer production.
3. Adequate CFM delivery is required to maintain dryer production and minimize degrade.
4. High fan system efficiency is required to minimize electrical costs.

To ensure safe designs, use only fans rated by the **actual fan manufacturer** for lumber kiln service. This is an issue that should be discussed thoroughly when purchasing a dryer because many dryer manufacturers do not actually manufacture the fans used in their dryers. In most cases, the fans installed in dryers are purchased from manufacturers whose specialty is the design and manufacturing of fans and blowers.

Selection of Metal Alloys for Fan Construction

Dryer fans are reversed numerous times each day, operate most of the time, are exposed to wide variances in temperature, and are exposed to potentially aggressive corrosive conditions. The selection of alloys for fan hubs, blades, and fasteners should match the type of dryer the fans will be used in. Durability, thermal expansion, alloy strength vs temperature, and the ability to resist corrosion are the key factors. If properly designed and maintained, dryer fans should last for decades.

The predominant metals used in the construction of wood dryer fans are mild steel and aluminum. Stainless steel fasteners are often used in drying corrosive species.

Mild Steel Sheet-Metal Fans

Mild steel sheet-metal fans have been used in thousands of dryers around the world. However, these fans have their pros and cons. First, they are typically heavy and difficult to both start rotating and reverse. The drives fitted to these fans have to be designed for the heavy inertia of these fans. Second, because sheet metal fans are not aerodynamic shapes, they do not have high aerodynamic efficiency at high backpressures. However, in many dryers for slow drying species, the backpressure on the fan wall is less than ¼″ water column and steel fans can perform at an acceptable level of efficiency. Steel fans are best applied to dryers requiring less than 200 feet/minute air velocity through the sticker openings.

Contrary to popular belief, modern complex aluminum airfoil-shaped fans may not be a better choice than steel fans at very low backpressures. Third, mild steel's strength can exceed the highest temperature expected in wood drying, making them suitable for ultra-high-temperature applications. Finally, if applied to dryers using E&C cycles, the corrosion of mild steel fans can be a problem.

Aluminum Fans

Aluminum fans are used in high-air-velocity wood dryers requiring either high aerodynamic fan blade efficiency or resistance to corrosion or both. Because aluminum castings are relatively inexpensive, complex hub and propeller designs are possible using this metal. Aluminum is also light compared to steel. Its use in propeller fan designs permits a wide range of tip speeds to maximize performance and fan efficiency. In most cases, the aluminum alloys used exceed the maximum expected temperatures inside lumber dryers. In all drying applications, including those approaching 300 degrees F, motor starting torque and fan rpm must be controlled to prevent structural failures. Numerous alloys of aluminum and heat treatments are available for improving both strength and durability.

Single Cast Versus Assembled Fans Depending on the aluminum alloy used, there is an upper limit to the size of the casting or mold in which propellers can be poured. Generally, 66″ diameter is the upper limit for single cast aluminum propeller fans. Some manufacturers limit their single castings at 60″ diameter. Some manufacturers have standardized on designs in which the hub and propellers are cast separately.

Fan Assembly Design Issues for Wood Dryers

Because dryer fans go through numerous thermal and reversing cycles, the following design issues must be addressed by the designer:

Cyclical Temperature Extremes

Wood dryer fans can experience thousands of temperature cycles each year. Every part of the fan assembly must be designed for these cycles. If steel fasteners or bushings are used with aluminum hubs, the design must consider the numerous differential thermal cycles the assembly will experience during its life. Standard engineering calculations for thermally induced strain need to be performed for every part of the fan assembly. If aluminum fan castings are used with steel fasteners, the choice of fastener alloy and cyclical strain fatigue must be reviewed.

Maximum Fan Shaft Torque

Depending on the design of the fan drive, the NEMA torque rating of the drive motor, and whether a soft start design is used, the fan will be subjected to a maximum level of shaft torque during its operation. This maximum torque will have to be known to determine the stresses subjected to the different parts of the fan system. Special attention must be given to the forces applied to keyways and tapered hubs.

Maximum Centrifugal Force on the Blades

The centrifugal force on each of the fan blades is determined by the rpm and distribution of mass along the fan blade. The centrifugal force is proportional to the square of the rpm. An increase of 20% in rpm will cause a 44% increase in centrifugal forces on the blades:

$$1.20 \times 1.20 = 1.44$$

Maximum Aerodynamic Forces on the Blade

The aerodynamic forces on each blade of a fan are determined by the number of blades on the fan, the CFM through the housing, the static head (backpressure) on the fan, the rpm (speed) of the fan shaft, and the density of the air stream. Aerodynamic forces on fan blades are a relatively small force when compared to the centrifugal forces acting on the blades.

Combined Reversing Stresses from Torque, Centrifugal, and Aerodynamic Forces

The fan's hub assembly including the hub's attachment method to the drive shaft must consider all the previously described forces and stresses in combined and cyclical modes. The blade shank, hub, hub bushing, fasteners, and all the keyways must be designed to prevent loosening, fatigue cracks, and failures.

Balancing the Fan Assembly

Each fan assembly should be precision balanced during the manufacturing process. The precision balance and manufacturing integrity of each fan assembly must absolutely be field checked after the final pitch is set and the fan is ready to be installed on the shaft. Any imbalance detected during a static roll test on hardened side rails must be corrected. Any manufacturing defects that create fan tip wobble, irregular blade surface paths, irregular pitch angles, etc., must be corrected before using the fan in a dryer. Either aerodynamic or blade mass fan imbalance will cause early failures of bearings and their support struts. It is best that a quality control rep for the buyer be present when fans are field inspected before installation. The reason is most fans are out of balance while still in their shipping crates and some fan manufacturers may fail to stand behind their product. This is the reason all fans to be installed in any dryer must be inspected months before being installed in any dryer. Doing this allows the buyer to resolve any manufacturing defects with the fan supplier before the planned startup date for the dryer.

Fan System Efficiency

Because fans in dryers essentially operate continuously, their overall electrical efficiency is a financial consideration in their selection. Fan system designs with low efficiencies should be avoided due to the high cost of electricity. To design the best system for the dryer, the following concepts and practices should be both understood and followed.

Fan Laws

The laws used in fan system designs are:

CFM is proportional to the first power of fan speed (RPM)
HEAD is proportional to the second power of fan speed
POWER is proportional to the third power of fan speed
HEAD is proportional to the first power of air density

System Curves

Every dryer has a specific relationship between the total CFM delivered by the fan system and the backpressure HEAD on the fan wall.

SYSTEM HEAD is proportional to the second power of the total fan CFM

The following system curve is for a high-temp, double-track kiln.

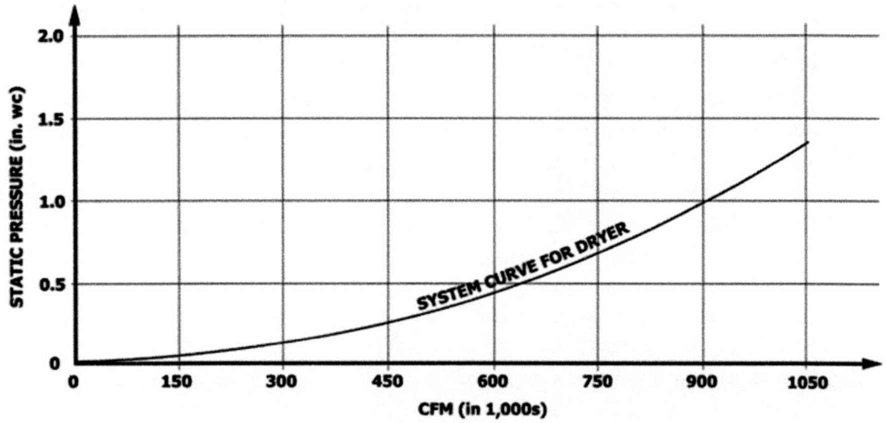

Changes in sticker thickness changes the dryer's system curve. A reduction in sticker thickness causes an increase in head at the fan wall, thus reducing the total CFM produced by the fans. An increase in sticker thickness decreases the head at the fan wall, thus increasing the total CFM produced by the fans.

Sticker Head @ Standard Air Conditions

The total pressure drop (head) across a lumber package can vary from as low as .005wc″ in low velocity hardwood dryers to as high as .150wc″ in high-air-velocity softwood dryers.

Total sticker head and cfm are dependent on the following variables:

The air velocity through the sticker opening
The sticker thicknesses
The actual width of the package
The package grid factor (sticker thickness/board thickness)
The relative roughness of the surfaces of the lumber
The number of lumber air gaps in the package width
The width of the air gaps between boards

See the ASHRAE Handbook for engineering methods for calculating air pressure drops in rectangular ducts in laminar flow, transition flow, and turbulent flow. You will notice that to use their calculations, so many assumptions have to be made that the reliability of the calculations will be extremely low.

I suggest that actual field data be collected at the plant where the dryer is to be installed before assigning a sticker head value for a specific board and sticker thickness and air velocity. You will need an accurate manometer and air velocity meter to collect the data. Locate both static probes 12″ from the outer sides of the packages and keep this same location during every test.

Collect head data for green lumber, with different levels of air velocities passing through the package using the actual board thickness and sticker thickness the new dryer will be using. Once the head data has been collected, you will need to plot a graph of head loss (wc″) versus air velocity (ft/min).

You will also notice that the sawing practices at the plant will have a dramatic effect on the roughness of the surfaces of the boards and thus head loss across the package. This is one reason why the performance of lumber dryers can be so elusive when trying to dry without an E&C cycle. A change in sawing practices or saw conditions will change the roughness of the surface of the boards causing significant shifts in sticker CFM from one dryer load to the next.

One problem with sticker head loss calculations and measurements is that below about 260 to 300 ft/min air velocity, the fluid dynamics for head loss changes. Above 300 ft/min, the fluid stream enters a region known as transition flow in which the boundary layers are not either laminar or turbulent.

In the transition region, unstable velocity pulsations are common and air movement acts in very strange ways, like a waving flag. These oscillations occur because of complex fluid dynamics too exhaustive to review in a book of this nature.

This is one of the reasons that vane type velocity meters bounce around so much when used to measure air velocities between 250 and 800 ft/min. in a lumber package. It is also the reason that a precision pitot tube and manometer is a better method for taking accurate velocity measurements. Do not use hot-wire velocity meters for taking readings. They are not reliable.

Fan Performance Curves

Depending on the design and size of a fan, the pitch of the blades, the fan's speed (RPM), and the tip clearance of the propellers, the fan will produce performance curves specific to those conditions. The following example fan curves are for one 72″ diameter propeller fan. Note that the graph includes a curve for CFM vs Static Pressure, and a curve for Brake Horsepower vs CFM.

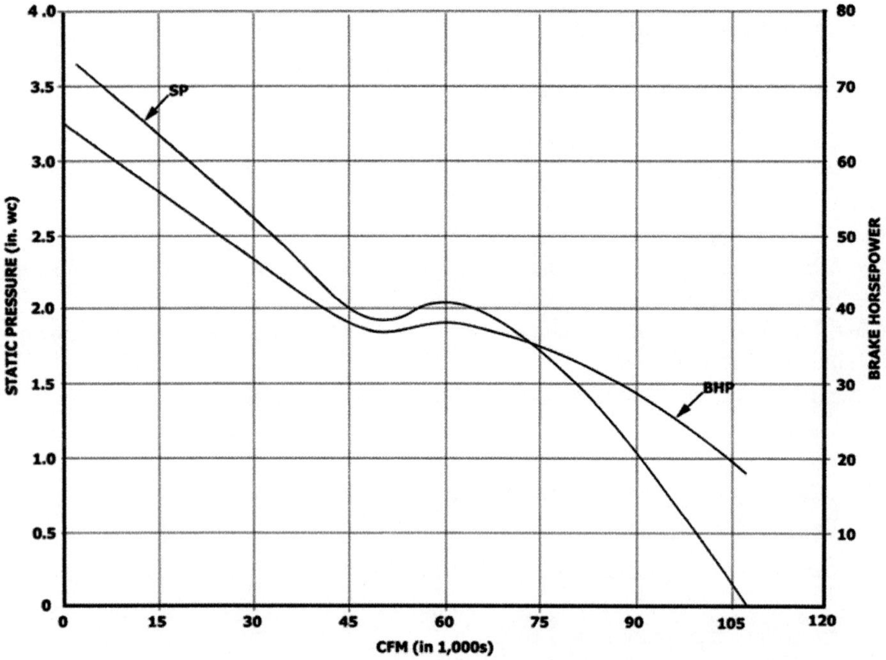

Fan performance curves are available from the fan manufacturer for different diameters, pitch settings, speeds, and number of blades/fans. Because the curves are a representation of the marketability, application, and performance of the fan, they are available at no cost to the party who will purchase the fan. Engineers must have copies of the fan curves for designing the fan system.

Methods for Rating Fan System Efficiency for Wood Dryers and to Avoid Cramming

There are three **methods** used to rate fan system efficiency in wood dryers.

1. The total kilowatts used per total CFM through **all** the lumber packages.

2. The electrical energy usage per pound of water removed during drying.
3. The kilowatts per unit of fan wall work where work is:

$$\text{WORK} = \text{CFM} \times \text{STATIC} \times \text{HEAD.}$$

The preceding term **cramming** is an energy term. Cramming is the act of putting excessive electrical horsepower into a convection dryer fan system to increase drying production rather than designing the dryer properly upfront.

During the early phase of a dryer design project, all three of the preceding listed methods should be studied to evaluate fan system designs in different dryer configurations such as: single-track, double-track, triple-track, quad-track, package type dryers, and low and high door heights.

Choosing the Right Fan Design for the Dryer

Fan systems for wood dryers are designed by going through a series of calculations.

First, the total required CFM through the kiln is specified.
Second, the HEAD at the fan wall, created by the CFM, is determined.
Next, the system curve for the dryer is plotted on a graph.
Next, the minimum allowable fan efficiency is established.
From the preceding text, and the type of dryer being designed, the optimum fan diameter is determined.
Next, the maximum possible number of fans is determined from the fan diameter and length of the dryer.
Comparisons are then made for different number of blades per fan.
Once a specific fan design is chosen, the blade pitch, rpm, and horsepower/fan are determined.
Next, the design is reviewed for the effects of changes in sticker and board thickness.
Finally, the CFM for a specific board and sticker thickness is established.

During this process, the engineer will superimpose the dryer's system curve onto the fan curve to determine the pitch setting, speed, and horsepower requirements. The procedure is repeated until an optimum design is reached. See the following graph that shows where the system curve intersects with the static pressure curve. The intersection point A is the operating point. In this example, the fan operates at 90,000 CFM at 1wc″.

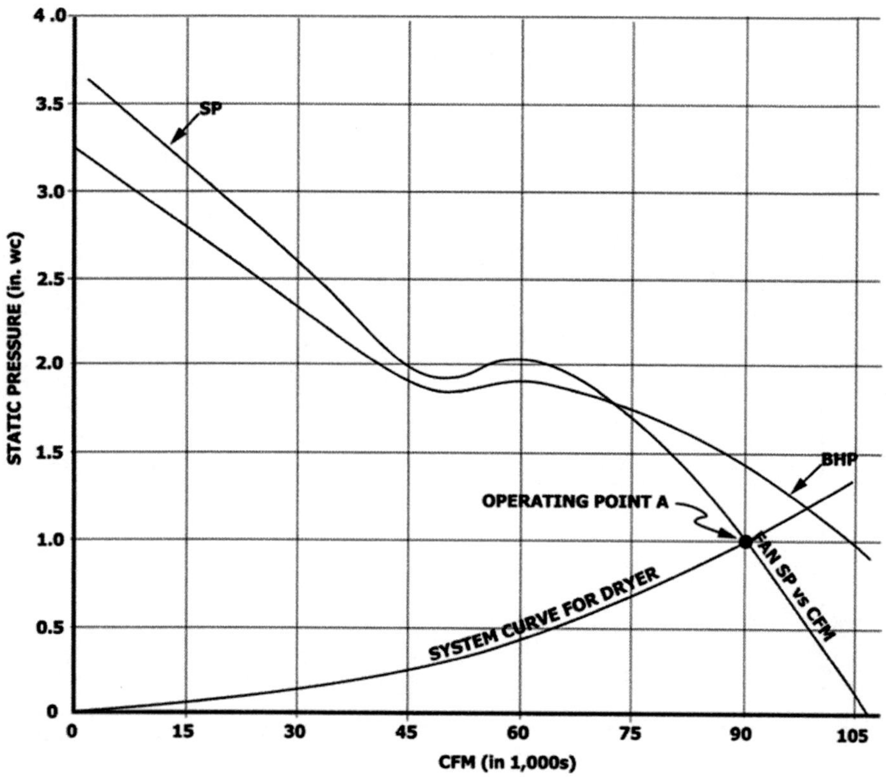

Preventing Surging

Fans in wood dryers can experience surging resulting in very low CFM delivery through the packages. Surging is caused by the system curve intersecting at a point too high on the fan curve and too close to the flat (stalling) portion of the curve. See the following graph. Point A is the operating point described in the previous example. Point B is a location on the curve above which surging occurs.

Surging occurs when the aerodynamic lift characteristics of the propellers begin to stall. To prevent surging from occurring, the fan's operating point A (at all dryer temperatures) should stay on the lower section of the fan curve. Careful analysis of the entire dryer is required to prevent surging from occurring.

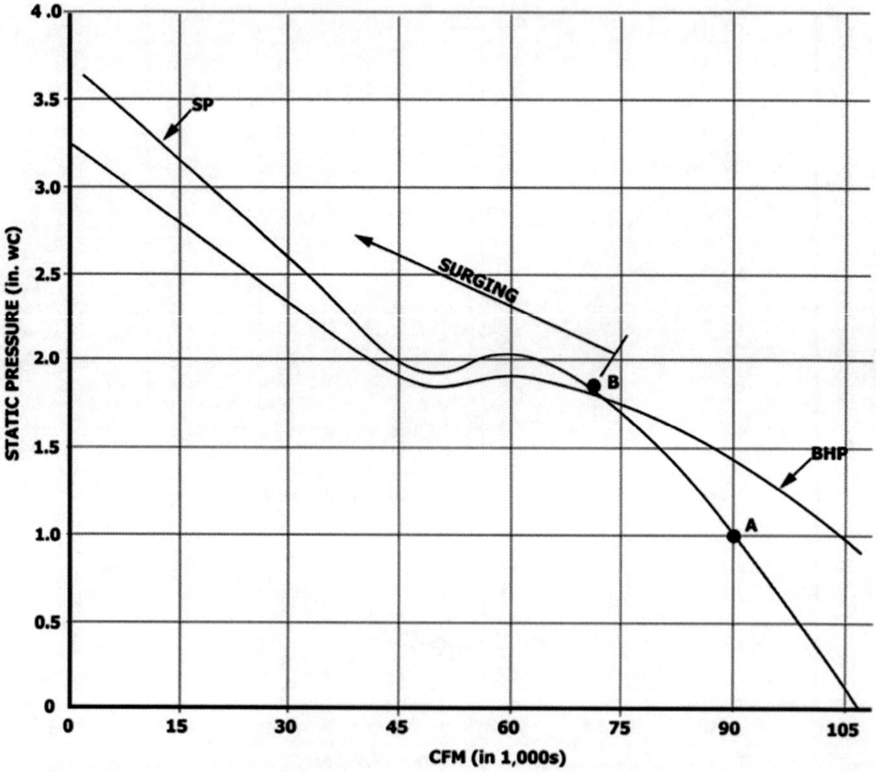

Surging in existing fan systems can be eliminated by either increasing the rpm of the fan and reducing the pitch or increasing the number of blades on the fan or both.

11.45 Designing Dryer Fan Systems for Maximum Electrical Energy Efficiency

Let us take a quick look at how internal dryer fans should be designed for maximum fan electrical efficiency. However, I want to caution you about the model in the following presentation. You need to remember while reading that this is a fluids dynamics airflow model and not an economic model.

Economic models look at front-end and operating costs. This model does not look at equipment cost, but instead only at electrical usage. Also keep in mind that the two objectives in this model are to **save** electricity and still **deliver** the minimum CFM determined by standard air model 74.

The other point is that this airflow model is based on fan designs in which the fan shafts are perpendicular to the lumber packages. Dryer manufacturers refer to this fan configuration as a cross-shaft design. Cross-shaft designs are used in dryers (such as high-temp track kilns) where both high electrical efficiency and CFM are needed.

Let us now look at an airflow model that keeps the fan system efficiency high by keeping dynamic pressure losses in the air stream to a minimum. To achieve this, the ratio of total fan area to total fan wall area should be kept as high as possible. This means the external fan housings' dimensions must be square, slightly larger than the fan diameter, and stacked tight against each other from one end of the fan wall to the other end.

First Thing to Do

Determine the fan wall length and **minimum** fan diameter.

1. The length of the fan wall is fixed for a specific dryer.

   For track kilns, the fan wall length is the inside length of the dryer
   For side loader (package) kilns, the fan wall length is the inside width of the dryer

2. The **minimum** fan diameter (MFD) is determined by the following model equation for minimizing dynamic pressure **energy losses** in the air stream at the fan wall:

$$\text{MFD} = \text{Ks} \times (\text{TSH} + \text{TCOH}) - \text{Where all units are in feet.}$$

   Where:
   Ks is a function of fan spacing and the type dryer.
   TSH is the total vertical sticker opening height in feet.
   TCOH is the total vertical cross-out opening height in feet.

Because the preceding formula was developed from both field-testing and fluid mechanics continuity laws, I did not include its proof in this book. Instead, I included a graph shown on the next page that demonstrates the results of the model.

Keep in mind that the two curves shown are energy-efficiency driven and do not address specific fan propeller performance issues such as number of blades, pitch settings, blade design, rpm, etc. These issues have to be addressed separately when selecting fans.

The graph shows linear (straight) curves, each with different slopes, for two types of dryers:

High-temp track kilns
Low-temp kilns, pre-dryers, and E&C chambers

The bottom scale is total sticker height (TSH) + total cross-out height (TCOH) in feet, and the vertical scale is minimum fan diameter in feet for achieving maximum electrical efficiency.

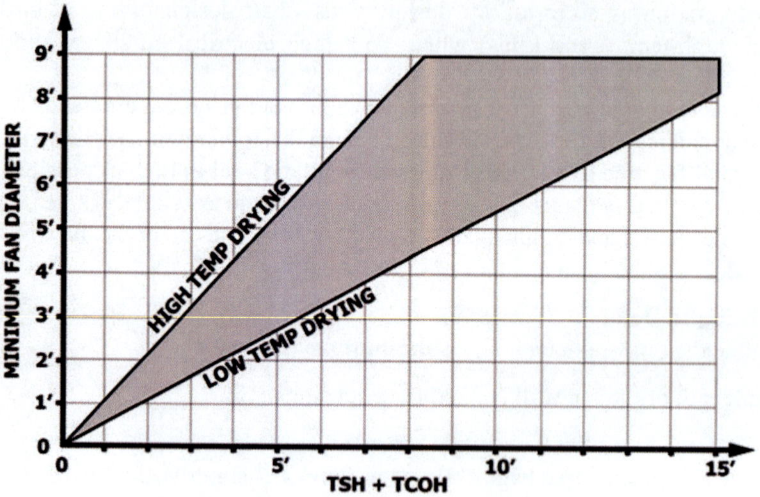

Looking at the graph, the two lines represent the **outer limits** of minimum fan diameters for all types of commercial dryers. Additional interior lines (within the gray area) could be drawn for accelerated and conventional dryers, but I will leave this task up to you.

Use this graph to plot where your dryers fit in this model and remember that your dryer fan housings are probably not stacked end-to-end in the fan wall. Calculate the percentage ratio of total fan area to total fan wall area in your dryers and compare your number to the ideal model percentage ratio of 60%. Next, using the fan diameter D in your dryer, calculate the percentage ratio again with the fans spaced at 1.14 x D. This exercise will give you an idea of how relatively efficient your fan system is.

EXAMPLE

Calculate TSH + TCOH for a specific track type dryer. Let us assume this calculation produced 5.5′. Go to the bottom scale and locate 5.5′. Next, go up vertically to a specific type of track dryer's line. From this point, go horizontally to the left (to the vertical axis) to locate the minimum diameter fan for maximum electrical efficiency.

Second Thing to Do

Determine the static head on the fan wall.

> Head values are highly dependent on the level of engineering put into dryers. Typical head values are shown in the following text to give you a starting point for selecting the correct fan.

Dryer Design and Application	*Static Head in Inches of Water Column (wc″)*
A. High-temp softwood track kilns	
Single-track designs – static heads can be from	.75 to .90
Double-track designs – static heads can be from	.90 to 1.10
Triple-track designs – static heads can be from	.95 to 1.25
B. Accelerated softwood track kilns	
Single-track designs – static heads can be from	.50 to .75
Double-track designs – static heads can be from	.75 to .90
Triple-track designs – static heads can be from	.85 to 1.0
C. Hardwood track kilns	
Single-track designs – static heads can be from	.20 to .30
Double-track designs – static heads can be from	.30 to .40
Triple-track designs – static heads can be from	.40 to .50
D. Package kilns	
Softwood designs – static heads can be from	0.50 to .60
Hardwood designs – static heads can be from	0.10 to .30

Third Thing to Do

Determine the number of fans that will fit inside the fan wall. Let us suppose your analysis for fan diameter came to 5.7′ and fan wall length came to 84′. Since the minimum fan diameter is 5.7′ and some standard fan sizes are: 60″, 66″, 72″, and 84″, we would use a 72″ diameter fan. For this diameter fan, we need to determine what the exterior width of the fan housings would be.

Most likely we would discover the following about fan housings:

Fan diameter	Fan housing width
48″	55″
54″	62″
60″	69″
66″	76″
72″	82″
84″	100″

Thus, in this example, we would choose the 82″ wide fan housing.

Now that we have selected a housing width 82″, the next thing to do is determine how many fans we could install in an 84′ long fan wall. Using the 82″ dimension for the fan housing width, we will discover that the maximum number of fan housings we could get in the dryer is:

$$N = (84 \times 12) / 82 = 12.29$$

Thus, we could install twelve 72″ diameter fans in the 84′ long fan wall.

Fourth Thing to Do

Locate a fan manufacturer who has the highest static efficiency rating for 12 fans operating at the total CFM predicted by standard air model 74 and at the static head selected from the previous table. See the following letter to a fan manufacturing company:

Dear Mr. Fan Manufacturer,

I am designing a lumber kiln and seeking propeller fans with the ***highest possible efficiency****.*

Please submit a quotation for 12 adjustable-pitch propeller fans for the following application:

Service: lumber drying kiln
Maximum dry-bulb temperature = 275 F
Starting torque – 1.75 × running torque
Reversing cycles – every 3 hours
Annual operating duty – .95 × 365 × 24 = 8,322 hours/year
Number of reversals/year – 8,322/3 = 2,774 reversals/year
Performance standards – 70 F @ 1 atm
Required CFM/fan = 39,500 @ 0.75wc″ wc static head.
Please state you most efficient fan's diameter, RPM, pitch setting, and horsepower consumption at the specified CFM and static head @ 70 F and 1 atm.
Please state your requirements for tip clearances and orifice shapes for your fans.

Thank you,
Mr. Dryer Designer

After you have gone through the preceding exercise in dryer fan selection, you have to ask yourself a question.

How do you speed up the process of finding the fan manufacture with the highest efficiency?

There is a very simple way to do this. First, as a dryer designer, you have all types of drying projects put before you. One day it may be a low-temp kiln for drying 8/4 white oak, and the next day it may be a high-temp kiln drying 4/4 pine stacked in 8′ wide packages, and the next day it may be a kiln drying ash.

In every case you need to set some ground rules for selecting dryer fans and discuss these with the buyer **before** the dryer design drawings leave the engineering department.

Rule #1 – Electricity is becoming very expensive and since dryers operate continuously, their total electrical costs can be significant over the life of the dryer. So, this rule is to establish in-house documents for presenting **valid** information to the buyers so that they are made aware of what their electrical costs are going to be during the life of the dryer.

Rule #2 – Minimum CFM delivery for quality lumber drying. Although high air velocities do produce better quality drying for most species, the downside of this is that high air velocities in a lumber dryer do not come cheap. It takes eight times as much fluid power to deliver 1000 ft/min of air velocity through a sticker opening as it does 500 ft/min. Such a dramatic relationship between electrical power usage and CFM means that both the dryer designer and the dryer buyer need to understand that at some point in increasing the CFM in a dryer, the electrical costs start to overtake the benefits in degrade reduction and dryer production.

Rule #3 – There is no one dryer fan design that is superior to all other dryer fan designs for every type of wood drying system. One manufacturer may offer a fan that outperforms all others at a certain head pressure but underperforms others at different head pressures.

Thus, you need to spend time searching the many fan manufacturers' **proven** performance data for different classes of head pressure and CFM. I suggest you build a table like the following to organize your search for the **most efficient** fan design using the fan **work** method.

	CFM / FAN							
STATIC HEAD	10,000	20,000	30,000	40,000	50,000	60,000	70,000	80,000
0.10wc″								
0.20wc″								
0.30wc″								
0.40wc″								
0.50wc″								
0.60wc″								
0.70wc″								
0.80wc″								
0.90wc″								
1.00wc″								
1.25wc″								

Once you do this, you will see (due to the fan laws) patterns emerging when you start getting information from the different manufacturers of fans. I will let you do this exercise so you can personally see the patterns emerge.

Blade Profile Versus Fan Efficiency

Because wood dryer fans have to reverse, their performance in both directions has to be the same. The fan performance curve must be the same no matter which direction they operate. To do this, the aerodynamic profile of the propellers must be symmetrical. Fan manufacturers have used different blade designs to achieve this. Some designs use flat or stamped sheet metal blades. Some designs use symmetrical flat aluminum castings. Some use cast-aluminum aerodynamic s-shapes.

Straight Versus Twisted Blades

Fan designs are offered in either a straight or twisted blade design. Of the two, the twisted design is far superior in both performance and efficiency. In the twisted design, the angle the blade makes with the air stream is an inverse function of the radius of the blade. In the twisted design, the velocity of the air stream leaving the blade (parallel with the fan shaft) does not change along the length of the blade. Such a design minimizes the dynamic pressure losses common in straight-blade fan designs, resulting in both increased CFM and higher efficiencies.

Since air dynamic pressure losses are a function of the air velocity squared, the ideal blade design keeps the exit air velocity uniform from the hub to the tip of the blade. The straight-blade design cannot do this and in addition may have reverse flow back around the blades next to the hub if the backpressure (static head) gets too high. For these two reasons, in applications where high static heads are encountered, twisted blade designs should always be used.

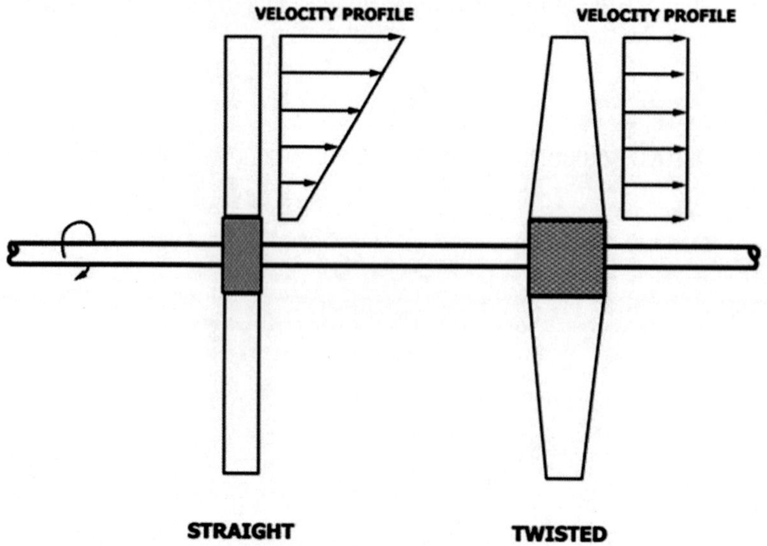

Custom Castings for Optimizing Fan Efficiency

Every casting of a specific twisted blade design has an **optimum** pitch setting at which the fan will produce its maximum efficiency.

To achieve maximum efficiency in a propeller fan, both the pitch and the width of the blades should be an indirect function of the radius of the blade. Additionally, in all blade designs the aerodynamic lift characteristics (performance) are also a function of approach velocity and fluid density.

If sufficient modeling and testing has been done on different **sections** of a blade, data may indicate that a slight **twitching** of the approach angle is needed for specific fluid conditions. Thus, a truly optimum blade design may appear on the surface to be twisted in an indirect linear relationship with the radius but may secretly be twitched to improve performance. In addition, the width of the blade at different radius may not follow a linear relationship.

Problems with All Cast Fan Blades

When a mold is built for a twisted blade design, a **specific** pitch has been selected for **that** casting such that the pitch does vary in an indirect fashion with the radius of the blade. Unfortunately, that one mold may be used by the fan manufacturer for a **range** of pitch settings above or below the **design** pitch setting.

If the fan is operated at pitch settings above or below its designed pitch setting, the **efficiency** of the fan will **drop** because the fan's pitch no longer follows an indirect linear relationship with the radius of the fan. Saying this differently, what efficiency exists with one pitch setting is not true for a different pitch setting.

To complicate matters worse, many published fan performance curves furnished by fan manufacturers are often projected curves calculated from fan modeling equations that do not consider the preceding effect described. Because of this, significant errors in published ratings may occur and not be known to the buyer, or the local fan company's sales rep.

If fan manufacturers offered custom castings for a specific pitch setting at a specific fan rpm, hub diameter, fan diameter, and fluid conditions, the efficiency and performance could be improved significantly over the current practices used in fan construction.

Looking to the Future

If there was a fan manufacturer who had the ability to model high efficiency twitched blade profiles for a specific dryer dry-bulb and wet-bulb temperature, fan head, fan horsepower, and RPM, and then custom manufacture blade casting molds for those conditions, this fan would be vastly superior to the fans now being used in lumber dryers.

Eventually, when electrical cost becomes very high, this method will be used for manufacturing many commercial fan propellers in all types of industries, not just lumber dryers. The days of wasted electrical energy due to inefficient fan designs are quickly coming to an end.

Comparing Fan Efficiency Between Manufacturers and Designs

You can use the following efficiency formula and guidelines to select the most efficient fan design using the fan **work** method:

$$\text{Electrical fan horsepower} = \text{Kf} \times \text{MCFM} \times \text{WC}''$$

Where:

Kf is an efficiency coefficient (horsepower/mcfm/wc″)
MCFM is fan CFM (in thousands)
WC″ is the static head at the fan (inches of water column)

Published test data I collected on numerous propeller fans revealed the following:

Kf	Work efficiency
0.295	Too low
0.285	Low
0.275	Good
0.265	High
0.255	Very high

Example: A propeller fan requires 10.0 horsepower to deliver 60,000 CFM at 0.5wc″. What is its Kf?

$$\begin{aligned} \text{Kf} &= \text{horsepower} / \text{mcfm} / \text{wc}'' \\ &= 10.0 / 60 / 0.5 \\ &= 0.333 \text{------------ This is a very inefficient fan.} \end{aligned}$$

One word of caution when calculating Kf from published fan performance tables and curves. Can you really believe what you are reading from published fan data?

Hub Losses and Hub Cones

The efficiency losses that occur around most fan hubs are a relatively minor design issue. Hubs are frontal areas where bearing supports are located, and their total cross-section area is usually small compared to the total face area of the fan. Aerodynamic cones can be added to streamline the hubs, but the front-end and maintenance costs usually outweigh the benefits. This money would be better spent on other areas of the fan system. To get the benefit of a hub cone for a reversible fan, the length of the cones on each side of the fan would have to be at least one fan diameter in length.

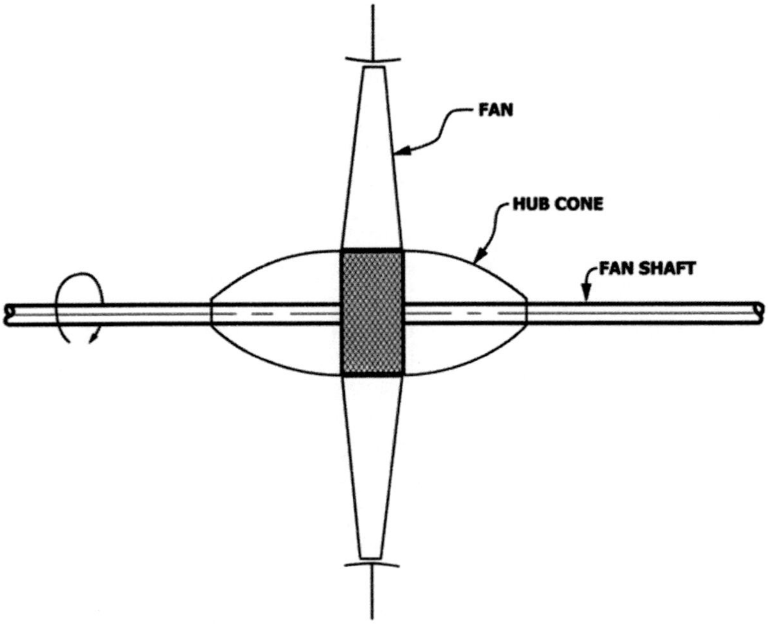

Fan Testing Standards and Laboratories

Reputable fan manufacturers have their fans tested by a certified independent testing laboratory to determine the performance of their fans under different operating conditions. The objective of these tests is to determine the actual delivered CFM and power requirements at different static heads, fan speeds, and pitch settings.

Air Movement & Control Association International, Inc. (AMCA) is the internationally recognized association for testing fan performance. Before any fan is used in a wood dryer, the fan manufacturer should submit proof that the fan has been tested by an AMCA accredited testing laboratory.

Contact the AMCA for copies of the following standards:

ANSI/AMCA Standard 210-99 (ANSI/ASHRAE 51-1999)

Laboratory Methods of Testing Fans for Aerodynamic Performance Rating

ANSI/AMCA Standard 230-99

Laboratory Method of Testing Air Circulator Fans for Rating

ANSI/AMCA Standard 204-96

Balancing Quality and Vibration

When comparing one fan manufacturer's design against others, I suggest sample fans from the different fan vendors be sent to the **same** testing lab for comparison. This will tell the truth about how one brand's fan performs against another brand. Do not be surprised if you discover that the results are quite different than the fan manufacturer's published curves. And even more important, the independent and neutral engineer that you retain to analyze the data has to understand how wood dryers and fans work.

I highly recommend this be done for dryer projects involved the purchasing of large numbers of fans. It should also be done by major wood product manufacturing companies who own and maintain large numbers of wood dryers. You may discover that your dryer designers know very little about wood dryer fans or fan engineering and they have cost your company millions of dollars in lost production, degrade, and electricity.

Fixed Versus Adjustable-Pitch Fan Designs
Fixed-pitch fans are less expensive and more durable than adjustable-pitch fans. If properly designed, they will give superior long-term performance than adjustable-pitch fans. If adjustable-pitch fans are used, expect problems to occur during the life of the dryer. In addition, the accuracy of the pitch setting of each blade is not to be taken lightly. Small variances in blade pitch can cause shaft vibrations and bearing failures. Precision pitch gages and hub scales should be used to verify that each blade in an assembly has the exact same pitch. A one-degree variance in blade pitch will cause vibration and problems with bearings.

Fan Blade Obstructions
All fans require an open space free of obstructions near the blades. Any object that is located close to a rotating fan assembly will reduce the performance of the fan and possibly set up unfriendly harmonics around the fan. Support struts for shaft bearings should be of such a design that as little aerodynamic cross-sectional area as possible exists in the area next to the fan where the shaft bearings are located. Standard aerodynamic calculations apply for determining their effect on the fan's performance. A drag coefficient of .8 is a conservative estimate for determining their effect.

11.46 Choosing the Best Drive for the Dryer Fans

Once a specific fan has been chosen for a dryer, the next task is to design the fan's drive. Cost and utility considerations dictate which type of drive is to be used. Reliability, maintenance requirements, and electrical costs are the major issues. The decision of whether to use internal or external motors is the first issue. If external motors are used, the decision of whether to use a line-shaft design or a cross-shaft design will have to be made. Each design has their merits and limitations. There is no single fan drive design that is superior to all others.

Internal Fan Motors

Internal kiln motors have developed a bad reputation throughout the wood dryer industry because designers have applied them to dryer applications they were not suited for, and dryer operators have subjected them to temperatures, humidity, and amperage loads above their design rating.

If properly applied, correctly manufactured, and not abused, internal kiln motors can provide excellent service. The motor manufacturer should rate the motor specifically for dry kiln service and variable frequency (speed) applications. The housing should be cast iron, open drip-proof (ODP) construction to maximize cooling of the windings. The stator and rotor should be epoxy coated to withstand 100% relative humidity in an acidic environment. Bearings should be heat stabilized. The motor's electrical leads should be heat resistant with a glass over braid construction and sufficiently long enough to make the electrical connections **outside** the dryer. The bearing lubricant should be rated for high-temperature applications.

The maximum temperature and amperage the motor will be subjected to will have a significant impact on the life of the insulation. Do not overload internal kiln motors by setting fan pitch too high. Kiln motors should only be subjected to maximum rated amperages when the kiln is at 70 degrees F, not at maximum kiln temperature.

Some motor manufacturers use two frame sizes over the standard NEMA size. The RPM should be kept as low as possible to extend bearing life and reduce vibration.

The motor controls should include a backup voltage system to automatically apply low voltage to the windings when the fans are off. This will prevent the windings from cooling down and collecting moisture.

Bearings should be inspected and/or replaced annually.

If internal motors are installed in common-wall kilns, expect numerous winding failures due to the humidity that can accumulate in the roof of the dryers during shutdowns.

External Fan Motors

If an external motor design is used, the orientation of the fan drive shaft relative to the airflow through the packages is the first decision to make. One design (the cross-shaft) keeps the fan shaft parallel with the airflow through the packages. The other design (the line-shaft) has the fan shaft perpendicular to the airflow through the packages. The following is a comparison of the two designs.

Line-Shaft Versus Cross-Shaft

Line-Shaft Designs

Least efficient design from an air-handling viewpoint
Lower maintenance problems than cross-shaft designs
Lower initial cost than cross-shaft designs

Applications

Low-temp drying of slow drying species using E&C cycles
Low air velocity requirements
Small effective package widths

Cross-Shaft Designs

Most efficient design from an air-handling viewpoint
More maintenance problems than line-shaft designs
E&C cycles will damage ball and roller bearings
Oil sleeve bearings should be used with E&C cycles

Highest initial costs of all designs

Applications

High CFM drying applications
Large effective package widths

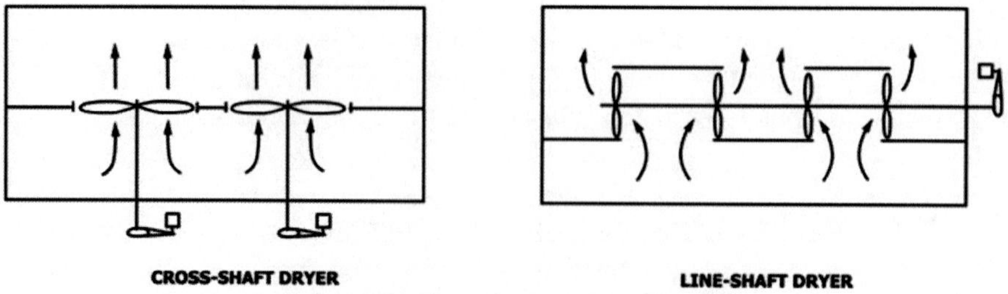

PLAN VIEW of TWO DIFFERENT FAN DESIGNS

Selection of the Fan Drive Motor

The selection of electrical motors for dryer fans plays a significant part in the success of the overall system. The following design criteria should be followed:

All dryer motors

1. Use only high efficiency motors to reduce electrical costs
2. Variable-frequency applications require motors suited for the drive
3. The standard NEMA rating B is used in virtually all dryers

Internal motors

1. The motor must be rated for lumber kiln service.
2. The housing design should be the open design for cooling of the windings.
3. The winding insulation should be rated for the maximum kiln temperature.
4. The insulation should be dipped and baked two times for moisture resistance.
5. The bearings should be a loose fit class rated for 275 degrees F service.
6. The type bearing lubrication should be rated for 275 degrees F service.

7. The motor's electrical leads should be of sufficient length to make the connections outside the kiln.
8. The motor control panel should include a low-voltage circuit for keeping the windings hot during periods when the dryer is shut down. This will prevent moisture from condensing on cold windings. It will also prevent thermal cracks from occurring in the winding's baked-on coating.
9. The frame size should be larger than the standard NEMA rating and depending on the rated full load amps. For high-temp kilns, it should be two sizes over the standard NEMA frame size.
10. The rpm should not exceed 875 in high-temp kilns and 1150 in low-temp kilns.

External motors

11. Use only TEFC designs to withstand water during rains
12. Use only heavy-duty, mill-and-chemical-grade motors

RPM Considerations for Fans, Motors, and Drive Shafts

Fan designs that operate at low speeds will increase bearing life, reduce vibration forces, noise, and centrifugal stresses on the fan assemblies. However, if the rpm is too low, stalling may occur resulting in poor performance.

Sizing the Fan Shaft

In most commercial wood dryers, the diameter of cross and line cold-rolled steel shafts varies from 1–3/16″ to 2–3/16″. The following table of minimum shaft diameters was developed from the allowable torsion stress created by shaft RPM and horsepower.

Contrary to popular belief, dryer fans shafts should not be oversized to improve longevity. Bearings are not always perfectly aligned and having some flex capability in the shaft can improve the life of bearings. Over-sizing the drive shaft also adds unnecessary cost to the dryer. It can also increase long-term maintenance problems and repair costs. If v-belt drives are used, pulleys should be located next to bearings. If oil sleeved bearings are used, diameters larger than the sizes listed in the following text may be required.

Type drive Shaft RPM Horsepower Minimum Shaft Diameter (inches)

Direct	875	30	1-11/16
Direct	875	20	1-7/16
Direct	875	10	1-3/16
V-belt	500	30	1-11/16
V-belt	500	20	1-9/16
V-belt	500	10	1-7/16

Fan Shaft Alloy

Most dryer fan shafts are made from 1020 cold-rolled steel. Some manufacturers use 1050 steel. During the last 50 years, I have never seen any viable justification for using 1050 steel. Typically, when drive shafts fail, the cause is something other than the yield strength of the shaft. If corrosion is a problem where the shaft passes through the wall of the dryer, there are several options for solving this problem. One is to install an insulation sleeve over the shaft located outside the dryer wall to reduce the amount of condensation on the shaft located inside the dryer. Another solution is to use a stainless shaft.

Fan Shaft Couplings

Avoid the use of shaft couplings in fan shafts. Couplings have a reputation of getting loose in reversing applications. Couplings should only be used in long line-shaft designs and even these should be fitted with soft start controls.

Preventing Fan and Bearing Drift

Dryer fans can become loose on their shafts due to the numerous reversals and temperature swings the fan experiences. For this reason, all fan shafts should be fitted with adjacent set collars and/or stop plates to keep the fan hub and its keys in place. Set collars should also be installed next to the shaft bearing that carries the thrust produced by the fan.

Belt Drive Versus Direct-Coupled Drives

Belt-driven fan shafts are the predominant design for external motors. Depending on the type of propeller used, a direct-coupled design can also be used to offer far less maintenance problems. V-belt drives are an ongoing maintenance problem requiring the replacement of the belts every year for reliable dryer performance. V-belt fan drives should be avoided if possible. If they are used, they should be a long poly design to reduce vibration and improve longevity.

V-Belt Versus Cog Belt Drives

Some dryer designers offer cog belts in place of v-belts for driving fan shafts. Due to the higher cost of the cog pulleys and belts, one would expect less problems and longer life from them. However, their use has not been widespread in the dryer industry.

Automatic Belt Tension

Whether using a v-belt or cog drive, an automatic belt tension device can be used to maintain the proper amount of belt tension to prevent excessive belt wear, jumped belts, and bearing failures.

The most common type of automatic belt tension device is a pneumatic air bag located under a hinged motor base. The motor base includes adjustable set bolts to prevent a total loss of belt tension in the event of a bag leak. An adjustable pressure regulator is used to control the bag pressure.

Smart controls can be added to operate the bag at a higher pressure during fan starts, and a lower pressure during normal operation. The bag pressure may also automatically adjust depending on the direction of rotation of the motor.

Fan Shaft Bearing Support Pedestals, Tubes, and Channels

Shafts for dryer fans must be supported by bearings that always stay in alignment. Bearing support pads and structures must be designed such that the bearings do not float, twist, or wobble. This is especially true when dryers expand and contract due to temperature changes. The solution is to use support structures that expand at the same rate and amount of the steel shaft being supported.

In long line-shaft designs, each of the shaft bearings are supported by a-framed pedestals mounted on a steel structure (the fan floor) that runs the entire length of the dryer. Proper alignment of the bearing pedestals is crucial to preventing bearing failures.

A long thin wire under high tension is used to check the alignment of the pedestals. If the dryer is experiencing foundation settlement, the alignment of the pedestals can be affected significantly. This will lead to both bearing and fan failures.

In cross-shaft designs, there are two methods for designing the bearing support system. One method, called the **fixed-cross-shaft** design, uses a steel tube, tee, or channel that runs the length of the shaft. Fixed bearing pads are mounted on the structure every 4′ to 6′. The tube, tee, or channel is attached to the fan housing and the external motor drive assembly (outside the kiln).

Of the bearings used on the assembly, only one is a non-expansion design. All the others are expansive types. The non-expansive bearing is always located at the drive motor end of the assembly. Properly designed and maintained, this system can provide decades of reliable service.

One fault of the fixed-cross-shaft design is its **propensity to corrode** at the point where it passes through the wall of the dryer. The corrosion can go unnoticed resulting in a sudden failure of the bearing support structure. Many fixed-shaft designs have been and will be removed from dryers because of this problem and replaced by either internal motors or the following floating-shaft design.

The **floating-cross-shaft** design is one in which the shaft is supported on hinged bearing pedestals mounted off the steel structure of the dryer.

In this design, there is no tube or channel adjacent to the steel shaft. All the bearings are rugged self-aligning non-expansive double-roller types. The bearing at the drive end is mounted off a hinged pedestal attached to the bottom of the motor base. The bearing at the fan housing takes all the thrust developed by the fan. The only part passing through the wall of the dryer is the shaft. Properly engineered and maintained, this design can provide decades of reliable service. Its only fault is the propensity to suffer misalignment if the foundation of the dryer settles.

Shaft speed should be kept as **low as possible** to minimize radial forces and vibration caused by out of balance fans. Never exceed 800 RPM with this design. Use only large diameter high efficiency fans and be sure to check the fan curve and system curve to make sure the system does not hunt. Fan balance is also crucial. Check and balance every fan assembly before installing them.

Many floating-cross-shaft designs, such as the one in the following text, have been in operation for more than 30 years.

11.47 Bearing Selection

Other than for ball bearings used in internal dryer motors, the two predominate types of bearings used for wood dryer fan shafts are the oil-lubricated sleeve bearing and the double-roller bearing.

Oil-Lubricated Journal Bearings

During the last century, dryer designers have successfully applied oil-lubricated journal bearings on wood dryer fan shafts operating at speeds less than 400 RPM. The lubricating oil is fed to the bearings by an external oil line connected to an oil cup mounted on the roof of the dryer. Frequent oiling is required of the dryer attendant. This is the only type of bearing that can withstand E&C cycles on a long-term basis. Many of these bearings have lasted for over 30 years in line-shaft applications.

Roller Bearings

During the last 40 years, self-aligning double-roller bearings have been successfully used for fan systems in which no E&C cycles are present. The roller bearing design can be designed with or without external grease lines. The selection of the bearings must include a self-aligning and an adjustable tapered inner locking race. The type of seals should be of all-metal construction. The initial adjustment of the bearing's running clearance must meet the bearing manufacturer's specifications for the RPM, load, and temperature.

Some dryer manufacturers use a split race roller-bearing design for ease of replacement. However, due to their nature, split roller bearings are not as durable as solid race designs, especially in applications in which large loads are placed on the bearing.

All ball and roller bearings are subject to high failure rates if used in dryers in which E&C cycles are used. Although frequent purging of the bearing with fresh grease slows down the corrosion of the precision rollers and races, it will not stop it. Wood acids will eventually pass through the seals and damage both the rollers and the races.

The rpm of all dryer fan shafts should be kept as low as possible to increase bearing life.

11.48 Fan Housings

The housings for dryer fans are one of the most important parts of a fan system. The following criteria should be followed in their design and manufacturing.

Material Selection

Because of the strength and workability of mild steel, this is the predominant material used for fan housings.

Aluminum and stainless steel have been used successfully for highly corrosive species in which E&C cycles are used.

Strength Requirements

All fan-housing materials should be of sufficient thickness to prevent unfriendly secondary harmonics produced by the motion of individual fan blades during the rotation of the fan assembly inside the housing. For fan diameters of 48″ or less, 16 gage or heavier materials should be used. For fan diameters over 48″, even heavier materials should be used. The outer frame of the housing should be of sufficient strength to maintain the overall structural integrity of the housing once it is bolted to the dryer's fan wall.

11.49 Fan Orifice Designs

Due to the increasing cost of electricity and energy for drying wood, the design of the orifice profile is becoming an issue. The selection of an improved orifice profile can affect the total air delivery capacity inside the dryer by as much as 5%. Although the effect is small, over several years of dryer operation, the accumulated savings can become significant. A comparison of cost versus drying benefits should be made to determine which orifice best fits the application. The face air velocity through the fan's cross-sectional area will determine which orifice should be used. The four orifice types are: simple, bell, tapered, and conical.

1. Simple Orifices

The simple orifice is a rolled flat bar that surrounds the outer tip profile of the fan's propellers. The width of the bar should be no less than two times the swept width of the propeller tips. The simple orifice is used for face velocities up to 1500 ft./min. The center of the fan tip should align with the center of the bar. The simple orifice is the lowest cost orifice to build, very strong, and rarely causes problems.

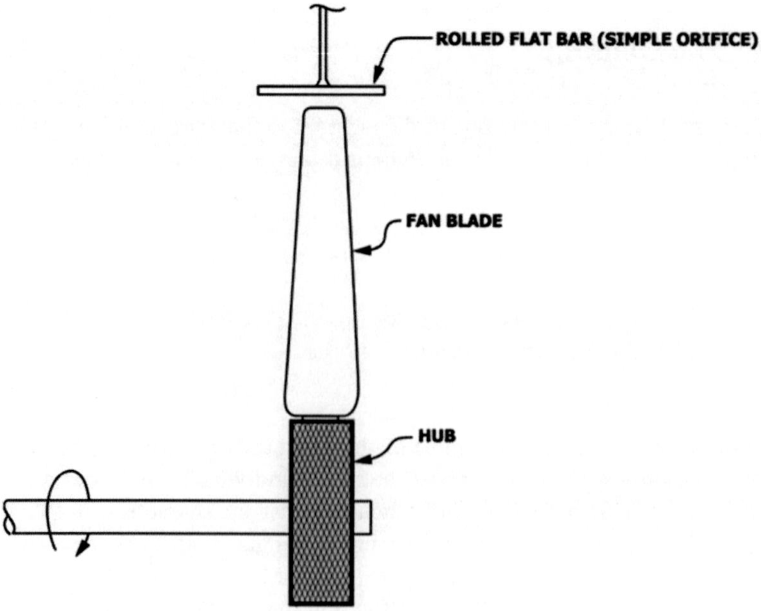

2. Bell Orifices

The bell orifice is a modified simple orifice. In this orifice, the outer edges of the rolled flat bar is fitted with curved surfaces to improve the airflow **entering** the orifice. The radius of the curved surfaces should be as large as possible, not interfering with other structures located near the fan housing. This design is the next to lowest cost orifice to build. It can be used for face velocities up to 2000 ft./min. The improvement in air delivery over a simple orifice is in the 1–2% range. Contrary to popular belief, the bell orifice does not have any significant effect on the air stream leaving the fan.

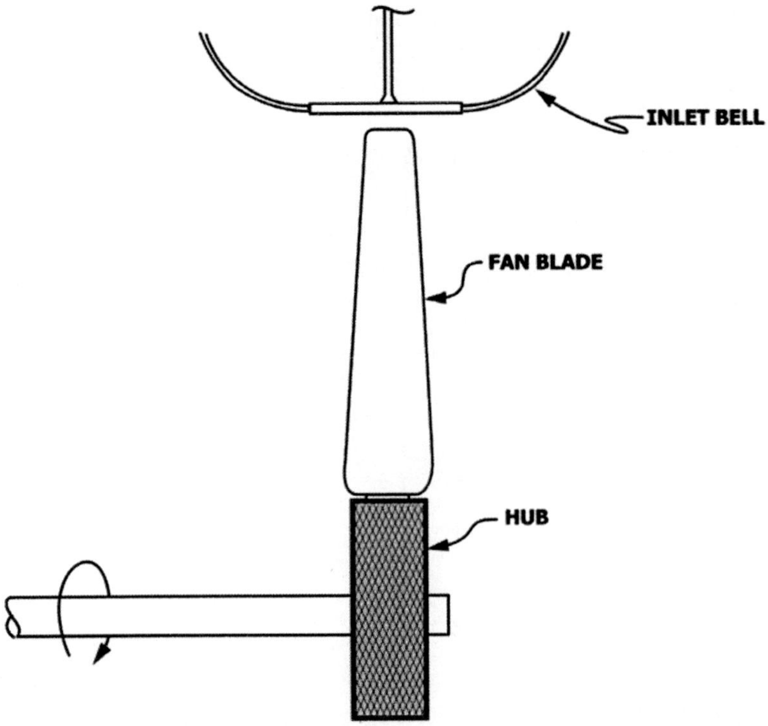

3. Tapered Orifices

In dryers that require high air velocity and high fan electrical efficiency, tapered orifices are the next step to improving air delivery. There are two types of tapered orifices.

The single-taper design in which the inlet and outlet of a simple orifice is fitted with tapered cones to improve the airflow entering the orifice.

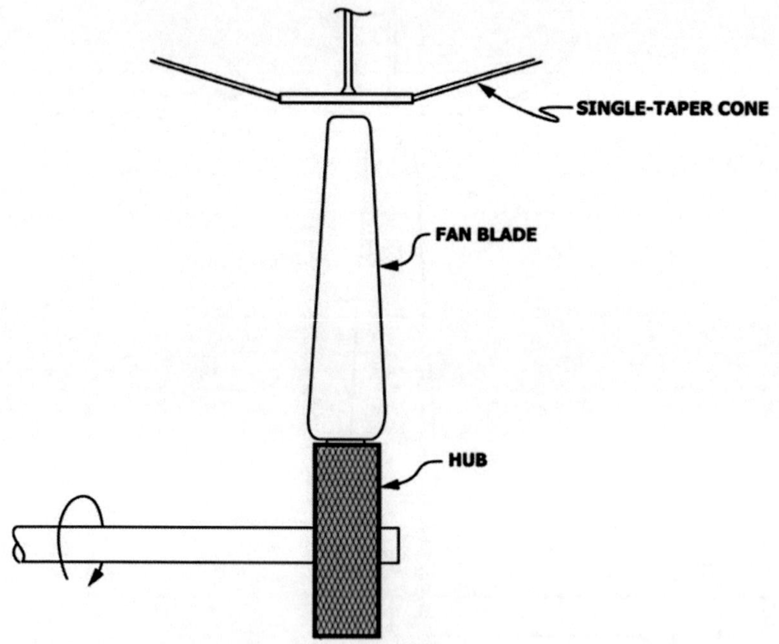

The dual-taper design in which the inlet and outlet of a single tapered orifice is fitted with a second tapered cone to improve the airflow entering the first cone.

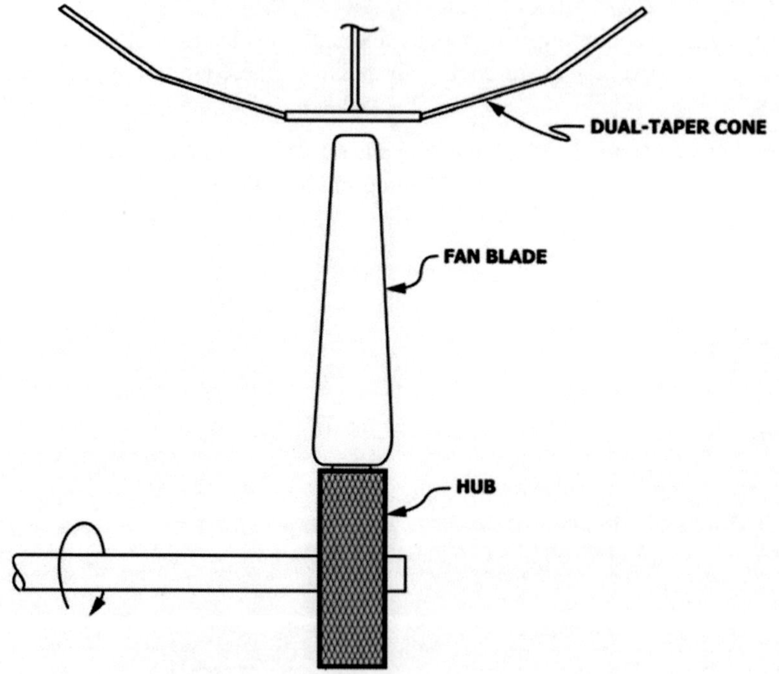

The tapered design is the next highest cost orifice now being put in lumber dryers. It should be used for face velocities over 2000 ft./min. The improvement in air delivery is dependent on the face velocity of the air leaving the fan, and the angle and length of the cones. The improvement in air delivery over the simple orifice can approach 5%. However, to reach this level of recovery, the total length of the two cones can exceed the diameter of the orifice. For fan drives that use external motors, fitting the large cones around the shaft support struts can be quite difficult and expensive. Their presence can also create unfriendly harmonics as well as significant obstructions for maintenance personnel attempting to work on the fan assemblies, motors, shafts, and bearings.

Parabolic Designs

Although none are in dryers as of the printing of this book, their use will appear in future dryer designs. In the coming decades, the cost of electricity for dryer fans will reach such a level that parabolic orifices will be promoted by dryer manufacturers.

Parabolic designs involve very long recovery cones designed to convert dynamic pressure into static pressure at a **constant recovery rate**. In some applications, these designs can result in recoveries of 10% over simple orifices. Parabolic fan orifices are very expensive and currently difficult to justify in wood dryers because of space limitations on both sides of the fan wall.

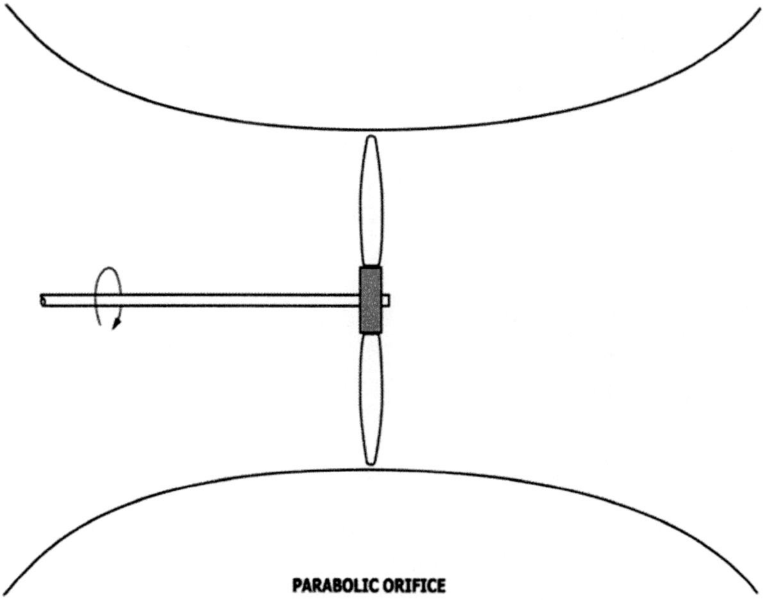

Sticker Thickness Versus Fan Orifice Design

Another way to view large expensive and complex fan orifices is to use a thicker stacking stick. In most cases, increasing the sticker thickness is a smarter way to improve CFM delivery through packages of lumber. In almost every wood dryer, a 1% increase in sticker thickness equates to slightly over a 2% increase in fan CFM delivery. To achieve the equivalent of a 5% increase in CFM from an upgraded orifice design, the sticker thicker needs only to be increased 2.5%. For a ¾″ sticker, this is only a .019″ increase in thickness.

Although it makes far better economics to design dryers for extra-thick stickers than to purchase and maintain expensive fan orifices, expect dryer salesmen to promote the expensive recovery orifices in the coming decades. Both dryer manufacturers and their salesmen will benefit from the initial sale of these expensive orifices and the sale of replacements when damaged by fan failures.

11.50 Fan Tip Clearance

The clearance between the end of a fan propeller tip and the inside of the orifice has a dramatic effect on the performance of a fan. Excessive clearance between a propeller and its housing causes air recirculation flows around the end of the blades. Recirculation causes both the delivered CFM and efficiency to drop.

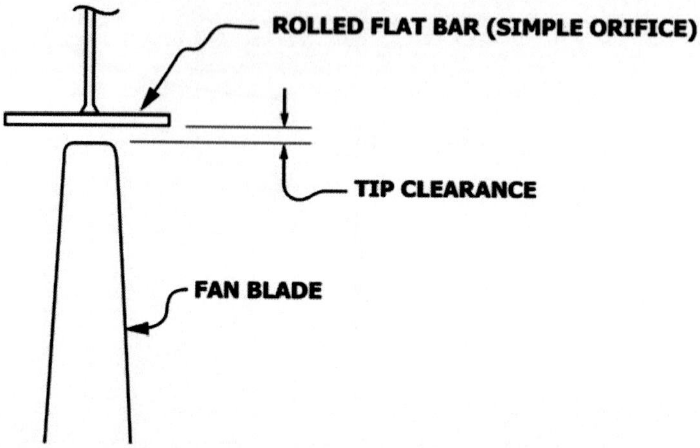

If excessive tip clearance is present, the delivered air velocity through lumber packages can be reduced by as much as 30% in extreme cases. This loss will lead to excessive drying times and moisture problems in all types of drying. The opposite of this (inadequate tip clearance) can lead to catastrophic fan failures when the tips of the propellers encounter the inside surface of the orifice. Sufficient clearance should exist to allow for thermal expansion and vibrations found in all types of dryer fan systems. The following tip clearances are a starting point for designing fan systems:

Fan diameter	Minimum tip clearance	Maximum tip clearance
84″	.42″	.52″
72″	.36″	.45″
66″	.33″	.41″
60″	.30″	.37″
54″	.27″	.34″
48″	.24″	.30″

In some cases, tip clearances far less than the preceding values can be achieved to temporarily improve fan performance. However, this may eventually lead to costly fan failures and potential safety problems. The degree of precision to which the fan assemblies are balanced and the rigidity of the entire bearing support structure including the fan wall has a strong effect on how small a tip clearance can be used. Discuss this issue thoroughly with the project engineer, the mill manager, the dryer operators, and the fan manufacturer before arriving at a final tip clearance.

11.51 Drive Attachment Pedestals and Struts

At the fan orifice, mechanical pedestals and struts are required to keep the center of the fan assembly shaft at the center of the orifice. Their presence in the fan's air stream has detrimental effects on both the performance and efficiency of the fan. The aerodynamic losses caused by their presence are a function of their cross-sectional frontal area exposed to the airflow. An aerodynamic drag coefficient of .8 is a conservative choice for calculating the effect.

The three predominate designs for struts and pedestals are:

The two-strut design in which two struts (located 90 degrees apart) are used to support the fan bearing platform. Although some dryers use this design, it has been my experience that these do not perform well with large-diameter, high-horsepower fans. They should only be used for shaft speeds below 400 RPM. A slight amount of imbalance in a fan assembly will cause these designs to fail.

The three-strut design in which two horizontal struts span the width of the fan housing, and one vertical strut supports the base of the bearing platform. This design has been used successfully on fan diameters up to 84″ and 30 horsepower.

The four-strut design in which each side of the bearing platform is supported by two struts. There are several variations of this design. Some designers use struts made of round tubing to reduce air drag. Some designs use ½″ thick steel plate aligned with the air stream. This is the strongest design capable of handling large high-horsepower fans.

11.52 Fan Walls

The walls that fans are mounted in have several features that require the attention of the designer. The first is that the wall must be sufficiently rigid to withstand the vibrations common in all fan systems. The second is that in the event of a fan failure, the wall will not be significantly damaged. I recommend fan failure guards be installed between fan housings to deflect failed fan parts away from adjacent fans thus preventing a domino effect.

In some fan designs, vortex breakers are installed in front of fans to stop the swirl exiting the fan. Vortex breakers have also been used in dryer buildings where the slope of the roof is too steep and too low to the fans. In such a situation, breakers are needed to break up the vortexes and the resultant pressure accumulation toward one end of the outer wall plenum.

Once a fan wall perimeter structure and the fan housings are in place, filler baffles are installed to prevent air leakage back through the fan wall. Baffles are installed at the top of the wall to seal against the roof, between each fan housing, and at the ends of the fan walls to seal against the two end walls. Any baffle that meets the inside surface of a prefabricated roof or wall panel should be formed to prevent rubbing tight against the soft inner skin of the panel causing a hole in the panel. Depending on the type of dryer, the baffles can be either bolted or field welded in place.

Once the final fan wall is in place, there should be very little air leakage back through in the wall.

11.53 Fan Decks: Air Baffling and Structural Requirements

Fan decks should be designed as both a barrier for leakage and a work platform for maintenance personnel.

The minimum structural design loading should be adequate for the most extreme concentrated load on one section of decking and its support structure. If internal motors are used inside the dryer, the deck design should consider the worst-case condition in which several motors and personnel may be present on one section of decking. Failure to do this could result in a collapse that could cause serious injury or death. The use of smooth heavy gage decking instead of light formed (rolled) decking is recommended. In all fan deck designs, the strength of the fan deck must be sufficient for the worst-case concentrated load.

The fan deck should be fitted with end-wall baffles to prevent air leakage at walls. The outer edges of the fan decks should be fitted with vertical rigid drop baffles for accompanying hinged ceiling load baffles used in the dryer. The gap between the bottom of the drop baffle and the top of the hinged baffle should be kept as small as possible to prevent air leakage between the two baffles. Although field-welded filler strips can be used to reduce the air leakage, I recommend the leakage problem be designed out of the baffle.

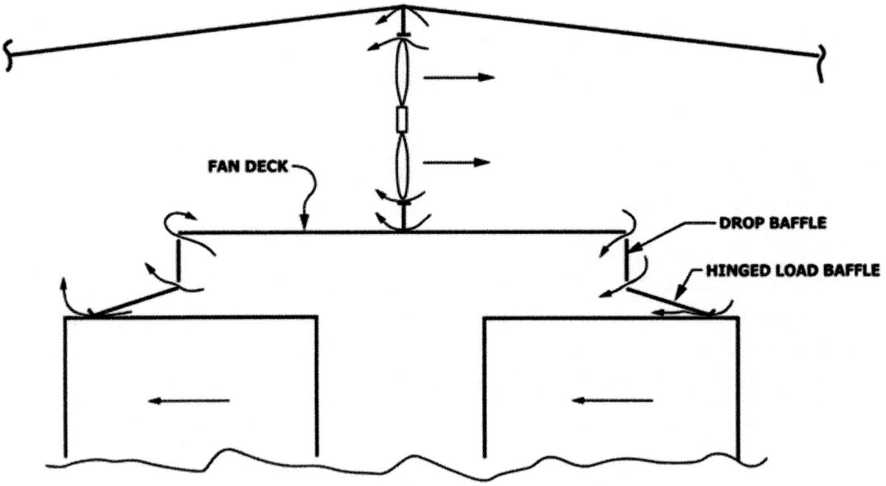

SOURCES OF AIR LEAKS IN FAN WALLS, DECKS, & BAFFLES

11.54 Propeller Fan Balancing, Installation, and Inspection

Read the following procedure **before** installing fans in a dryer. This procedure is mandatory for preventing costly fan system failures and lost dryer production.

#1. Remove the fans from their shipping crates. Assemble the fans per the manufacturer's instructions. Preset the pitch to the dryer's specifications.

#2. Check the true radius of the fan blades. Locate the longest and the shortest blades. You will have to mount the fan on a rotating shaft to do this. If the differences in blade radius is over .080″ contact the fan manufacturer about replacing the fan or how to trim the blades.

#3. Check the static and aerodynamic balance of each assembled fan. A bench roll test is recommended. Any imbalance should be corrected. Consult with the manufacturer for the correct procedure to balance the fans. Do not let any person, including the fan salesman, convince you that some imbalance is normal.

Balance the fan before you put it in a wood dryer! If you put an unbalanced fan in a wood dryer, you will have failures!

#4. Check the fan drives **before** installing the fans.

A. Inspect all bearing and motor fasteners for proper assembly and torque.
B. Inspect all v-belt drives for proper alignment and tension.
C. Inspect all couplings for proper alignment and assembly.
D. Add lubrication to the bearings.
E. Run the motors in forward to check rotation.
F. Run the motors in reverse to check rotation.

G. Inspect all the bearings for noise, vibration, and excessive heat buildup.
H. Run the motors for 30 minutes.
I. Record the amperage and voltage of each phase of each motor (with no load on the motor).
J. Inspect each bearing a second time for excessive heat buildup.
K. Store all your notes in a secure file so you can refer to them the next time you will be doing maintenance or repairs on the fans.

If properly installed, motors, couplings, and fan shafts should run very quiet. Any noise that comes from either a motor, coupling, or bearing is suspect. All v-belt drives produce a certain amount of noise, but these noises can be distinguished from bearing noises.

#5. Install the fans and set collars on the fan shafts. Do not use oils or anti-seize compounds on tapered bushings or fasteners. These will create problems.

#6. Check the fan's tip clearance with machined feeler gages.

Locate the longest fan blade. Using the feeler gage, rotate the longest fan blade 360 degrees around inside the fan orifice. Use the following recommended tip clearances or what the fan manufacturer requires.

Fan diameter	Minimum tip clearance	Maximum tip clearance
84″	.42″	.52″
72″	.36″	.45″
66″	.33″	.41″
60″	.30″	.37″
54″	.27″	.34″
48″	.24″	.30″

If the above tip clearances are not possible, consult with the dryer manufacturer before proceeding. Do not under any circumstances proceed if the correct minimum tip clearance is not achieved.

#7. Cold test

A. Test run each fan and listen for sounds coming from the fan.
B. Start all the fans and listen for sounds coming from the fans.
C. Make repairs and adjustments as needed to correct the problem.
D. Start all the fans.
E. Record the amperage and voltage of each phase of each motor.
F. Compare the no-load amperages to the loaded amperages for each phase of each motor.

Significant differences between the no-load and loaded amperages between fans should be investigated by a qualified electrician or engineer or both. Often, this problem is caused by poor electrical connections or inadequate field wiring size for the motors.

#8. Hot test.

A. Close the kiln doors and roof vents with no lumber in the kiln.
B. Start the fans and heating system.
C. Heat the kiln up to its maximum dry-bulb temperature.
D. Listen for sounds of the fans rubbing the fan housings as the dryer is heating up.
E. Run the kiln at the maximum dry-bulb temperature for 30 minutes.
F. Shut off the heat, open the roof vents and doors, and let the kiln cool down.
G. Listen for fan noises while the fans are running, and the kiln is cooling down.

#9. Final inspection after the kiln has cooled down.

A. Inspect each fan orifice for evidence of rubbing.
B. Adjust the fans and orifices as needed.
C. If no problems exist with the tip clearance, re-torque all the fan's fasteners to the manufacturer's specifications.
D. Re-torque all set collars to prevent fan and bearing drift.

#10. Adjusting the fan pitch.

A. Load the kiln with cold lumber and place the baffles against the lumber.
B. Start the fans.
C. Measure the amperage on each fan motor.
The cold amperage should be at the nameplate amperage for the motor.
D. If the amperage is too high or too low, adjust the pitch on all the fans.
A small change in pitch setting (one degree) will have a large effect on amperage.
E. Re-torque the fan bolts and nuts to the manufacturer's specifications.

After you are **sure** you have **followed** all the preceding procedures, then the fans are now ready for putting the dryer in operation.

#11. Fine-tuning the pitch settings.

A. Load the dryer with lumber and start the fans and heating system.
B. Measure the amperage on each phase of each motor.
Depending on the type motors and motor drives, the fan pitch can be adjusted to settings higher than the cold setting but do so with caution. Excessive pitch settings cause excessive motor amperages, and costly motor failures. An increase of one degree in pitch setting can lead to motor problems, and the benefits in drying may not be significant.

11.55 Baffles in Convection Wood Dryers

This is a part of lumber drying that rarely receives the attention it deserves. Baffles are one of the most important parts of a lumber drying system.

Baffles are needed in lumber dryers to direct the air produced by the fans to the lumber packages. Baffles are also required to keep both the temperature and quantity of air flowing through each package the same no matter where the package is in the dryer.

The failure to design and use baffles properly can make the difference between a successful drying system and a drying disaster. The effect that proper baffling has on dryer performance is directly related to the drying rate of the lumber being dried.

The following table demonstrates the qualitative importance of baffling for different drying processes.

Lumber drying system	*Drying rate (%MC/hour)*	*Baffling requirements*
Softwoods	4–6	Extremely high
Softwoods	2–4	Very high
Softwoods	1–2	High
Softwoods and hardwoods	.5–1	Moderate
Hardwoods	.1–.5	Low

If the species or thickness is very slow drying and subject to checking, the installation of baffles in a dryer may cause additional degrade. In many older hardwood kilns, the accuracy of the temperature controllers is insufficient to maintain very small depressions. In these situations, increasing both the temperature and air velocity through lumber packages can cause degrade problems due to the additional heat transfer rates created by the increased heating capacity entering the packages.

An analysis of the dryer fan system, heating system, board thickness, sticker thickness, and the schedules used should be performed before making significant changes to the dryer's baffles.

Steam Coil Baffles

Overhead steam coils should be fitted with perimeter baffles to cover any exposed width more than 4″.

Reheat coils located between packages should be fitted with perimeter baffles to cover any exposed width more than 2″. The only exception is at the area above the top row of reheat fin pipe. If baffles are installed in this area, it may increase the amount of degrade that occurs on the top layers of lumber.

Fan Deck Baffles

Fan decks are large baffles for directing the air from the fans to the outer plenums of the dryer. The design of all fan decks should prevent air leakage through the deck.

In cheap side-loader dryers, fan deck baffles may be completely left out. Do not ever purchase one of these dryers. Such a design is totally defective and will cause major drying problems.

Fan Deck Drop Baffles

Fan deck drop baffles are used in kilns with large vertical distances between the top of the lumber and the fan deck. They are used for supporting the upper-hinged side of baffles located above the lumber. They should be designed to minimize air leakage. In some lumber kilns, automatic powered drop baffles, fitted with top package loaders can both provide a tight air seal and an automatic top load on the lumber providing a significant increase in dryer production and lumber warp control. I highly recommend this type of baffling system.

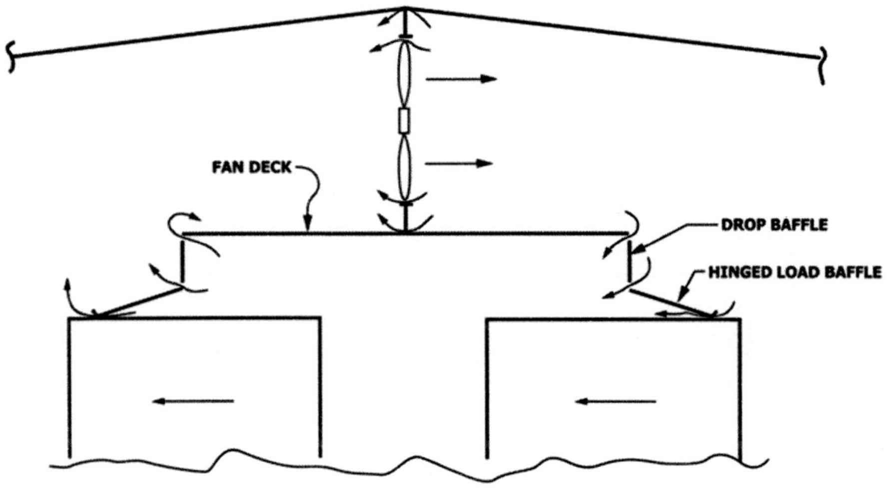

SOURCES OF AIR LEAKS IN FAN WALLS, DECKS, & BAFFLES

Plenum and Splitter Baffles

Plenum and splitter baffles first appeared in wood dryers over a century ago to correct undersized plenum widths. Although their use reflected poor dryer designs that led to uneven air flows through lumber packages, they can still be used to fine-tune air distribution in dryers with tall load heights, insufficient sticker thickness, inadequate fan CFM, and/or poor stacking and loading practices.

Because there are so many different types of dryer configurations and plenum widths, the design of plenum and splitter baffles must be custom designed for each dryer. Field-testing is required to do this right.

Fixed plenum baffles are located in the plenum space between the lumber packages and the outer walls of the dryer. They run the entire length of the dryer plenum. They are designed to redirect airflows in the plenum space to improve the air velocity profile passing through the packages of lumber.

Hinged plenum baffles are the same as the fixed designs except they are provided with adjustable hinges.

Splitter baffles are used to split and deliver the air coming from fans to plenum baffles.

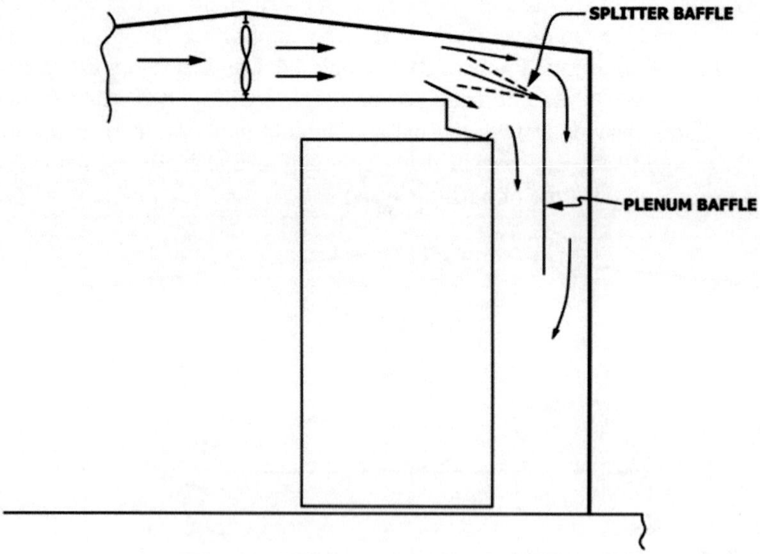

Return-Air Plenum Baffles

A large amount of return-air CFM leaving a direct-fired dryer can create problems for some types of dryers. When the dryer's internal fans reverse, the effect of the return air CFM can upset the airflow patterns through lumber packages. Lumber located close to the inlet of a large return-air duct can experience erratic drying problems.

Return-air plenum baffles are installed in kilns to **distribute** the return airflow throughout the length of the dryer. The best return air baffle design runs the entire length of the dryer and draws air from the **roof** of the dryer.

In some direct-fired kilns, two return ducts (one for each side of the kiln) have been used to minimize the effect return-air has on the air velocity through the lumber.

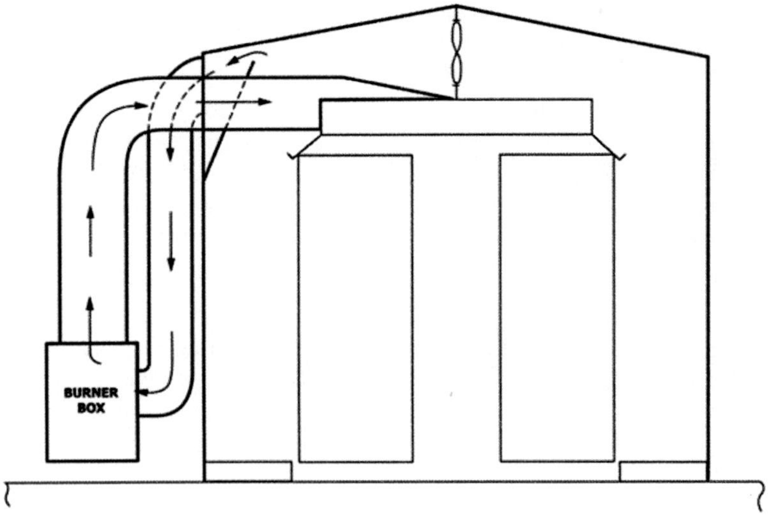

In some kilns, the return-air baffle's inlet is located at the **floor** of the dryer. Although this design is vastly superior to the **conventional** wall-mounted inlet duct with no plenum baffle, it will not perform as well as the roof-inlet design.

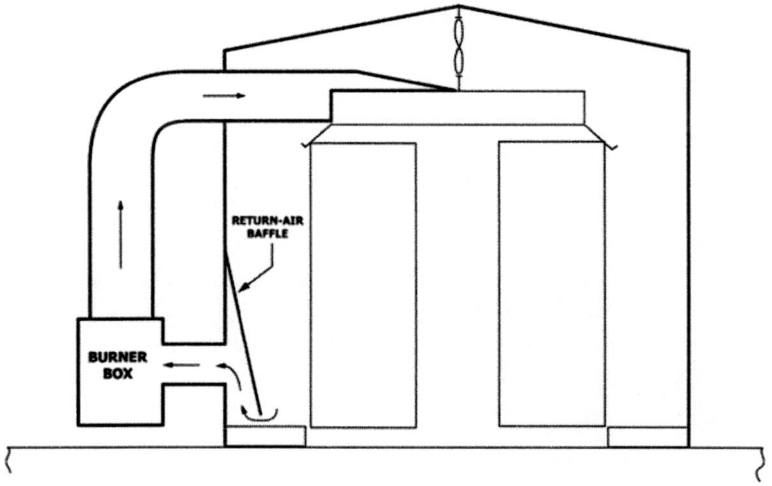

I do not recommend any return-air inlet being in the plenum space between the lumber and the exterior walls. Of the three designs, the roof-inlet design is vastly superior to the other two. This design essentially eliminates the large variances in pressure drops across the lumber packages due to direct-fired designs.

Adjustable slots are required to balance the distribution of return air entering the baffle. In some short dryer designs, a portion of the return airflow can be drawn into the baffle at the ends of the dryer, but this flow should be kept to a minimum.

My experience with return-air plenum baffles is that if they are designed and adjusted properly, the return air effect on the lumber drying can be reduced to an acceptable level or eliminated in roof-inlet designs. In most direct-fired dryers, a smart fan-reversing staging scheme can also be used to enhance the performance of the baffle system as well as increase the drying rate of the dryer.

Ceiling-Load Baffles

Ceiling load baffles are located at the top of the lumber to prevent air from passing over the top of the packages. They are a major source of air leaks if they do not rest fully on top of the packages of lumber. If the height of the packages in the dryer varies by as much as one course of lumber, very large air leaks will occur.

All types of wood dryers require ceiling-load baffles to force the air from the fans through the packages of lumber and not around them. The failure to install and use them properly will result in extended drying times and moisture problems. Never purchase a dryer that does not have them. If your dryer does not have them, see that they are installed and used properly.

Ceiling baffles are prone to damage when dryer operators fail to lift them before loading or unloading a dryer. The cables used for lifting the baffles can either be individual cables operated by hand, or by a remote-operated cable winch system. Considerable thought should go into the design of both the baffles and the cable arrangements.

The remote cable system is used in many wood-fired direct-fired dryers where fires are common. The remote system allows the baffles to be lifted (from outside the dryer) so the load of burning lumber can be quickly pushed outside the dryer with a fork truck. If a remote cable system is used, each baffle cable should be fitted with small pre-load weights to keep the individual cables from twisting and kinking on the cable spools.

Some dryers have used remote-powered operators to lift entire banks of large ceiling baffles at one time. This is especially useful in large, tall side-loader dryers, and track kilns in which drying capacity is limited. Remote-powered baffle systems allow dryer operators to remove a dried load of lumber and reload a fresh load in a small amount of time.

Door-Front Load Baffles

Door-front baffles are located at the door-fronts of track kilns. They are installed to prevent air from passing around the ends of the packages located next to the kiln's main doors.

Two strategies are used in door-front load baffles. One is to baffle the side of the packages (**side baffles**), and the other is to baffle the end of the packages (**whole-door baffles**).

Side baffles are hinged off a post that is located flush with the inside of the dryer's main door while the door is in the closed position. The side baffle can be either a dual-hinged or a triple-hinged design. Of the two, the dual design is the most common.

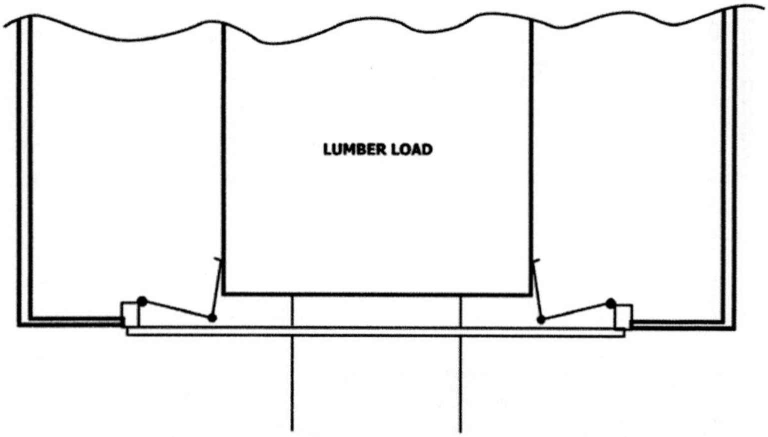

PLAN VIEW of SINGLE TRACK KILN WITH DOOR FRONT SIDE BAFFLES

The whole-door baffle is an idea I developed several decades ago for track kilns in which the total length of lumber inside the dryer consistently does not match the length of the dryer. The baffle is hinged off the same doorpost but covers the **ends** of the packages instead of the sides. After the main kiln door is opened, the whole-door baffle swings outside the dryer to allow sufficient clearances for changing loads. The whole-door design uses an 8′-wide baffle that rotates in a frame assembly hinged off a doorpost. The rotating whole-door baffle is especially useful when the door-to-door length of the dryer is not an integer of the average package length. The design can also be installed on just one end of a dryer if every kiln load is consistently pushed to the same far end door. This arrangement allows for effective door-front baffling even when entire loads end up being 1–7′ short of the dryer length. The baffle is secured in place by a large spring cable with a hook. The spring can also be fitted with a sister slider-lock cable for dryers with high air velocities.

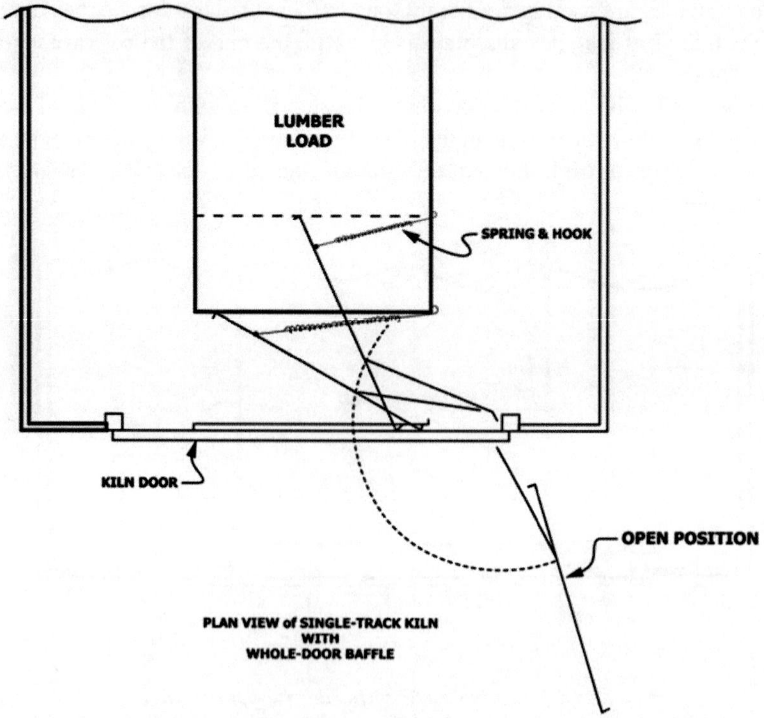

The selection of which door-front baffle designs to use is based on the length of the kiln and the effective total length of the packages loaded in the kiln. Both side and whole-door baffles can be installed in the same kiln to further reduce air bypassing the ends of lumber packages.

Floor Baffles

There are three ways to baffle the bottom of packages. One way is to cast a concrete baffle adjacent to the bottom of the packages. Another way is to hinge metal or wood baffles off the floor of the dryer. Another way is to roll a large metal sheet metal duct next to the package after the kiln is loaded. The pipe simply rests on the floor of the kiln and rolls next to the packages. It can be held in place with a concrete block.

Side-Loader (Package) Kiln Baffles

Almost any baffle idea that works well in a track kiln should work well in a package kiln. The problem is that package kilns involve fork truck traffic, and this can be disastrous for any baffle located in areas where packages are handled. Eventually, baffles will be destroyed by packages catching them and ripping them loose from their hinges.

The loading of most package kilns is haphazard leading to a high frequency of incomplete and misplaced loads inside the kiln. The rule for all package kiln baffles is to locate them behind and above the reach of moving packages.

Rear plenum baffles located behind loads of lumber should include fixed upper baffles to prevent air from passing over the top of the lumber loads. The rear sidewalls next to the packages should include fixed baffles. All side loader dryers should be fitted with ceiling load baffles like track kilns.

Unfortunately, the use of baffles in a package kiln is often an exercise in futility. Due to the nature of package kiln designs and fork trucks, no baffle system really works well or lasts very long.

There is something about lumber kilns and fork trucks that just will not mix. There are many cases of fork trucks ramming large packages of lumber through the rear walls of package kilns.

Not only will the rear baffles have to be replaced but the entire rear wall and the fork truck driver.

11.56 Heating Systems for Wood Dryers

At some point in the design of every drying system, the source of heat energy for the drying process will have to be chosen. In addition to the energy source, all the equipment for converting and delivering the energy source to the air flow through the packages of lumber will have to be purchased, designed, manufactured, installed, operated, and maintained. This includes:

Pre-dryers, kilns, E&C chambers, and dry storage facilities
Steam boilers, steam systems, and condensate systems
Burners and furnaces for the system
Fuel storage and handling systems

In addition to the preceding, the operation of all types of energy-conversion and drying processes involves safety, health, and environmental issues.

Passive Versus Reactive Energy Systems

If no chemical reaction for converting heat energy occurs **inside** the plant's property line, the drying process is classified as passive. Examples of passive systems are electrical, wind, and solar heating systems. However, even these are not truly passive. Fossil fuels are needed for fork trucks that operate in passive systems. If chemical reactions are the predominate energy conversion process, then the drying system is classified as reactive. The combustion of fossil and wood fuels are examples of reactive processes. The levels of passivity and reactivity will dictate how much capital costs and safety, health, and environmental risks exist in the plant.

Energy Cost Factors

The factors that determine energy costs for drying wood are:

1. The types of drying equipment used at the plant.
   - Lumber handling equipment
   - Indirect-heated systems
   - Steam-heated systems
   - Hot water/glycol heated systems
   - Hot oil heated system
   - Dehumidifiers with or without heat backup
   - Solar-heated hot-water designs without heat backup or with heat backup with dehumidification
   - Direct-heated systems
     - Wood fired
     - Gas fired
     - Oil fired
     - High frequency
2. The drying schedule used in the dryers.
   - Low-temperature schedules
   - Conventional schedules
   - Accelerated schedules
   - High-temperature schedules
   - Hyper-temperature schedules
3. The average specific gravity of the wood.
4. The initial average moisture content of the lumber.
5. The final average moisture content of the lumber, before E&C.
6. E&C losses.
7. The effective R-value of the dryer buildings.
8. Weather conditions.
   - Average ambient air temperature and humidity
   - Average rainfall
9. The dryer's venting losses.
   - Low temp – high venting losses especially in cold climates
   - Conventional temp – less venting losses
   - Accelerated temp – less venting losses
   - High temp – minimal venting losses
   - Hyper temp – minimal venting losses
10. The electrical usage by fans, blowers, conveyors, and machinery.
    - Internal propeller fans
    - Blowers for direct-fired kilns
    - Dehumidifier (heat-pumps)
    - Air recirculation blowers for heat pumps
    - Recirculation pumps for hot-liquid heating systems
    - Blowers for powered exhaust vents
    - Electric motors for support machinery, conveyors, etc.

Saturated Moisture Content (SMC): Calculating Maximum Drying Energy Requirements

When logs enter a lumber mill, the weight of the logs that is water and the weight that is dry wood fiber should be known for determining the **maximum** drying energy requirements. For one cubic foot of green log, there is a theoretical maximum weight of water that can exist in that one cubic foot. This water mass is determined by the specific gravity of the wood fibers in the cell walls. For a cell-wall fiber specific gravity of 1.51, water specific gravity = 62.43, and log specific gravity (**SGL**) = (dry weight of 1.0 cu. ft. of green log volume)/62.43, the following equation can be used: – Note: The geometrical proof (by the author) of this formula is not shown in this book.

$$\mathbf{SMC\% = 100 \times \left(\left(1.0 / SGL \right) - 0.6622 \right)}$$

SGL	SMC %
.30	267.1
.40	183.8
.50	133.8
.60	100.4
.70	76.6
.80	58.8

Standard MC and SG formulas should be used to determine the actual weights of both water and wood fiber in the saturated log. Once you collect data for a specific plant, you will discover that the actual initial moisture content of the logs will be less than the values calculated by the preceding SMC equation. However, I have seen several cases where the actual MC was very close to SMC.

During extremely cold wet weather, the actual moisture content of fresh logs entering a sawing operation can approach the values predicted by the preceding SMC equation. This is the time when the total energy demand on the drying system will be at its highest level.

Warning All green wood drying systems should be designed for initial moisture contents equal to SMC. Failure to do so could possibly result in large capital outlays if the drying system does not meet the terms of the contract. For all dryers to be designed properly, the dryer designer absolutely must know the average specific gravity and average moisture content of the wood entering the dryers.

11.57 Temperature Control Zones

Because temperature control loops and heat control parts are expensive and have to be maintained, there is an optimum economic number of zones for every dryer design. If an insufficient number of zones are installed, moisture control and degrade problems can occur, especially in poorly designed track kilns. If too many zones exist, front-end and long-term maintenance costs will be excessive. Standard accounting practices should be applied to the following three factors found in dryer zoning:

1. The degrade capacity function
2. The total front-end costs of zoning equipment
3. The total cost of maintaining zoning equipment

Degrade Capacity Function (**DCF**)

DCF is a term I developed during the early 1970s for wood drying that applies to the fiscal capacity of a drying system to waste product value and reduce profits at the plant level. Thus, DCF is an accounting term that refers to all losses associated with lumber degrading. The term includes all fiber, labor, and overhead costs associated with degrade issues.

Degrade in wood drying can also be categorized as **static** or **dynamic**. Static degrade is associated with final moisture content alone. Dynamic degrade is associated with drying rates alone. Both types are present in every wood drying system.

Furthermore, both types of degrade involve **rebound** dynamics. The term "rebound" refers to degrade **shifts** that occur after the fiber leaves the dryer proper.

Static degrade includes value losses due solely to fiber shrinkage during drying. The term includes longitudinal, radial, and tangential shrinkage losses.

One example of static degrade economics is that in the drying of southern pine construction lumber. The southern pine industry changed moisture standards several decades ago from KD15 to KD19 specifically to allow southern pine mills to ***gain*** *an economic* ***advantage*** *over competitor lumber from the west coast and Canadian mills. Since the amount of static degrade in construction lumber increases as final moisture content decreases, the southern pine industry raised its moisture standard 4% for kiln dried lumber to reduce static degrade losses at the plant level.*

Dynamic degrade includes value losses due solely to drying rates. Dynamic losses occur significantly in the drying of hardwoods subject to checking, splits, and honeycomb when excessive drying rates are used.

Rebound degrade shifts occur after the fiber leaves the dryer proper. Moisture changes after drying causes wood fibers to either expand or shrink resulting in grade shifts.

Front-end zoning cost includes the total front-end installed costs of heat and temperature control equipment associated with zoning the dryer, excluding temperature safety devices used for fire protection purposes.

Maintenance costs includes all labor and materials for maintaining temperature systems such as controllers, sensors, control valves, field wiring, etc.

Once sufficient information has been collected for a dryer project, the optimum number of temperature zones can be determined for a specific final target moisture content.

If the dryer is to be a direct-fired design, the number of temperature control zones will be 1, 2, or possibly 3 in long track kilns. The number of zones in track dryers heated with steam or hot liquids can vary from 1 to as high as 24.

The level of engineering put in a dryer has a dramatic effect on the required number of temperature control zones. Poorly engineered low-end dryers require large numbers of temperature control zones to correct design deficiencies. Properly engineered dryers require low numbers of temperature control zones.

One common myth is that the more zones any dryer has, the better the dryer is. Unfortunately, there is an economic limit to the return on investments in paying for and maintaining any control system with large numbers of control loops.

The following demonstrative graph shows the effect of zoning in single-track kilns. In multiple track designs the number of zones will increase proportionally.

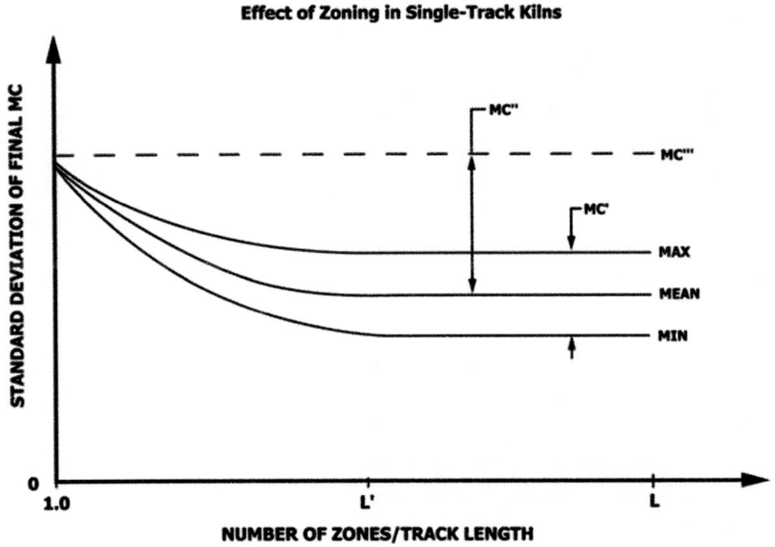

Where:

L is track length
PL is the minimum package length
L’ is the maximum effective number of zones (a function of L/PL)
MC’ is the incoming moisture content variance
MC” is the effect that zoning has on moisture content standard deviation
MC”’ is the moisture content standard deviation with one zone

The same type of analysis also applies to vertical zoning (top-to-bottom) of packages.

Cascade Zone Control

Cascade control schemes can be used to target final moisture contents in each zone of a dryer using statistical feedback schemes, and by using in-kiln moisture sensing devices. In such a control system, each zone is monitored during the drying cycle and adjusted to approach its targeted setpoint "against all other zones."

The degree of success of such a control scheme is dependent on the accuracy of the sensors against the final average moisture content of each zone. There are several ways to do this, involving control valve positions, temperature drops across loads, active moisture sensors, etc. During the drying cycle, the zones may transfer from temperature control schemes to moisture control schemes.

The number of temperature control zones in a lumber dryer is also dependent on whether an E&C cycle follows the drying cycle. The following two **demonstrative** curves demonstrate the difference between the required number of zones in a dryer with and without E&C. The bias and slope of both curves are dependent on the level of engineering put into the dryer, excluding temperature controls. Linear curves are shown in this example to simplify the concept when the curves are actually stepped.

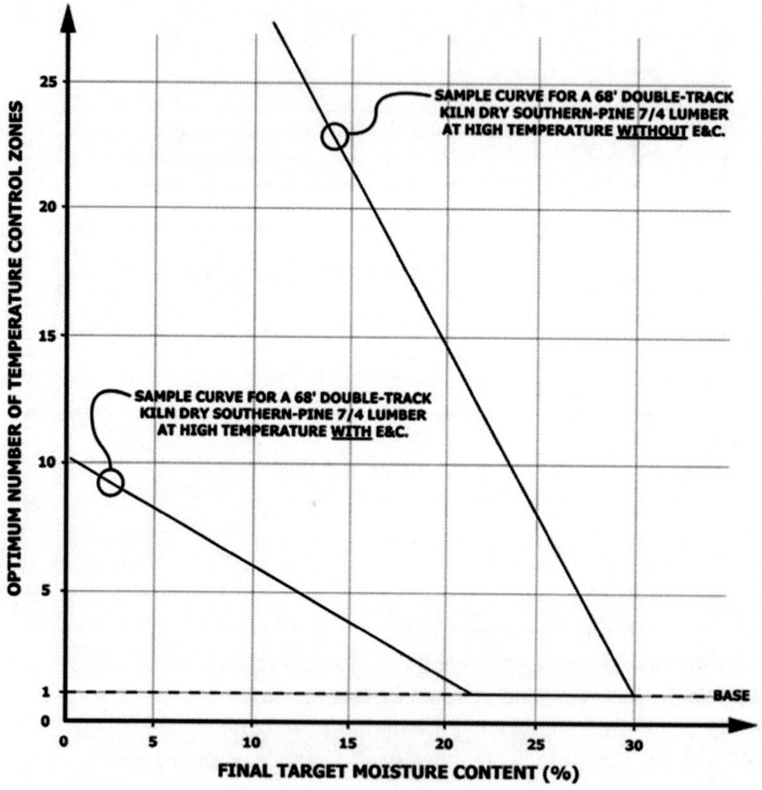

Dryer manufacturers typically oversell a large number of temperature control zones to lumber manufacturing operations under the guise of leading-edge technology when, in reality, this technology is frequently used to cover up (in politically acceptable ways) poor plant drying practices.

1. The initial design of the drying system was inferior leading to poor drying quality
2. The maintenance of the dryer was poor leading to poor drying quality
3. The drying system did not include E&C following the drying cycle
4. The drying operation was lacking in quality control training and practices

This leads us to the next issue. If all lumber dryers included an E&C cycle following drying, what effect does the final target moisture content have on the economic benefits of E&C? The answer to this question involves the same proportional dynamics found in economic zoning. As the final target moisture content approaches FSP, the economic benefits of E&C (for controlling degrade) drop.

Determine the optimum economic number of zones in a lumber dryer at your plant for:

A final target moisture content of 5%
A final target moisture content of 10%
A final target moisture content of 20%
A final target moisture content of 30%

Plot the optimum economic number of temperature control zones vs final target moisture content on the following graph to see the effect that final moisture content has on the required number of zones in a dryer.

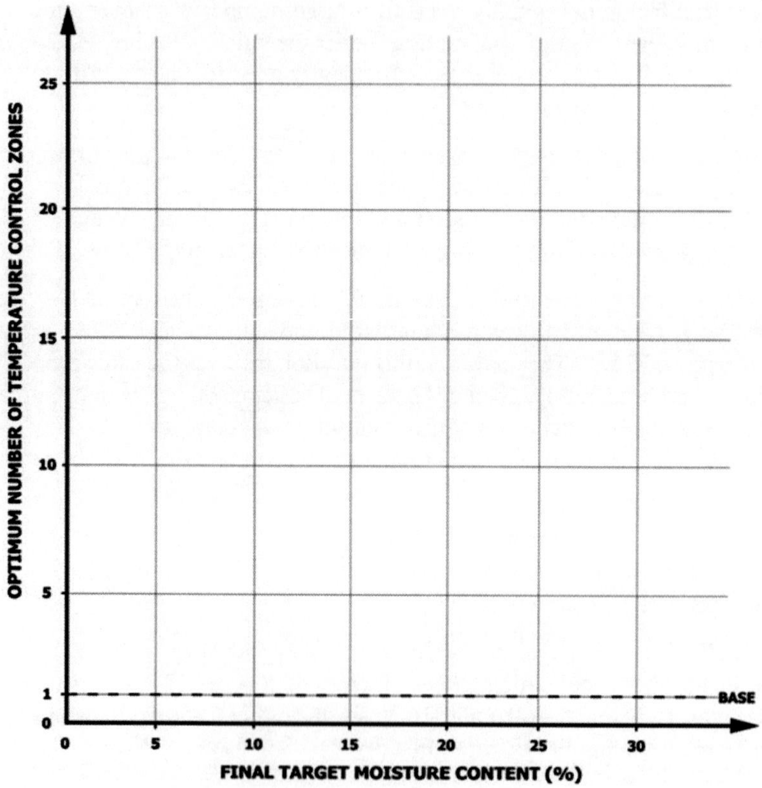

11.58 Energy Studies for Commercial Drying Operations

The purpose of an energy study is to determine how one drying operation compares to others and from this develop the best drying methods with the highest rate of returns for reducing energy costs. Energy studies should include **all** the parts of a drying system to determine the **total** energy costs/pound of water removed in the drying process. Energy from fuels and electricity should be combined to determine the total BTUs per board foot of lumber dried, and the costs of all the different types of energy consumed.

Green stacking equipment electrical usage
Green storage yards (lighting) electrical usage
Lift truck fuel usage
Transfer car electrical usage
Pre-dryers, kilns, and steaming chambers fuel and electrical usage
Rough-dry-storage fuel and electrical usage
Boilers, steam, and condensate systems fuel and electrical usage

Burners and furnaces for dryer equipment fuel and electrical usage
Fuel storage and handling systems electrical usage

When analyzing the different electrical users in a drying system, a list of motors should be prepared. The running amps and run time per week for each motor should be tracked.

When analyzing fuels, the heating value of each fuel type should be determined by laboratory testing.

Some common energy terms and conversion factors are listed in the following text:

BTU = 778 foot-pounds of mechanical work
BTU = .000293 KW
BTU = .000393 horsepower-hour
horsepower = 550 foot-pounds/second
horsepower = 33,000 foot-pounds/minute
horsepower = 1,980,000 foot-pounds/hour
1 horsepower = .707 BTU/second
1 horsepower = 42.42 BTU/minute
1 horsepower = 2545 BTU/hour
1 horsepower = .746 kilowatts
1 kilowatt = 3413 BTU/hour
1 kilowatt = 1.341 horsepower
1 kilowatt = 1000 watts
1 megawatt = 1000 kilowatts
1 gigawatt = 1000 megawatts

As mentioned earlier, wood dryer system rated energy efficiency (REF) is defined as follows:

$$\text{REF} = \left(1054.3 / \text{TDE}\right) \times 100\%$$

And

TDE = total energy used by the drying system/total pounds of water removed

Every drying operation should determine their TDE and REF.

Example #1:
An 84′ long, high-temp, double-track southern pine kiln has nine 25 horsepower fans. What is the BTU/hour of energy that these motors contribute to the drying process if the motors are 100% efficient?

$$\text{BTU / hour} = 9 \times 25 \times 2545 = 572{,}625$$

If the lumber dries in 19 hours, what is the total energy contributed by the fan motors to the total drying energy load if the motors are kept fully loaded during the drying?

$$\text{BTU} = 19 \times 572{,}625 = 10{,}879{,}875$$

If the kiln dried 135,000 board feet of lumber that requires 3000 BTU/board feet to dry, what is the percentage of the total energy load that the fans contributed to?

$$\% = \left(10{,}879{,}875 / \left(135{,}000 \times 3000\right)\right) \times 100 = 2.69$$

Example #2:
A lumber mill has a 1.0-megawatt electrical generator operating off the plant's boiler. How much equivalent horsepower does the generator produce when operating at its rated capacity?

$$\text{Mechanical Horsepower} = 1.0 \times \left(1{,}000{,}000 / 746\right) = 1340$$

Example #3:

How many 30-horsepower lumber kiln fan motors could the 1.0 megawatt generator supply electricity for, assuming the kiln fan motors were rated at 100% efficiency?

$$\text{Number of motors} = \left(1340 / 30\right) = 44.67$$

The following table demonstrates the typical annual cost of electricity for drying system fan and blower motors that are 95% efficient and operate 90% of the time during each year.

$$\$ / \text{yr} = 365 \times 24 \times .90 \times 1.0526 \times .746 \times \text{HP} \times \$ / \text{kwh} = 6{,}191 \times \text{HP} \times \$ / \text{kwh}$$

Total horsepower	***Electrical cost per kilowatt-hour ($)***			
	.08	**.10**	**.12**	**.14**
1	$495	$619	$742	$867
100	$49,500	$61,900	$74,200	$86,700

What does this say about using lumber dryer designs that require lots of electrical horsepower?

What does this say about using lumber dryer designs that use fan motors with low operating efficiencies?

What does this say about using lumber dryer designs that use inefficient fan and blower designs?

What does this say about using variable frequency drives for reducing electrical usage?

If a lumber dryer could be designed to use a thicker stacking stick that would use 25% less fan electrical energy, what does this say about doing a thorough analysis of the dryer before it is purchased?

If proper baffling is required to achieve efficient air utilization in a lumber dryer, what does this say about doing a thorough analysis of the baffles?

11.59 Heating System Design and Operation

One of the first decisions to make in the design of a wood drying system is the fuel source and the type heating system to use. The next decision is whether to use an indirect-fired or a direct-fired design. Each has their advantages and disadvantages. This section discusses the fundamental issues for heating systems. Only predominate dryer designs are discussed.

Health, Safety, Environmental, Legal, and Insurance Issues
Before you purchase a new heating system, you need to find out what local, state, and federal laws apply to the project. Not only are new systems affected, but upgrades and modifications can also be. The issues involve:

Construction permits
  Local and federal environmental laws and permits
OSHA regulations
  Fire and explosion prevention codes and standards
Insurance coverage issues
Business liability issues

You will need to contact the proper local, state, and federal agencies, as well as your insurance companies about both new equipment and upgrades of existing heating equipment.

Some of the many issues are:

Environmental permits for air, liquid, and solids discharges
Construction permits
Fire and explosion prevention and protection programs
Employee health and safety regulations
Insurance requirements (fire, property, business liability, business interruption, boiler and machinery, workers comp insurance, health insurance, life insurance, etc.)
Equipment design and manufacturing safety code requirements
Equipment inspection code and insurance requirements
Operation and maintenance requirements
Maintenance of records

Because the laws in each country and state are different, have your safety manager and attorney coordinate these activities. Each government agency and insurance company will require sufficient information about the project (type fuel, furnace size, annual fuel usage, etc.) so they can get the proper legal documents to you.

I also recommend that only engineers with considerable experience in wood dryer design be used in both designing and upgrading heating systems. There are numerous national safety code and standards issues in heating systems that, if not followed, can lead to environmental and health issues, accidents, and lawsuits.

Indirect-Heated Designs

Indirect-heated dryers use heat exchangers to separate the heating medium from the dryer's internal air space. The heat exchangers prevent the primary source of the heat energy from entering the interior space of the dryer where the lumber is located.

Predominant types of indirect-heated dryer designs are listed in the following text.

- Combustion systems using all-metal heat exchangers
    - Steam-heated systems
    - Hot-water-heated systems
    - Hot-oil-heated systems
    - Hot-air heated systems (very rare)
- Electrical systems
    - Dehumidification (heat pump) systems
    - Electrical strip heaters
    - Radio frequency systems
- Solar systems
    - Solar-heated hot-water systems
    - Solar translucent building panel systems

Finned-pipe heating coils are used in steam, hot water, and hot oil heated dryers to transfer the heat energy from the hot fluid to the dryer's internal airflow.

Plate type heat exchangers have been used in hot-air heated wood dryers, but these are rare. The hot air is produced by natural gas, oil, or wood combustion systems.

In dehumidifier dryers, the heat transfer process involves both heating and cooling. In these systems, a refrigerate is circulated through a closed heat pump loop involving a compressor, a hot condensing coil, expansion valve, and cold expansion coil.

From the dryer's perspective, the objective of every heating system design is to provide a controlled method for heating the air stream entering the packages of lumber inside the dryer. This may seem to be a simple task, but it is not. The air heating system must be able to react quickly to different conditions inside the dryer and maintain the proper dry-bulb temperature at each of the lumber packages. The air temperature control scheme at the packages can be entering, exiting, or averaging temperature. The selection of which control scheme to use is based on the design of the dryer, the species, the temperature schedule used, and the level of experience of the dryer operators.

The advantages that indirect-heated designs have over direct-heated (direct-fired) designs are:

1. Their superior ability to distribute heat energy and control temperatures inside the dryer
2. Higher wet-bulb temperatures are possible without the use of sprays
3. Less lumber degrade problems because of 1 and 2
4. Longer building life due to less internal pressure buildup

The disadvantages of indirect-heated designs over direct-heated designs are:

1 The total capital cost is much higher due to the additional cost of heat source and heat-exchange equipment
2 The total maintenance costs are much higher, same reason as 1
3 The total labor costs are much higher, same reason as 1
4 The overall energy efficiency is much lower, same reason as 1

In manufacturing complexes in which a central heating system, such as a steam boiler, is present for serving other processes besides drying, the use of indirect-heated dryers is the predominate method.

Steam Heating System Design

The first issue for designing a steam heating system is the difference between the steam temperature and the air temperature inside the dryer. Because heat transfer is directly related to temperature difference, the total surface area of the heating coils inside the dryer will determine the final total installed capacity of the heating system. The following table demonstrates how steam temperature and dryer air temperature effects the temperature difference at the heating coils.

Table of Temperature Difference

Steam pressure (psig)	Steam temp. (F)	Dryer air temperature (F)				
		240	180	140	100	50
150	366	126	186	226	266	316
100	338	98	158	198	238	288
60	307	67	127	167	207	257
30	274	34	94	134	174	224
15	250	10	70	110	150	200
5	227	–	47	87	127	177
0	212	–	32	72	112	162

Look at the temperature difference for 150 psig steam and 240 F dryer air temperature. The temperature difference is 126 F. Now look at the temperature difference for 60 psig steam and 240 F dryer air temperature. The temperature difference is 67 F. From these two examples, we can see that the ratio of the two is 126/67 = 1.88. This means that if we were to reduce the steam pressure going to a kiln from 150 psig to down to 60 psig, we would have to have 88% more heat transfer effective surface area to keep the same drying rate.

The next factor is the heat demand rate (BTU/hour) for the dryer. Once the temperature difference and the heat demand rate are known, the design of the heat exchanger coils can begin. Depending on the use of the dryer (species, board thickness, drying rate, etc.) the amount of required airflow inside the dryer will also change.

The total CFM of airflow inside the dryer will determine the ability of the heating coil surfaces to transfer heat to the air stream. In most wood dryers, the effective heat-transfer convection coefficient is an approximate .5 power function of the air velocity over the heating coils. This means that for every 2% increase in airflow (CFM) through the heating coils, there is a 1% increase in heating capacity.

If the fan CFM increases 2%, the heat transfer coefficient at the heating coils will increase 1%.

If the fan CFM increases 10%, the heat transfer coefficient at the heating coils will increase 5%.

If the fan CFM increases 20%, the heat transfer coefficient at the heating coils will increase 10%.

Peak Demand for Convective Heating Surface Area

In wood dryers operating at constant-temperature set points, the ratio of the heat-demand rate (for heating coils) to the available steam-to-air temperature difference will change during the drying cycle. When hot steam enters the heating coils, the air temperature inside the dryer will try to reach the steam temperature inside the heating coils. Thus, for every constant-temperature, steam-heated convection dryer, there is a **peak demand** for convective heating surface area during every kiln charge.

The following graph demonstrates a typical batch kiln charge in which the **demand** for heating coil surface area starts at a low value, increases to a peak value (the peak demand), and then decays to lower levels as the lumber dries. In most batch steam kilns, the peak demand for surface area will be reached within 1/4 of the drying cycle.

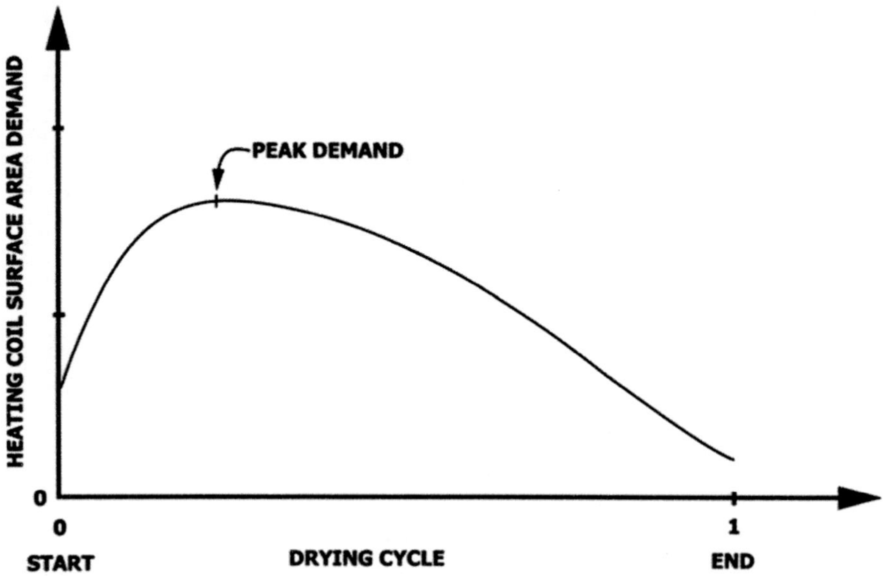

In dryers with insufficient heating coil surface area, the heat demand on the steam coils exceeds the total surface area built into the heating system. If this happens the lumber drying rate will, for a period of time, be determined by the total surface area of the heating coils. During this event that can last for many hours in some dryers, the lumber drying rate is "heat-limited" instead of "air-flow-limited."

All steam-heated wood dryers should be designed with no less than 15–20% excess heating coil surface area to prevent heat-limiting from occurring. However, there are some dryer manufacturers who believe heat limiting is a good way to design a batch steam-heated dryer. The most common spin is that such a design does not shock the lumber or the boiler.

The problem with heat-limiting in many steam-heated dryers is that the dryer operator, in most cases, has no control over the steam supply pressure (and thus steam temperature) going to the dryer. In the real world, wild swings in steam pressure can cause erratic drying, a host of control problems, and frustrated dryer operators. Heat-limiting is also a problem if load shedding schemes are installed on several dryers. The inadequate heating system will not respond properly to control system demands (such as a process computer). In addition to this, condensate flashing, and subsequent boiler makeup water and energy losses will be excessive during heat-limiting. If you add extra heating capacity to a steam dryer, the dryer will produce less flash steam and thus reduce energy usage, boiler water, and chemical treatment costs.

In low-temperature dryers with small heating systems, designers often add significant excess heating capacity for heating the dryer up during cold weather conditions. It is common for these dryers to have banks of coils that can be closed off manually by the kiln operator to maintain tight control of the dry-bulb temperature inside the dryer.

Minimum Steam Pressures

There are many thousands of different steam-heated wood dryer designs in existence. Many of these designs have oversized steam heating systems allowing them to operate on low steam pressures, and many have undersized steam systems requiring they operate on high steam pressures. The following is an overview of the ranges of steam pressures used for different wood dryer designs.

Drying operation	Steam pressure (psig)	Minimum temp. diff.
Hardwood operations	5–15	87–110
Hardwoods and softwood	15–60	70–127
Low-temp softwood operations	30–60	94–127
High-temp softwood operations	60–150	67–126

The preceding table is based on conventional industry practices in steam kiln designs. Using modern computer technology, all wood dryer heating systems can be designed by CFD computer modeling programs to operate on temperature differences much less than those shown earlier resulting in less fuel and electrical energy usage.

This brings us to the next subject in this book.

High- Versus Low-Pressure Steam: Which One Is the Better Choice?

The biggest and most common mistake made during the mating of wood dryers with steam boilers is to undersize the steam coil surface area in the dryer and purchase a high-pressure boiler. Both are done to reduce the total installed front-end cost of the drying system.

If the steam pressure is excessive, the entire dryer system will require additional long-term fuel and electrical energy, boiler water chemical treatment costs, and boiler and steam system maintenance costs over that for a low-pressure steam design.

Furthermore, high-pressure steam heating systems are vastly more prone to create hot and cold spots inside lumber kilns. For this reason alone, the use of high-pressure steam in lumber dryers should be avoided.

Every steam-heated wood-drying system should have a study done to determine the optimum economic boiler pressure. Standard accounting practices apply for calculating payback.

And there is a big difference between optimum economic boiler-pressure for new construction projects versus existing systems.

Steam Usage Calculations

The steam flow (pounds/hour) to a dryer is be determined by the specific enthalpy (heat energy) of the steam going to the dryer, and the dryer's energy demand rate.

The following table shows the specific enthalpy of saturated steam for different pressures (From Keenan and Keyes "Thermodynamic Properties of Steam"):

Steam pressure (psig)	**Steam temp. (F)**	**Specific enthalpy (BTU/pound of steam)**
150	366	857
100	338	881
60	307	905
30	274	929
15	250	945
5	227	961
0	212	970

In the preceding table (for 100 psig steam), it takes 881 BTUs of heat energy to boil 1 pound of water into 1 pound of steam. This same energy is released when the steam is condensed inside a heat exchanger in a wood dryer.

Example #1

A low-temp popular dryer has a 5,000,000 BTU/hour heat demand rate. The steam supply pressure is 30 psig. What is the pounds/hour of steam flow going to the dryer?

$$\text{Pounds / hour} = 5{,}000{,}000 / 929 = 5{,}382$$

Example #2

A high-temp pine dryer has a heat demand rate of 25,000,000 BTU/hour. The steam supply pressure is 150 psig. What is the pounds/hour of steam flow going to the dryer?

$$\text{Pounds / hour} = 25{,}000{,}000 / 857 = 29{,}171$$

Total Steam Usage

During the time it takes to dry a charge of lumber in a batch dryer, the steam demand will change depending on fan reversals, dryer temperature, lumber moisture content, etc. To determine what the total steam usage is for drying a charge of lumber, an integrating steam flow meter will be needed.

The accuracy of the flow meter and its integrator is crucial for accurate energy studies and dryer upgrades. I suggest a shedding-vortex flow meter be used. Consult with an established flow meter manufacturer, not a boiler or dryer supplier, for prices, accuracy, and other technical information. Do not use an orifice plate type flow meter. They are not as accurate unless they include pressure-compensation capability.

Heating Coil Unit Design

The design of heating coils for wood dryers has changed very little during the last century. Front-end cost is a big factor in most wood dryer prices, and this limits how much true engineering is put into their designs. Up until the mid-1970s, most steam coil units used in wood dryers had very little true engineering involved in their designs. Because of the management philosophies of dryer manufacturers, very little attention was given to improving coil designs. Today, advanced heat-transfer computer models are available for engineers to optimize coil designs. Discuss this with dryer manufacturers before buying their product.

Location of Coil Units

One decision to make when designing steam systems is the location of coil units. There are several places in a dryer for locating coil units. One is in the upper plenum space next to the exterior walls of the dryer, and the other is in the reheat space between the tracks of lumber in track kilns. In some dryer designs, heating coils may be located above the fan deck. This configuration should be studied carefully before doing this. Such a design makes it very difficult to replace the coils if they become corroded or develop leaks. Plenum coils should be installed under the elevation of the horizontal support steel for the fan deck so that their replacement is not laborious. Floor-level heat and reheat coils should be supported in such a manner allowing for their easy replacement.

Fin Pipe Orientation

Horizontal fin pipe has a higher heat transfer rate than vertical fin pipe because of the flow pattern of condensate inside the pipe. In vertical pipe, all the condensate flows down the entire length of the pipe. At the lower end of the pipe, the thick stream of condensate forms a barrier to the transfer of heat at the wall's boundary layer.

Research has shown that this situation can reduce the internal heat transfer coefficient at the lower end of a 20′ long fin pipe by as much as 70%. This situation can cause more than three times as much heat release at the top end of the coil unit then at the lower end. If tall vertical fin pipe is used for reheating between tracks of lumber, the result will be uneven lumber drying from top to bottom. Horizontal fin pipe is not affected in this manner because the distribution of condensate on the interior walls of the pipe is uniform from one end of the coil to the other.

One downside of using a horizontal design is that all the condensates will stream down a narrow line on the lower inside of the pipe and cut a trough if carbolic acid is present. Under the same corrosive environment, the inside of vertical pipe will last over three times as long as horizontal pipe before it has to be replaced. Another difference is that dryer manufacturers use long horizontal coil units to reduce the cost of manufacturing the coil units. There are fewer headers to manufacture, less holes to drill in headers, less welds to make, and less pipe connections to make.

Another downside of using horizontal coils is that long sections of fin pipe tend to sag. Support frames will be required to prevent sagging. Some of these support frames can also get to be rather complex and their presence in the air stream hinders heat transfer at the fin pipe.

There are many dryer applications in which a vertical fin pipe is a better choice than a horizontal design. In low-steam-pressure dryers requiring small amounts of fin pipe, short vertical coil designs can be easier to install, last longer, and be far easier to replace. If the height of the coil is not over 10′, the loss of heat transfer from top to bottom will not be significant.

We can summarize the **best fin-pipe orientation** in a dryer-application table like the following: However, the reader should understand that the total front-end and long-term costs must be evaluated carefully when making the decision to go from one orientation to the other. In the following table, a coil is classified horizontal if the slope of its fin-pipe is less than 90%.

Type Dryer	*Type Boiler Feedwater*	
	Alkaline	Acidic
Low-temp hardwood dryers	Vertical	Vertical
Conventional dryers	Vertical	Vertical
Accelerated dryers	Vertical	Vertical
High-temp dryer upper coils	Horizontal	Horizontal
High-temp dryer reheat coils	Horizontal	Horizontal
Hyper-temp veneer dryers	Horizontal	Horizontal

Some reheat coil designs (located between packages) use multiple staggered vertical coils. These are by far the best of all designs. However, they cost more. Quality always does in heat-transfer equipment.

Types of Fin Pipe

Pipe Size

For wood dryers, the pipe diameter ranges from 1″ to 3″. Both schedule 40 and 80 pipe are used. The thicker schedule is used to increase the life of the pipe when internal corrosion is a problem.

Metallurgy of the Pipe and Headers

The metallurgy will depend on the maximum working pressure and temperature of the heating medium inside the pipe. A53 and A106 are common pipe materials used. In low pressure steam systems (5–15 psig), A-120 has been used. However, in all steam coils, each coil unit must be pressure tested at no less than 2 times its maximum expected working pressure after the coil unit is manufactured.

Metallurgy of the Fin Material

Fin material will depend on the maximum temperature the fin pipe will be exposed to and the required heat-transfer coefficient. Mild steel, stainless steel, aluminum, and copper are common materials used for fin pipe construction.

Attachment Method for the Fin

The fins can be rolled tight onto the pipe without welding, rolled and welded, or stress fitted by expanding the pipe inside the fins.

Fin Geometry

The spacing of the fins, the outside diameter of the fins, and the shape of the fins may vary depending on the application. Some fins are crimped and then rounded in a spiral fashion onto the pipe and some fins are square for press fits. Some spiral fins are serrated giving a porcupine appearance. The optimum economic fin geometry can be predicted by modern heat-transfer computer models. The air velocity over the fins and the temperature difference from the heating medium to the air stream will determine the best fin geometry for a specific dryer.

Types of Headers

Either round pipe or square tubing can be used for headers.

Configuration of Coils for Drainage and Uniform Air Heating

Horizontal coil units should be sloped no less than 2% for condensate drainage.

The degree of slope should also overcome any sagging of the fin pipe.

Temperature Rise Across the Coils (TRAC)

The temperature rise (TRAC) of an air steam passing through a heating coil is the measure of how much energy is being transmitted to the air flow through the coil.

Venting Tramp Air Out of Coil Units

Steam boiler deaerators do not get all the air out of the makeup water that enters boilers. Because of this, air will carry over to the dryer and collect in steam coils. When this occurs, large cold spots (air locks) can occur causing reduced heating capacity and uneven heating. If large heating coil units are used in a dryer, large cold air pockets will collect inside the coils.

To get the air out of the coil, two methods are used. One is the use of thermostatic vents mounted on the coil units, and the other is the use of a steam trap design that will vent the air through the trap.

Of the two, the coil-mounted thermostatic vent is a superior method. Steam traps for condensate removal are designed for handling condensate. They do not do a good job of removing air from large coil units because the trap receives condensate from one drain point on the coil, and trapped air may or may not exit the coil at this point. Thermostatic air vents can be mounted directly on coil headers. On very large coil units, the use of several air vents may be needed.

Thermostatic air vents are a small type of steam trap. The device has a small valve inside it attached to a bellows. The bellows is filled with a mixture of water and antifreeze. The end of the bellows opposite from the valve is attached to a rigid mounting. The bellows is designed such that a certain amount of clearance exists at the valve seat. When the coil unit and the bellows are cold, the valve is open. As steam enters the coil unit, the air inside the coil starts escaping through the valve. After the air is vented, hot steam starts passing through the valve. This causes the bellows to heat up. Once the temperature of the liquid inside the bellows reaches a predetermined level, the liquid inside the bellows starts boiling. When the liquid boils, the pressure inside the bellows increases. The bellows then expands causing the valve to close. The relationship between the spring constant of the bellows and the initial clearance of the valve will cause the valve to prevent pure steam from passing through the valve. When operating properly, mostly air, not steam, escapes through a thermostatic air vent.

In some dryer installations, the exhaust of each thermostatic vent is plumbed to the exterior wall of the dryer with small copper tubing. This allows the dryer operator to check each vent after the dryer is started up.

The only problem with thermostatic vents is that their bellows eventually fail, and they then vent steam constantly. Because of the large number of thermostatic vent failures that can occur in a dryer, they must be constantly checked for proper operation. However, very little attention is given to them, and when they start failing in large numbers, dryer operators too often see them as a pesky problem.

Dryer operators have numerous tasks to do when operating lumber dryers and often pay little attention to the numerous thermostatic air vents inside dryers. Even with the external vent lines, the operator has to inspect them regularly. Many plants use them. Some plants remove them after they fail. I recommend their use if heating capacity is a problem in a dryer, and the plant's production is limited by drying capacity. If drying capacity is not a major issue, the justification for using thermostatic air vents in a properly designed coil system is precarious.

Corrosion of Fin Pipe

Internal corrosion is caused by an inoperative deaerator at the boiler or caustic boiler makeup water or both. In plants using highly alkaline makeup water to the boiler, wood dryer coil units can be destroyed in less than 1 year of use.

External corrosion of fin pipe is caused by attack from wood acids in dryers using E&C cycles. Dryer designs using E&C cycles will suffer significant corrosion to coil units. All E&C dryers should include coil designs for easy replacement. Coating a corroded heating coil with a mixture of diesel fuel and graphite does not increase the life of the coil significantly. This practice is dangerous and should never be done. It may cause a fire or explosion.

Required Heat Transfer Capacity of Steam Coils

A properly designed steam heating system has sufficient heat transfer capacity to respond to the dryer's dry-bulb temperature controller. When the controller calls for heat, the heating system should respond to the demand at such a rate the temperature does not overshoot the set point of the controller. If too much heating capacity exists in the steam coils, overshooting will occur. If inadequate heating capacity exists, the dryer's temperature may require hours to reach set point. In severe cases, the temperature may never reach the set point for days because of insufficient heat transfer surface area. Ideally, for control purposes, the steam heating coils should allow quick response with no overshoot.

Hardwood dryers may be fitted with several banks of coil units, or an adjustable main steam pressure regulator, or both. Heating capacity can be reduced or increased by adjusting the regulator or closing off or opening banks of coils. This allows the dryer operator to fine tune the heating capacity to match the drying conditions and maintain tight control over the dry-bulb temperature inside the dryer.

Softwood dryers may include a main steam pressure regulator, but most do not. Instead, the heating capacity relies on the temperature controllers and steam coil control valves to control the temperature inside the dryer. Softwood dryers are usually designed with sufficient heating capacity to prevent long periods of the control valves staying wide open. If a softwood dryer has insufficient steam heating capacity, this can cause erratic drying, excessive condensate flashing, wasted energy, and lost drying capacity.

Fin pipe footage in a dryer is referred to as L, meaning length of fin pipe. To calculate the proper amount of fin pipe footage, the following variables have to be known:

1. The peak heat demand of the dryer (British thermal unit (BTU)/hour)
2. The conditions inside the dryer when the peak demand occurs.
3. The difference between the coil steam and air stream temperatures.
4. The air stream velocity approaching the fin pipe.
5. The temperature and humidity ratio of the air stream.
6. The number of banks of coil units.
7. The geometric orientation of the fin pipe and coil units.

We can write the following equation for calculating L:

$$\mathrm{BTU} / \mathrm{hour} = \mathrm{H} \times \mathrm{L} \times (\mathrm{Ts} - \mathrm{Tf})$$

Or,

$$\mathrm{L} = \mathrm{BTH} / \mathrm{hour} / (\mathrm{H} \times (\mathrm{Ts} - \mathrm{Tf}))$$

Ts is the steam temperature and **Tf** is the temperature of the air stream approaching the coil unit. The heat transfer coefficient H is affected approximately by the square root of the approaching air velocity, the number of coil banks, geometric factors, and the heat transfer properties of the air/steam fluid stream approaching the coils. Due to wide variances in air stream velocities, banking, geometric factors, and fluid stream properties, the coefficient H can vary from 2.5 to 20.0 in commercial wood dryers.

Case Study

A southern pine double track kiln has 850 ft/min of air velocity through 7/8″ stickers in 8′ wide packages. The heating system in the dryer has 5500′ of fin pipe. The dryer dries 2 × 4 lumber in 25 hours. Tests have shown that the initial steam demand of the dryer is 4.33 × 5500 = 23,815 pounds per hour when the average air temperature approaching the steam coils is 50 degrees F. The steam supply pressure is 150 psig. What is the H coefficient for this dryer?

Solution:

150 psig steam has a temperature of 366 F. Thus Ts = 366.

The air temperature approaching the coils is 50 degrees F. Thus Tf = 50. The length of the fin pipe is 5500′. Thus L=5500.

The BTU/hour is calculated as follows:

$$\begin{aligned} \text{BTU / hour} &= 23{,}815\,\text{pounds steam / hour} \times \text{the specific enthalpy of steam at 150 psig} \\ &= 23{,}815 \times 857 \\ &= 20{,}409{,}455 \end{aligned}$$

The equation presented earlier is:

	BTU/hour	= H × L × (Ts − Tf)
Thus,	H	= BTH/hour/(L × (Ts −Tf))
Therefore,	H	= 20,409,455/(5500 × (366-50)) = 11.74

Use of the H Coefficient to Evaluate a Convection Dryer Design

The H coefficient is a useful tool in comparing similar dryer designs operating at the same stream flow conditions (air flow, temperature, humidity ratio, etc.). However, to use it for comparing different coil designs, the total pressure drop through the airside of the coil units has to be known. If coil units are tightly packed to increase heat transfer, this will create additional backpressure on the dryer fans. Any additional backpressure on the fans will result in a reduction in the total CFM output from the fans, drying problems, and unnecessary additional electrical and fuel costs. Such situations should be avoided because the losses go on from year to year after the dryer has been installed. Eventually, the collective electrical costs can exceed the total cost of the coils in the dryer. The term "coil pressure drop" CDP is:

$$\text{CDP} = \text{H / DP}'' \text{ where DP}'' \text{ is the air flow pressure drop through the coils}$$

In this equation, the unit for DP is inches of water column. In the preceding case study, tests revealed that the total pressure drop through all of the steam coils in the dryer was 0.23″ water column.

Thus,

$$\text{CDP} = 11.74 / 0.23 = 51.04$$

For all types of commercial wood dryers, CDP varies over a wide range. However, when comparing one heating system design against another for a specific application operating under the same conditions, it is a simple method to use. CDP is also an **indicator** of how **energy efficient** the heating coil design is.

Only by modern CFD computer software designed specifically for heat transfer processes can the **optimum** CDP be determined for a specific dryer application. If you have a dryer project that requires high performing efficient heating coils, you should retain a mechanical engineer familiar with heat transfer technology to work with both your dryer contractor and the fin pipe manufacturer.

Steam Control Valves

The selection of a steam control valve is determined by the following:

1. The maximum steam pressure and temperature the valve will be subjected to
2. The maximum steam flow rate to pass through the valve
3. The pressure drop through the valve at the maximum steam flow rate
4. The maximum shut off pressure
5. The type of actuator used on the valve
6. The type of trim characteristics required for proper temperature control

Control valves used in most commercial wood dryers are in the range of 1–4″ pipe size. Either a cast iron or carbon steel body is used. Both flanged and screwed bodies can be used.

Standard pressure and temperature ratings of cast iron valve bodies are listed in the following text for valve sizes of 1–12″.

Body Rating	Pressure-Temperature Rating		
	Pressure (psig)	at	Temperature (F)
125 lb. cast iron	175		150
125 lb. cast iron	125		353 (saturated steam)
250 lb. cast iron	400		150
250 lb. cast iron	250		406 (saturated steam)

Carbon steel valve bodies are rated by ANSI 16.5. They are offered in 150 lb, 300 lb, and higher ratings. In wood dryers using saturated steam less than 175 psig, the use of the more expensive carbon steel body over a cast iron body cannot be justified.

For saturated steam pressures of 125 psig or less, the 125 lb. cast iron body is sufficient. For saturated steam pressures between 125 psig and 250 psig, the 250 lb. cast iron body should be used.

Once the valve's body material has been selected, the required flow capacity Cv, and trim characteristics of the valve will have to be determined. The Cv will dictate the final size of the valve body. The trim characteristics will depend on the valve's control requirements.

Trim characteristics	**Applications**
Non-modulating (on-off)	Small dryers, small valves, large batteries of dryers, usually hardwood dryers
Modulating	Dryers receiving steam from water-tube boilers
Linear trim	Dryers for wood species subject to heavy checking
Equal-percentage trim	Softwood dryers, high-temp dryers

Valve manufacturers offer a wide range of valve sizes, types, and designs. The following approach is recommended for selecting a valve size based on flow capacity.

In the following example, published data from a control valve manufacturer is used.

One zone of a high-temp softwood dryer's steam heating system has 750′ of fin pipe. The pressure at the steam header is 145 psig. The condensing rate of the steam inside the fin pipe and total steam flow rate has been predicted by computer analysis to be the following:

Temperature Approaching the Fin Pipe	Condensing Rate (#/hr/foot fin pipe)	Flow Rate (#/hr)
70 F	6.47	4,852
260 F	1.87	1,402

From the preceding, the control valve has to pass 1,402 pounds/hour of 145 psig steam when the dryer reaches set point. It also has to pass 4852 pounds/hour of steam at startup. Let us take a quick look at the required valve Cv for both cases. Assume a 10% pressure drop across the valve in both cases.

Looking at the control valve manual, the following Cv data is found:

Inlet pressure	Pressure drop across the valve	Required Cv/10,000 #/hour flow
160 psia	15 psi	72

Thus,

For a steam flow rate of 1402, the Cv = 72 × (1402/10,000) = 10.09
For a steam flow rate of 4852, the Cv = 72 × (4852/10,000) = 43.93

The average Cv = (10.09 + 43.93)/2.0 = 22.51

Looking at the control valve manual again, the following data on Cv is:

Body size	Trim	Cv
¾, 1	Stem	10.0
1	Stem	17.0
1½	Cage	34.0

For this application, the use of the 1″ valve with the Cv of 17 is the best choice. There would be more than a 10% drop across the valve at startup, but the small trim would allow better control of the temperature once the dryer reaches set point. The 1½ valve with the Cv of 34 could also be used but may operate too close to its closed position near the end of the drying cycle. Looking at the ¾, 1″ valve with the Cv of 10.0, this is too small a valve but would work fine for slow drying schedules. The choice would have to be made based on the dryer designer's experience with the types of wood being dried.

Control Valve Actuators

Every control valve must have means by which the valve's position is controlled. The two classes of operators are pneumatic and electric. Most control valve operators used on wood dryers use pneumatic. Electric operators are also used, but they have a reputation for high failure rates due to the high humidity corrosive environment that surrounds these valves. However, there is a trend for going all digital in wood drying control systems requiring the use of electric control valve operators.

Pneumatic operators operate off a 3–15 psig control signal range from a pneumatic controller or converter. The pressure the valve operator will experience will be 20 psig when the signal from the controller is at its maximum output. The lowest pressure the valve operator will experience is zero when the temperature controller is at its minimum output. In both cases, an increasing air signal to the operator causes the valve to open. When the signal is 0–3 psig, the valve is closed. At 3.1 psig, the valve is barely open. When the signal is 9 psig, the valve travel is half open. When the signal is 15 psig, the valve is 100% open.

Steam Pressure Reducing Stations

Pressure regulators are installed in the main steam supply line going to dryers to keep the steam header pressure at the dryers at a certain level. A steady header pressure improves temperature control inside the dryers. It is also a means by which the steam demand to the boiler can be limited in the event the dryers demand more than the boiler is rated for.

Numerous control schemes have been used for automatically changing the setting of header pressure regulators to prevent overloading boilers. Their operating principle is to reduce the header pressure once the total steam flow at the boiler reaches a certain level. Some control schemes use the control valves in the dryers to control header pressure. Some use the flow rate of steam produced at the boiler for controlling the header pressure. Some schemes use both header pressure and boiler steaming rate.

Steam pressure regulators can be either a manual design or a remote operated design. The manual design is adjusted by a spring wrench at the regulator. The remote design uses a pneumatic operator controlled by a pressure controller. The pressure controller can be located next to or a considerable distance from the regulator.

The required Cv of the pressure regulator can be significant in softwood dryers. Often, a 4″ or 6″ valve body is required. In some cases, an 8″ body is required.

Principles of Condensate Removal

There are two types of forces involved in the movement of condensate. One force (gravity) pulls down on the liquid causing it to drain and drip down the surface it condensed on. The other is differential-pressure forces.

Condensate Production

Steam produces condensate when it loses energy through heat transfer. The condensing process is referred to as a change of state. Before the loss of energy, the water is in the form of a gas (steam). After the loss of energy, the water is in the form of a liquid (condensate). The amount of energy released per pound of steam condensed is the specific enthalpy. The specific enthalpy changes with absolute saturated steam pressure.

Steam pressure (psia)	Steam temp. (F)	Specific enthalpy (BTU/pound of steam)
300	417	809
200	382	843
160	364	859
140	353	868
120	341	878
100	328	889
80	312	901
60	293	916
40	267	934
30	250	945
20	228	960
14.696	212	970
5	162	1001

Atmospheric Condensate Systems

As hot condensate drains or drips down from a cool surface, it eventually encounters a physical object such as a collection pan, shroud, bucket, etc., in which it collects. Once the hot condensate accumulates inside the collector, the task of handling the condensate presents itself. If properly designed, the collector is sufficiently sloped such that the condensate flows freely to a drain opening located at the lowest point in the collector. The drain opening is connected to a sloped drainpipe that directs the hot condensate to a trapping device. The purpose of the trap is to permit only liquid condensate to pass. In atmospheric condensate handling systems, float-operated valves are used for traps.

Pressurized Condensate Systems

When saturated steam collects inside a cooled pressurized vessel, such as a fin pipe coil unit, it reaches a temperature dependent on the pressure inside the vessel. While the steam is collecting inside the coil, the steam velocities inside the coil unit can vary over a wide range. At the inlet to the coil unit, velocities can approach several hundred miles per hour. At the far end of the coil unit near the drain connection, the velocity is very low, even being zero for long periods of time. Depending on the opening of the steam control valve for the coil unit, and the operation of the condensate trap for the coil unit, the velocities, pressures, and temperatures inside the coil unit can change dramatically.

These pressure and velocity changes create quite a task for any trapping system used to remove condensate collecting inside the coil unit. If insufficient pressure exists inside the coil to force the condensate through the pipe going to the trap, the flow of the condensate to the trap will drop below the condensing rate inside the coil. When this occurs, condensate will start collecting at the bottom side of the coil unit.

Under certain conditions, the level of condensate inside the coil can rise to such a point that the coil will be partially full of condensate instead of hot steam. If the condensing rate of the partially flooded coil matches the drain rate from the coil, the condensate level in the coil will stabilize. If the coil becomes full of condensate, the heating capacity of the coil unit will drop to zero. If the coil unit is half full, the coil's heating capacity will also be half.

Under certain conditions, the pressure inside the coil unit can drop below atmospheric pressure causing a vacuum inside the coil unit. If the heat exchanger is in an environment less than 212 degrees F, and the steam supply valve closes, the steam inside the coil will keep condensing until a partial vacuum is created inside the coil. This vacuum will cause a backflow of condensate from the traps installed on the coil. If the discharge of the trap piping in the condensate receiver tank is not designed properly, cold air at the tank may be sucked back into the coil unit.

To ensure reliable operation of the coil unit at different operating pressures (and temperatures), the condensate piping, the trap, and the condensate receiver tank should be designed such that any condensate produced inside the coil unit is promptly removed, and not allowed to flow back into the coil. If the entire system is not designed for all possible conditions, the coil unit will not operate properly.

Air Contamination in Steam Coils

Another problem that plagues steam coils is air. If air enters a coil unit, it will tend to settle down to the lower portion of the coil due to its heavier density than steam. However, internal steam flow patterns can lift the air sending it to different areas inside the coil. Wherever the air is, its presence acts like a blanket between the steam and the interior surface of the coil. The air causes less steam condensation on the interior wall of the coil resulting in less heat transfer, and thus a cold spot. For this reason alone, air should be kept out of steam heating coils.

Objectives of Steam Trap Designs

There are three main objectives in designing a condensate trapping system for steam-heated coil units.

1 Maintain the rated capacity of the heat exchanger by proper air and condensate removal
2 Make allowances for different heating coil operating conditions
3 Use trap designs that are rugged, cost effective, and easy to maintain

Selecting the Trap

The factors that go into trap selection are:
The maximum flow rate of condensate entering the trap
The maximum inlet pressure the trap will be subjected to
The maximum backpressure the trap will be subjected to
The range of inlet and outlet pressures that will affect the operation of the trap
The ability to remove air from the heating coils

When selecting a trap, operating **curves** should be constructed to determine the required trap capacity for every inlet and outlet pressure expected to occur during the operation of the heating coil. Two processes are occurring at one time, one is the condensing rate inside the coil, and the other is the condensate flow dynamics through the trap. Both must be reviewed to choose the right trap.

Types of Traps

There are six types of traps used in wood dryers. Each has their pros and cons.

Thermodynamic disk
Thermostatic bellows
Float
Inverted bucket
Bimetallic
Orifice

Thermodynamic disk traps operate by the impulse principle. A flat disk, mounted on top of a valve seat, opens when a slug of condensate impacts the disk. The condensate then passes under the disk and out the trap. Thermostatic traps are the smallest and lowest cost of all designs. They can be used successfully on dryers if they are inspected on a regular basis. Because of their small size, they can be installed close to the coil units inside the dryer or located outside close to the condensate collection header.

Thermostatic bellow traps use a liquid-filled bellows to operate a valve. The expansion of the bellows opens and closes the valve when the pressure and temperature inside the bellows reaches the saturation curve for steam. The failure of the bellows is the biggest problem with these traps. Like the disk, they are small, and easily replaced.

Float traps are in effect a float connected to a valve using a mechanical linkage. Float traps must be piped to the heating coil in such a manner that condensate will drain freely from the coil unit to the trap. Float traps are very large and heavy requiring plenty of space for their installation and maintenance. They are also the most expensive of all types of traps. Their use on wood dryers is rare because of their high cost and large size.

Inverted-bucket traps are like float traps except they use an inverted bucket for operating a valve. Although they are large and heavy, they are efficient and their use on wood dryers is common. The internal valve seats and linkages require annual inspections to maintain their efficiency. Worn valves and valve seats in inverted bucket traps will cause significant wasted steam if not replaced promptly.

Bimetallic traps use a bimetallic spring to operate a valve. These traps operate at temperatures below the saturated steam temperature. Bimetallic traps are used to save energy and reduce boiler makeup water. Because the condensate entering the trap is colder than the saturated steam temperature in the coil, less flashing occurs downstream of the trap than occurs with all other types of traps.

Bimetallic traps can cause significant problems with cold spots inside heating coils, as well as reduced heating capacity. If used, the **heating coils** must be **designed specifically** for bimetallic traps. Aggressive corrosion, numerous cracks, and leaks in heating coils may occur if bimetallic traps are used. This is an especially serious problem with horizontal coil units receiving steam from a boiler with highly alkaline makeup water. I have seen new dryer coils **destroyed** in 6 months due to these conditions.

Orifice traps are either holes drilled inside a plate, or adjustable throttling-type needle valves. Plate orifices may have one or many holes drilled inside it to allow the passing of liquid condensate. These are very easy to build. Simply drill holes in a plate and install the plate between two pipe flanges. The number and sizes of the holes can be changed to get the desired result. Orifice traps are best applied to systems in which the flash downstream of the trap is being recovered for other steam users.

Adjustable needle valve traps are often used on continuous driers such as veneer dryers. The small amount of steam that passes through the valve collects in a pressurized flash recovery tank and is not wasted. The valve is field adjusted to maintain an inlet pipe temperature just below the steam temperature inside the coil. Adjusting these valves is a rather simple task. First, the valve is opened 100%. The valve temperature is then observed. As the valve is closed, its inlet pipe temperature will rise slightly. As the valve continues to close, the temperature will reach a maximum value. Then the valve is slowly closed more. Once the inlet pipe temperature starts dropping, the valve is locked in place. An accurate thermometer, mounted in a well in the inlet pipe, is required to adjust the valve properly. These simple traps work best on steam systems in which the pressure inside the coils is constant. The valve and the seat must be hardened to withstand the high velocity of the condensate and steam. The final settings of these traps should be agreed on by all involved plant people and documented.

Trap Location

There are two philosophies about where to locate steam traps on dryers. One is to locate them outside the dryer so they can be tested while the dryer is in operation. The other is to locate them inside the dryer so they will be protected from freeze damage and to reduce piping costs. Because this is a matter of personal preference, the high probability exists that either design will work if properly maintained and will not work if not maintained. If you put them inside the dryer, locate them close to floor level so they can be easily tested and replaced.

Freeze Protection of Traps and Pipes

All traps and their lines should be designed such that they drain freely when the dryer is shut off. If a bucket trap is used, there will be a certain amount of residual water inside the trap when the dryer is shut off. The main condensate line going to the flash tank should be fitted with a drain valve.

Trap Strainers and Blow-Down Valves

Strainers are used with traps to prevent rust and scale from entering and plugging or damaging the trap. Some traps are built with an internal screen. Some dryer manufacturers do not install strainers with traps. I prefer to see every trap have a strainer and a blow down valve. The strainer should be blown down once a week. While blowing down the strainer, watch the discharge stream for evidence of rust or dirt.

Trap Discharge Piping

The piping exiting a steam trap must be of sufficient size to prevent excessive backpressure on the trap. In all steam traps, when hot condensate passes through a valve seat, there is a sudden pressure drop that causes the condensate to flash. The flashing causes a large rise in the pressure inside the piping. The trap's discharge piping should be of sufficient size to prevent the discharge pressure from reducing the capacity of the trap. This problem works differently for different trap designs. Inverted bucket traps are not affected as much by high backpressure as other traps are. If the trap's discharge pipe is the same pipe size as the trap, but not over twenty times the pipe size in length, the backpressure will not affect the rated capacity of the trap. However, beyond this pipe length, the size of the pipe should be increased to the point where it collects only in the main condensate header. Ideally, steam traps should be located as near the condensate return header as possible to prevent excessive backpressure. The main condensate return header should have a minimum inside cross-section area equal to three times the total cross-section of all the trap lines attached to the header. All condensate headers should have a pressure gage mounted on their topside so the pressure inside the header can be observed.

If multiple dryers are connected to one condensate flash tank, the design of the return lines and isolation blocking valves should allow for maintenance to be performed on one dryer while the other dryers are in operation.

Testing the Steam Trap for Proper Operation and Wear

A trap design that works well will test well. In addition, a trap design that does not work well may also test well. Knowing why both test well but one does not work well is the key to understanding how steam traps work.

MYTH #1 – A trap that is hot is working.
MYTH #2 – A trap that is cycling is working.

A maintenance person marks a trap with a welder's crayon and watches the mark melt. Then he listens to the trap with a stethoscope to see if the trap is cycling. The trap is hot and cycling. Is the trap operating at maximum efficiency? Probably not if the trap is over a year old.

To test a steam trap properly, two measurements are required. First, the trap must be working. It must be cycling if it is a cycling design. Next, the temperature of the condensate entering the trap should be just below the temperature inside the coil unit it is serving. If designed and operating properly, the trap will have an entering temperature slightly lower than the saturated steam temperature entering the coil. To measure this, thermo-wells are needed at both the coil unit's inlet header and the pipe at the entrance to the trap.

When testing the trap for all possible load conditions, the same thermometer, or thermometer meter, should be used to measure the temperature at both the coil and the inlet pipe to the trap. The first thermo-well should be mounted on the coil's feed header. The thermo-well at the trap should be mounted within 5′ of the inlet of the trap. The type of trap will determine the allowable (maximum) difference in temperature reading. Discuss this with your trap manufacturer's representative.

Once temperature data has been collected on a specific trap design serving one coil in a dryer, then the proper trap testing temperature can be established for a specific steam pressure test. If high temperatures are found at a trap, this may be a sign of worn valve seats. Annual inspections should be performed on all traps to locate worn valves, seats, and linkages.

Startup Steam Demands Created by Lumber Kilns

This is a subject that involves corporate and plant politics, and environmental, safety, and legal issues.

To keep yourself out of danger and possibly serious legal trouble, never load a steam boiler beyond its design rating. Willfully overloading a steam boiler is something that only an ignorant or totally irresponsible person would do. Furthermore, if you think you know more about steam boilers and steam boiler safety codes than the engineers who design them, you could find yourself in a lot of legal trouble or dead.

Steam boiler designs are based on a host of safety codes that were specifically developed by very smart mechanical engineers to prevent people from doing dumb things that can get themselves and a lot of other people killed. Furthermore, most steam boilers in operation today are required to meet federal emissions standards based on permitted steaming rates. If a boiler exceeds its permitted steaming rate, federal law is being violated. No steam boiler should ever be pushed beyond its design rating. No drying system should ever demand more of a steam boiler than what the boiler was rated or permitted for.

Uncontrolled Steam Demand

Uncontrolled steam demand is what happens when a steam system goes out of control and creates a steam demand that is excessive or changing too fast.

Controlled Steam Demand

Controlled steam demand is what happens when steam demand is controlled, not excessive or changes too quickly.

Controlled Steam Demand Schemes

Wood dryers can have numerous steam control valves opening or closing at the same time. During fan reversals, the heat transfer of steam coils drops to essentially zero when the air flow in the dryer stops during the reversals. During the startup of a cold dryer, the steam demand in large softwood dryers can exceed 40,000 pounds per hour for a short period of time. Special controls may be needed to prevent large dynamic steam demands.

During the past century, dryer operators manually closed and opened gate valves to control the demands on boilers. Today, many wood dryers have automatic process-control schemes for controlling steam demand.

Header-pressure control loops
Header-temperature control loops
Steam-flow-limiting control loops
Boiler furnace firing-rate control loops
Load sharing between multiple dryers

At this point I want to remind the reader that control systems applied to steam boilers must only be designed by engineers who are property trained and experienced in both steam boiler and process-control technology.

Header-Pressure Control Scheme

This control system monitors the saturated steam pressure at the main steam header at the dryers and sends this signal to a pressure controller. The pressure controller sends a signal to a low-selector relay at the dryers. If the steam header pressure never drops to the set point of the pressure controller, the dryers are not affected. If the header pressure drops down to the controller's set point, the control valves in the dryers start closing. The closing of the dryer control valves causes the steam demand to decrease. The decrease in demand causes the header pressure to rise. The set point of the header pressure controller is adjusted slightly below the setting of the boiler master (pressure controller) at the boiler.

Because this is a pressure-control system, there may be times when the boiler is overloaded for long periods of time beyond its rated steaming capacity.

Header-Temperature Control Scheme

This control system monitors the saturated steam temperature (instead of pressure) at the main steam header at the dryer and sends this signal to a temperature controller. The temperature controller sends a signal to a low-selector relay at the dryers. If the steam header temperature never drops to the set point of the temperature controller, the dryers are not affected. If the header temperature drops down to the controller's set point, the control valves in the dryers start closing. The closing of the dryer control valves causes the steam demand to decrease. The decrease in demand causes the header temperature to rise. The set point of the header temperature controller is adjusted slightly below the setting of the saturated temperature in the boiler's steam drum. The boiler master (pressure controller) at the boiler indirectly controls this temperature.

Because this is a temperature-control system, there may be times when the boiler is overloaded for long periods of time beyond its rated steaming capacity.

High-Steam-Flow Limiting Controllers

This control system monitors the steam flow rate being produced by the boiler and sends this signal to a steam-flow controller. The flow controller sends a signal to a low-selector relay at the dryers. If the steam flow at the boiler never rises to the set point of the flow controller, the dryers are not affected. If the boiler steam flow rises to the flow controller's set point, the control valves feeding the dryers start closing. The closing of the dryer control valves causes the steam flow at the boiler to fall. The set point of the flow controller (located in the boiler control room) is the rated steaming capacity of the boiler, or a lower level determined by the boiler operator. **Do not ever set this controller's set point above the rated capacity of the boiler.**

Because this is a flow control system, there will never be times when the boiler is overloaded beyond its rated steaming capacity. The accuracy of the boiler's steam flow meter is crucial to the success of this system.

In some versions of this system, a large steam control valve is installed in the main steam line at the boiler instead of using the valves at the dryers to control the steam flow. I prefer this design because of its simplicity. This design allows boiler operators to quickly reduce steam demands on the boiler when they see the need to do so. This capability at the boiler house leads to a smoother boiler operation and less tension between the boiler and dryer operators.

Boiler Furnace Firing Rate Schemes

There are occasions when the steaming rate of a boiler can be limited by limiting the combustion rate inside the boiler's furnace. If a boiler is experiencing large uncontrollable steam demands, the solution is to control the demand, not limit the combustion of fuel at the boiler. This type of control system may cause wide swings in boiler pressure. This should be avoided. Repetitive wide swings in boiler pressure can damage boilers.

Demand Priority Schemes

There are cases when numerous steam boilers, dryers, and steam users exist at one plant. There may be times when the total steaming capacity of the boilers is not adequate for all the steam users. In such a situation, priority control systems can be used to both maximize production and protect the boilers and the drying system's equipment.

Priority control systems use the lead-lag principle for controlling either boiler pressure or steam flow. Of the two, steam flow is the safest. It also produces superior steam utilization than a system that uses steam pressure or steam temperature.

A master index scale is developed for the maximum available steam flow. This scale is used to adjust a second scale based on steam demand. The different steam users are given set points within the demand scale. The steam user with the highest priority is at the top of the scale. The second user is at another set point just below the first, and so on. The dryer operator changes the priority given to each steam user. If careful design of the control algorithm exists, the system can work properly. It can safely load a boiler or boilers up to their maximum design or permitted rating and hold them there for long periods of time.

The same control principle can be applied to large biomass-fueled boilers with a gas-fired backup boiler.

Steam and Condensate Systems

Code Requirements

All steam and condensate systems should comply with the ASME safety code. Final designs should be approved by an engineer who has experience in both steam systems and the code.

Steam Pressure Requirements for Wood Dryers

Steam pressures encountered in wood dryers vary from 3 psig to 350 psig.

Drying Operation	Saturated Steam Pressure (psig)
Hardwood operations	3–15
Hardwood and softwood operations	15–60
Low-temp softwood operations	20–60
High-temp softwood operations	40–150
Veneer driers	200–350

Minimum Pipe Size for Steam Lines

Steam velocities in lines should not exceed 100 feet per second. The allowable pressure drop through the entire piping system will be dictated by the sensitivity of the steam users to the delivered saturated steam temperature and pressure.

Pipe Slope and Supports

All steam and condensate piping should have no less than 1% slope to allow condensate to drain downstream. Pipe supports should be spaced according to the ASME piping code.

Condensate Removal Traps

Steam lines and steam headers located outside of the dryer should be fitted with drip traps such that condensate does not collect inside the pipe and cause water hammer. Each trap should be fitted with a blocking valve to allow replacement of the trap during the operation of the steam line. Each trap should be fitted with a strainer and blow-down valve. Traps should be located such that they can be drained during cold weather.

Thermal Expansion

All pipe and supports should be designed for allowances in thermal expansion. Stress calculations should be performed for each bend, ell, and branch to determine the percentage of yield stress produced by thermal expansion. The manufacturers of flex-bellows expansion joints should be a nationally recognized brand.

Insulation

All pipe, flanges and valves should be insulated to protect against freeze damage, reduce energy losses, and water hammer. The minimum R value of the insulation will be determined by the steam temperature and the unit steam production costs, which includes fuel costs. Products containing asbestos should not be used.

Condensate Return Lines

Lines that return condensate back to the boiler should be schedule 80. The allowable pressure drop in the return line will be dictated by the design of the condensate pumping system.

Condensate Flash Tanks

Flash tanks should comply with ASME pressure vessel codes. An engineer should review the design.

Open Flash-Tank Designs

An open flash tank is a non-coded tank in which sufficient venting capacity exists to prevent any significant accumulation of pressure inside the tank. Condensate will flash when it enters the piping downstream of steam traps. The flashing will produce a predictable volume of flash. The flash will exit the tank through the tank's vent stack. The internal tank pressure is a function of the dynamic pressure created when flash steam passes through the stack. In most cases, a discharge coefficient of .5 can be used to calculate the tank pressure caused by the flow of flash steam out the stack. To keep the pressure of the tank at a low level, the inside diameter of the vent stack should be sized for the maximum expected flash steam flow rate passing through the vent. Steam velocity passing through the stack should not exceed 100 feet/second.

Minimum Tank Size

Flash tanks should be designed such that liquid condensate does not exit through the tank's vent stack. The size and configuration of the tank, the condensate feed line, and the stack must be designed accordingly. In some tank designs, the feed line is pointed toward the end of the tank away from the vent stack such that the liquid condensate settles down preventing liquid condensate from passing out the vent.

Closed Flash Tank Designs

In all cases, a pressurized coded flash tank design is used for heat recovery purposes. They are also used to reduce the amount of boiler water lost during flashing, resulting in the boiler requiring less makeup water and chemical treatment. Designs should comply with the ASME non-fired pressure vessel code.

In these designs, a portion of the flash is recovered and piped to other steam users in the plant. The design may also include a steam makeup valve to maintain the pressure inside the tank when the flashing is low. The tank should be fitted with safety valves designed for the pressure rating of the tank. The tank should be designed for the maximum pressure the tank can experience. Both the capacity of the safety valves and all sources of flow into the tank must be considered.

Condensate Return Pumps

The method of returning condensate back to a boiler can be either a mechanical pump design or a residual pressure design.

Mechanical Condensate Return Pumps

Mechanical pumps are used with either atmospheric or pressurized flash tanks. The pump must be designed to deliver sufficient head to overcome the piping losses and any backpressure at the condensate receiver vessel. The pump capacity should be sufficient to handle the rated capacity of the boiler plus 20%. The pump's shaft seal should be a leak-free design. The range of all operating conditions at the pump should be reviewed to prevent cavitation (bubbles and shock waves) at the pump's inlet.

Residual-Pressure Condensate Return Systems

Residual pressure designs are often used in place of mechanical pumps. There are differences in how these systems are designed based on the type of dryer they are used on.

In some small low-steam-pressure dryers located close to the boiler, the condensate return system only consists of the piping between the dryer traps and the feed water receiver tank at the boiler. In this design, the flash provides the pressure for pumping the liquid condensate back to the tank.

In dryers that operate above 15 psig steam pressure, the condensate leaving the dryer's traps is fed into a coded receiver tank fitted with a set of inlet and outlet valves controlled by floats and level switches. In this design, the liquid flash is pumped back to the boiler using pressurized flash that collects in the top of the receiver tank. Some of these designs include a steam makeup valve to keep the tank pressurized during low condensate flow rates.

Duplex Pumps

When using mechanical pumps, it is wise to use two pumps controlled by a level controller in the flash tank. The controls are designed to cycle between pumps to keep the wear on both pumps the same. A low-level safety float switch should be installed on the tank to prevent the pumps from running dry.

Condensate Flash-Energy-Recovery Systems

Energy conservation and flash recovery are significant issues in wood products plants. One method to do both is to install flash recovery heat exchangers at flash tanks. The heat exchanger can be cooled by air or makeup water to the boiler, or both. There are many versions of flash-energy-recovery systems. The selection of which design to use is dependent on the design of the entire drying system and fuel costs.

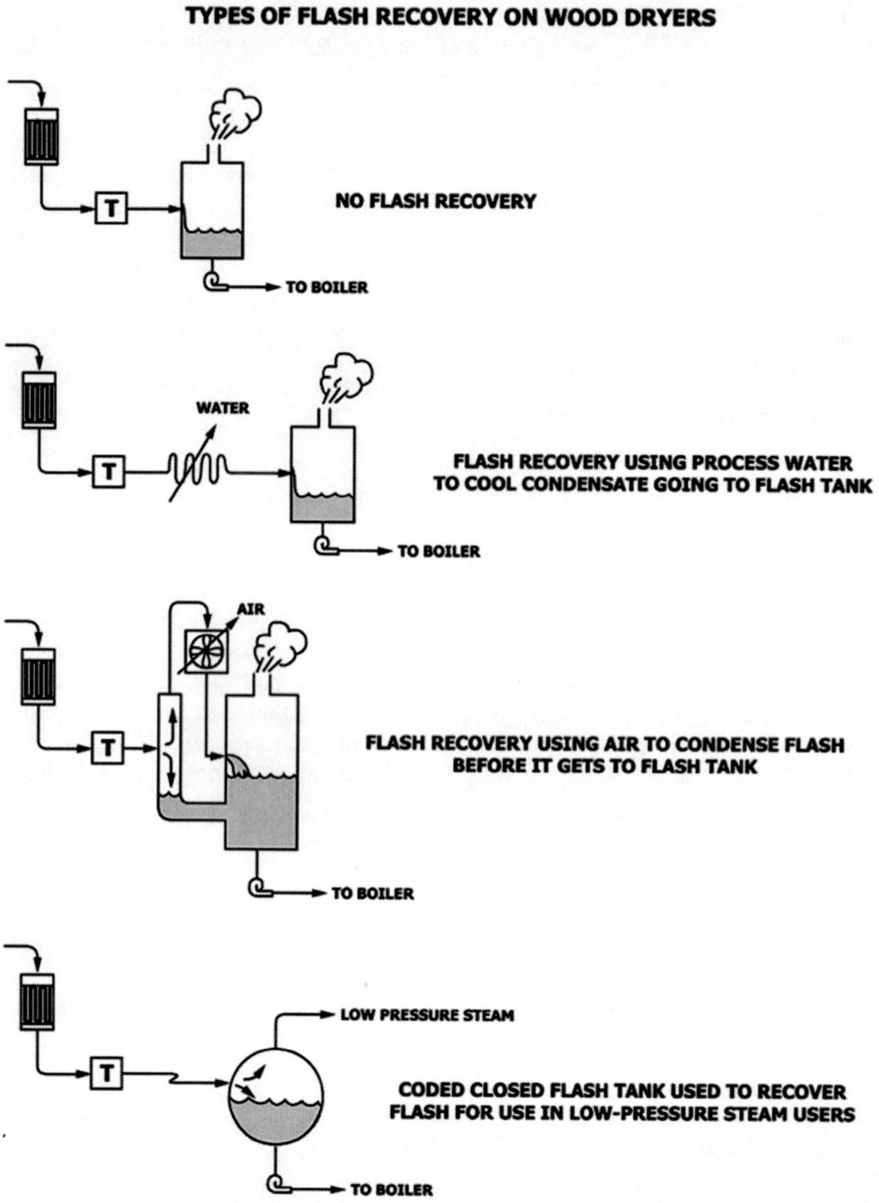

Hot Liquid Heating Systems for Wood Dryers

Some wood dryers use hot liquids (thermal oil or water and glycol) as heating mediums instead of steam. The advantages that hot liquids have over steam are:

The savings in fuel costs due to their higher thermal efficiency
The elimination of boiler water treatment costs
The savings in labor required for operating the system

Unfortunately, large hot-liquid heating systems can be very expensive. Most of their extra cost is due to the large size of their heat exchangers including the numerous recirculation pumps and valves required to make them perform on a level with steam.

Hot-liquid heat exchangers, when used in wood dryers, require both heat-delivery and recirculation pumps to keep the hot liquid always circulating through the coils. Considerable fluid velocity is always required inside the dryer's coils to maintain (1) a sufficient heat transfer coefficient inside the coil unit, and (2) the distribution of heat energy inside the coil unit.

If the number of recirculation pumps is reduced to save front-end costs, the heating coils can have cold spots. If recirculation pumps are left off a coil, the coil should be a two-pass design.

Hot-liquid heating systems can perform as well as steam designs in wood dryers, but in many cases, the level of engineering, pumping, piping, and valves are reduced to such an extent they do not. In most wood dryer applications, a properly designed hot-liquid heating system could have a difficult time competing with a properly designed steam system. The same design principles apply to hot-oil and water systems.

Hot-liquid heating systems are available using gas, oil, or wood heating fuels. In each case, a central heat exchanger and hot-liquid storage tank are needed to transfer and store the heat energy produced by the combustion system. The hot-liquid storage tank is kept at a constant temperature by a temperature controller that controls the firing rate of the burner.

If you decide to use hot liquids in a wood dryer, you would be wise to choose a highly reputable design build firm with decades of experience in hot-liquid heating systems. Proper equipment selection is a big factor in having a hot-liquid system that can operate reliably for decades.

If the dryer will be drying corrosive species of wood, coil leaks, replacement cost, and dryer downtime can be significant issues for hot-liquid heating systems.

11.60 Dehumidification (Heat-Pump) Systems for Drying Wood

Back in 1971 I built my first experimental dehumidification dryer. It consisted of a one-ton window air conditioner unit and one very large pasteboard box. I placed the box on a table, put the air conditioner unit inside the box, attached a garden hose to the condensate drain line, and ran the power cord through the wall of the box to a 220-volt outlet. Then I hung a water-soaked bathroom towel in front of the air conditioner's exhaust air vents. Then I closed the top of the pasteboard box, taped all the openings in the box, and wrapped the box with several cotton blankets to insulate the system. Then I turned the power on. Within minutes water started draining out of the garden hose. After about 30 minutes, the water stopped. Then I opened the box and found a dry bath towel.

Dehumidifier wood dryers are in effect no different than a residential air conditioner placed inside a large, insulated moisture-proof box. They are simply heat pumps driven by electric motors. This technology has been used for decades to cool, heat, and dehumidify buildings. They are totally driven by electricity, very reliable, low-maintenance, and efficient.

All heat-pump-type dehumidification wood dryer systems use a simple process control scheme. Provide a tight highly insulated wood product storage building surrounded by a tight moisture barrier. Then after the moist (green) wood products are placed inside the building and the main door is closed, the relative humidity inside the dryer's forced but slow-moving air stream will approach 100%. Theoretically, the relative humidity never reaches 100% because of tramp heat energy from fan motors and building losses or gains but can get very close to 100%. Thus, during the initial stage of wood product drying with a low stacking sticker air velocity, the wood product fiber equilibrium moisture content is kept close to fiber saturation point (29% moisture content) and thus, very little fiber shrinkage exists and thus, excessive drying stresses will not occur in the wood products. Then start the heat pump air recirculating blower and compressor while controlling the temperature of the fins on the refrigerate expansion coils to ensure adequate condensing of the moisture exiting the wood occurs. Then control the amount of heat released by the high-side coils located inside the building to ensure the required internal circulated dryer air temperature depression (dry bulb – wet bulb) for the specific species and thickness of wood to be maintained during the entire drying cycle. Then, once the drying starts, these drying systems will produce a discharge stream of condensate that is the same water being removed from the lumber. And, by measuring the flow rate (pounds/hour) of this condensate stream, the dryer operator, or automatic control systems, can quickly determine if the wood product is being dried too fast or too slow and adjust the heat pump components and dryer air handling system controls accordingly. Inside the dryer building, both the dry-bulb and wet-bulb temperatures are monitored and controlled for each phase of the specific wood species drying schedule. The temperature of the fins on the moisture condensing coil is kept below the dew point temperature of the internal dryer air stream by a combination of refrigerate flow rate and air velocity approaching the fins. Because dehumidification drying systems are classified as heat-limiting, these drying systems produce very little product degrading when compared to other types of convection drying systems. Additionally, the economic performance of these drying systems is heavily dependent on how much attention is given to the design of the process control systems.

In some locations, getting an environmental permit for the liquid discharge may be required. It is not the contents of the liquid water stream that is the problem, but instead the quantity of the discharge. The discharge stream from these drying systems is not significantly different than rainwater runoff from a forest during a rain.

Heat pump dehumidification is an old technology that has been around for many decades. The maximum dryer temperature determines which refrigerate is to be used in the system. The electrical usage per pound of water removed from the wood is determined by the dryer's temperature, the temperature depression for relative humidity control, the type refrigerate used, and the thermal efficiency of the heat pump loop.

Virtually all dehumidification dryer buildings are loaded and unloaded using fork trucks. The physical sizes of these buildings can vary from as small as 40′ × 40′ floor area to larger than 100′ × 100′. Door heights can vary from 10′ to over 20′. Inside the dryer, stacked package heights with cross-outs can exceed 20′. Detail time and motion studies are required in selecting the number of and the building sizes to match the species, product sizes, package sizes and product drying times. Do not try to save money up front by trying to dry different wood products in one large dryer building. It is always better to match the number of and sizes of the buildings to the maximum range of different products to be dried. Avoid designs that require fork trucks to have to constantly enter these dryers to load or unload products because such designs waste large amounts of electricity and dryer production capacity.

The dryer building should have an insulation R-value of at least 30. Many of these dryer buildings are built of wood because their internal temperatures rarely exceed 140 F. If dryer temperatures exceed 140F, the building should be of prefabricated metal construction.

It is imperative that all wood products dehumidifier dryer buildings, concrete floors, and doors include tight moisture barriers to prevent tramp humidity entering the dryer from the environment outside the dryer building or the soils under the building. All doors must include overhead rain, ice, and snow guards, and soft tight-sealing gaskets to prevent air and moisture leaks. Failure to do this will cause product drying problems, lost dryer production, and additional wasted electricity.

The internal dryer air circulating systems require uniform air distribution like in all types of convection dryers, but at much lower sticker air velocities. Internal-dryer, air-circulation, dead-spots are to be avoided. Both downflow and crossflow air-handling systems should be examined for specific species, product sizes, and thus drying rates. In the larger buildings, both types of air handling systems may be required to prevent dead spots.

The moisture condensing coils must be resistant to the corrosive water vapor coming from the lumber. If you purchase one of these systems, get it from a company that has decades of experience in designing heat pumps specifically for wood drying. See the ASHRAE Handbook for a detailed analysis of heat pump design and analysis.

The final heat pump configuration for drying hardwoods and softwoods will be different because of the different drying temperature schedules for limiting drying degrade.

Information about approved commercial heat pump refrigerants can be found at ASHREA.org or www.wikipedia.org/wiki/list_of_refrigerants, and are sorted by ASHRAE designated numbers.

High-Temperature Dehumidification Systems Although research is being done to increase maximum dryer air temperatures for dehumidification dryers (to speed up wood drying rates), when drying large amounts of lumber, current heat-pump wood-drying systems cannot compete with convection-type drying systems using fossil-fueled heating systems such as wood waste, fuel oils or natural gas, especially if high-temp drying schedules are being used.

11.61 High Frequency Heating Systems

HF technology is commercially available for drying both small and large amounts of wood. RF systems involve plates that produce a field of electromagnetic radiation in which the wood is placed. The system is totally driven by electricity. The electrical field heats the water inside the wood causing it to be forced to the surface of the wood. RF technology can produce minimal lumber degrade caused by dynamic moisture gradients.

11.62 Direct-Heated (Direct-Fired) Kiln Designs

A direct-fired dryer is one in which the heat source is released directly into the interior of the dryer through a chemical-energy-conversion process. The conversion process is the combustion of natural gas, #2 fuel oil, or wood or combinations of the three.

In addition to the internal kiln fans that circulate the heated air through the lumber packages, a recirculation blower is required to circulate the hot combustion products through air-handling ducts located inside the dryer. The recirculation blower runs at one speed even through the burner operates at different firing rates.

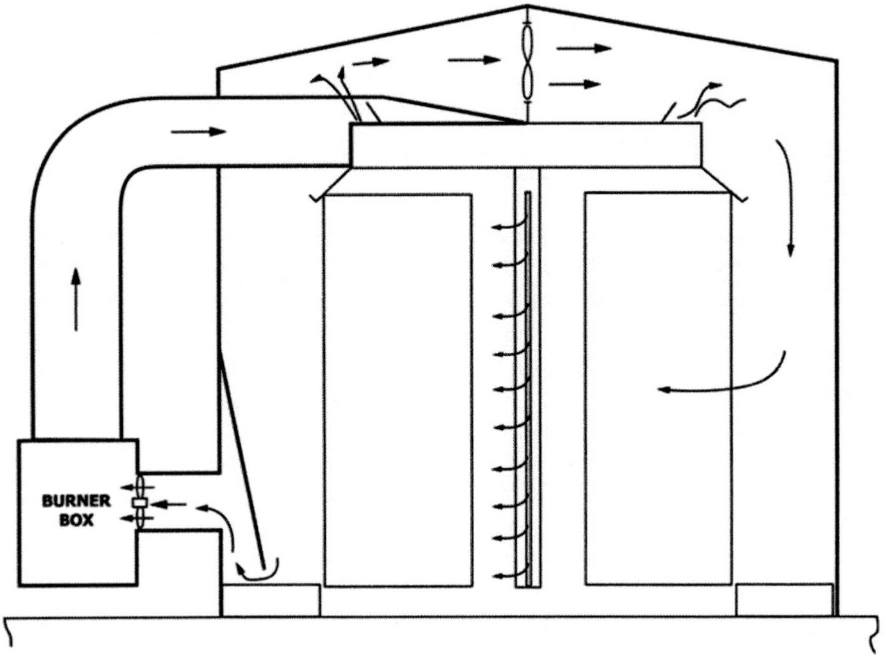

Direct-fired kiln designs have been around for over a century. Up until about 1970, most direct-fired kilns had very little true engineering put into their designs. They were considered by many people in the industry as being on the low end of the dryer industry and only used as a low-cost way of drying wood. Even today, with modern computer controls, better fan and burner designs, direct-fired dryers are still considered by many people to be an inferior way to dry wood when compared to indirect-heated dryers. And that is certainly true for drying hardwoods.

However, when one examines the total costs and benefits of the two, a different picture can emerge depending on the size and species of wood being dried. For some applications, they are the best design.

Pros and Cons of Direct-Heated Drying Systems Versus Indirect-Heated Drying Systems Using the Same Type of Heating Fuel (Comparison includes the dryer and the heat source equipment.)

Pros

1. Direct-fired drying systems are more energy efficient than indirect-fired systems.

2. The total installed cost of direct-fired designs is considerably less than indirect-fired designs when all the drying equipment and support equipment is included.
3. There is less equipment in direct-fired systems to maintain and operate than indirect-fired systems.

Cons

1. Direct-fired designs do not have the same capability of controlling heat distribution in dryers that indirect-fired designs, such as a steam system, offers.
2. Because of their lower wet-bulb temperature and poor temperature control, direct-fired dryers cause more degrade and moisture problems than indirect-fired designs.
3. Direct-fired dryers are more prone to having fires than indirect-fired designs. There is an especially high risk with wood-fired dryers.
4. Wood-fueled, direct-fired dryers produce far more harmful, toxic, and cancer-causing emissions in the vicinity of the dryer than all other types of commercial wood dryers. However, these dryers can be fitted with negative-air venting systems to collect air-born emissions and transport them to a safe point for either treatment or discharge.

Energy Usage of Direct-Fired Dryers

To size the heating equipment in a direct-fired heating system, the required heat energy for drying the wood will have to be determined. The factors are:

1. The drying schedule used in the dryers.

   Low-temperature schedules
   Conventional schedules
   Accelerated schedules
   High-temperature schedules
   Hyper-temperature schedules

2. The initial moisture content of the lumber
3. The final target moisture content of the lumber, before E&C
4. The specific gravity of the wood
5. The effective R value of the dryer building
6. Weather conditions

   Average ambient air temperature
   Average rainfall

7. The dryer's venting losses

   Low temp – high venting losses especially in cold climates
   Conventional temp – less venting losses
   High temp – minimal venting losses
   Hyper temp – minimal venting losses

8. The electrical usage by fans and blowers

   Internal propeller fans
   Blowers for direct-fired kilns

All the preceding energy components go into the total energy requirements for drying. In high-temp lumber dryers, the total heat energy usage is approximately 1300 BTUs per pound of water removed. In some low-temp drying operations in extremely cold weather, the usage can exceed this value by as much as 400%. Experience with the specific dryer design and species is required to establish a benchmark for the heating system.

Calculating the Average Drying Cycle BTU/HR

Once the total energy requirement is determined for a load of lumber, then the average drying cycle BTU/HR is calculated by dividing the total energy consumption by the drying time in hours.

Example:

A dryer uses 2700 BTU/board foot. The dryer holds 116,000 board feet. The drying time is 72 hours. What is the average drying cycle BTU/hr.?

$$\begin{aligned}\text{Average drying cycle BTU / hr} &= (2700 \times 116{,}000) / 72 \\ &= 4{,}350{,}000\end{aligned}$$

Calculating the Required Size of the Burner

There are several ways to size a burner for a specific dryer in a specific drying operation. Each method is affected by the temperature schedule used during the earliest stage of drying.

For direct-fired dryers, the ratio of the required size of the burner (BTU/hr) to the average drying cycle BTU/hr is called the heat peak ratio (PR).

$$\text{PR} = \text{required burner size BTU / hr / average cycle BTU / hr}$$

The heat peak ratio is determined by species, board thickness, and the drying schedule. The following table lists common heat peak ratios alongside the air peak ratios presented earlier in this book.

Drying Schedule	Air Peak Ratio	Heat Peak Ratio
Constant temperature schedule	2.80	2.00
Stepped temperature schedule	2.00	1.40
Constant-rising schedule	1.75	1.22
Heat-limited schedules	1.50	1.00

Calculating the Required Burner Turndown Ratio

All burners have a maximum firing rate and a minimum firing rate. The two extremes are determined by combustion air patterns in the burner, fuel particle dynamics, flame front velocities, emission controls, etc. For a specific burner design, the ratio of the burner's maximum continuous output to its lowest continuous output is called the turndown ratio (TDR).

$$\text{TDR} = \text{max.burner BTU / hr / min.burner BTU / hr}$$

In many burners, such as cyclonic double-vortex wood-dust burners and wet-wood burners, the temperature of the refractory inside the burner has a dramatic impact on the burner's ability to operate properly. If the refractory is hot and reflecting heat energy back into the flame, the TDR is one value. If the refractory cools down, the TDR will drop. If the temperature of the refractory drops too far, the ability of the burner to operate at any level will be affected. In some burner designs, a low-temperature sensor (TDR safety) is embedded in the refractory wall to shut off the burner or turn on a gas-fired support burner.

When a burner is installed on a wood dryer, there are times when the burner is operating at its maximum rated capacity, and there are times when it is operating at its minimum rated capacity. If the burner is fully capable of responding to the demands of the dryer's temperature controller, the temperature inside the dryer is under control. The dryer temperature does not lag the controller excessively or overshoot the controller's set point. This is the main objective of a burner on a dryer.

The drying schedule will affect the required TDR of a burner. When temperature set points for a dryer are changed, the heating demands on the burner will change accordingly. Following is an overview of TDR for different drying operations:

TDR Standard Practice (one species and one board thickness)

TDR = 3:1 to 5:1 (softwoods)
TDR = 2:1 to 3:1 (hardwoods)
TDR = 1:1 to 1.5:1 (heat-limited drying)

Effect of board thickness (BT)

Low-temp drying

$$\text{TDR} = \text{k} \times (\text{BT} \times \text{BT})$$

High-temp drying

$$\text{TDR} = \text{k} \times \text{BT}$$

Burner TDR should be based on experience with the species and board thickness required of the dryer. If **different** board thicknesses are dried in the **same** dryer, the required TDR could be as high as 30:1. Such a design would require multiple burners, or a special low-fire burner control system now discussed.

Automatic Low-Fire Cycle Controls

In many burner designs, controls are included that automatically turn the burner off during periods of low heat demands, and automatically turn it back on when the demand increases. A time-delay relay is connected to the burner's low-fire sensor. This sensor is mounted on the main shaft of the combustion-air damper assembly. This sensor is referred to as the low-fire safety switch.

After a period (usually 3–10 minutes) while the burner is operating on low fire, the time-delay relay trips and turns the pilot burner on and the fuel supply to the main burner off. Following this event, the dryer continues to evaporate water from the lumber. Within a minute, the temperature inside the dryer will start to drop. It then drops below the set point of the dryer's temperature controller. The controller then calls for heat from the burner. However, the burner does not immediately respond to the demands of the temperature controller. Instead, a second time delay relay has been installed to wait for a period (usually 1–3 minutes) to see if the temperature controller is still calling for heat. If the controller is still calling for heat after the time has elapsed, the main burner is turned back on and the pilot burner is turned off. The pilot is kept on during each of these events to prevent the burner from having to go back through its purge cycle. The controls are also setup such that the burner does not leave its low-fire position unless the main burner is on.

These designs must meet national burner safety code criteria and should only be designed by a qualified engineer familiar with the applicable codes. Review these controls with the burner manufacturer before using them.

Blend Box Design

Burners are mounted on box-like structures installed in the recirculation air stream. These structures are called blend boxes.

Negative- Versus Positive-Blend-Box Designs

If the blend-box is mounted on the suction side of the main recirculation blower, it is called a negative pressure blend box. If the blend box is mounted on the discharge side of the blower, it is called a positive pressure blend box. Both designs have their pros and cons. The negative design is the less expensive to manufacture but requires a larger more-expensive recirculation blower. If the blend-box is mounted on the discharge side of the recirculation blower, the blower will handle colder air, and will not have to be as large.

Positive-pressure blend boxes must be designed such that heated air does not leak out of the box. Of the two designs, the negative-air design does a better job of mixing the hot gases from the burner with the recirculation air stream. Of the two designs, I prefer the positive-pressure design. If properly designed, it will have fewer long-term maintenance problems.

Mounting the Burner on the Blend Box

Because burners can be heavy, the blend box must be structurally adequate for mounting the burner. Hot spots can occur next to the burner flange if hot gases flow back through the mounting. If a positive box design is used, all leaks around the burner mounting must be sealed off.

Protecting the Burner Flame

Every burner produces a hot flame that can damage mild steel and insulation materials. The burner flame requires a protected volume (cross-sectional area and length) for the combustion to occur. Stainless steel, or refractory-lined flame shields, should be used to protect gas and oil flames before they reach the cold humid air stream coming from the dryer.

Sizing the Blend Box

The blend box must be of sufficient size such that the presence of the burners and flame shields do not interfere with the recirculation air stream. The flow direction and velocities of the air stream inside the blend box should prevent flame wrapping and quenching. In some blend boxes, turning vanes are used to turn the incoming air stream, making the circulated airflow parallel with the burner flame.

Return Air Ducts

When recirculated air is pulled out of a dryer for reheating, it passes through a duct that connects to the burner blend box or the recirculation blower. The duct through which the air travels is called the return-air duct. The shape, size, and length of the duct can take many forms. In ground-level control rooms, the return-air duct is mounted in the wall of the kiln next to the control room. In over-head control rooms, used with track kilns, the return air duct can either connect to the gable end wall or the roof of the kiln. In some designs, the return-air duct runs the entire length of the dryer and is fitted with inlet slots. In some designs, the duct is connected to an underground concrete duct or an overhead duct with inlet connections and automatic dampers on both sides of the lumber.

Water can be a problem for underground return-air ducts. Water can collect in underground ducts if the dryer heats the lumber too fast during the startup of a cold charge of green lumber. If the kiln is built over an area where the water table is high, the duct can collect water. Sump pumps located around the perimeter of the dryer may be needed to keep water out of underground ducts.

The return-air duct should be sized such that excessive velocities do not occur. Avoid high air velocities in these ducts. As a rule, the velocity in every part of a return-air duct system should be no more than 75% of the velocity at the inlet flange of the recirculation blower used on the heating system.

The inlets of return ducts must be fitted with safety guards to prevent people from being sucked into the duct while the blower is running. Do not use small mesh screens for safety guards. These will create an unnecessary pressure drop that will reduce the performance of the heating system.

Main-Recirculation-Air Blowers

An air-handling blower is required to deliver heat from a burner to the dryer. The blower must deliver sufficient CFM of air to prevent excessively high temperatures from entering the dryer. High supply-duct temperatures are a cause of fires in lumber dryers. Supply duct temperatures should be kept as low as possible. To properly size the air delivery system for the burner's rated heat output, air mass flow and energy calculations should be made by a qualified engineer for the maximum supply duct temperature with a return air duct temperature equal to the dryer's maximum dry-bulb operating temperature.

Maximum Allowable Heat Supply Duct Temperature

The maximum allowable supply duct temperature is a function of how the dryer and its heat distribution system are designed. The key fire-safety factor is the distance between the supply duct **heated air outlets** and the lumber. This is especially true for track kilns. The following maximum operating temperatures are recommended for heat supply ducts in wood dryers:

Distance from heated air outlets to lumber (feet) Maximum heat supply duct temperature (F)

Over 10′ – 550 F,	Over 7′ – 500 F,	5.1′ to 7′ – 450 F,	4′ to 5′ – 400 F

No heated air outlet should ever be located less than 4′ from a lumber package. All heat supply duct **surfaces** located within 4′ of lumber should be covered with an insulated metal jacket that prevents the exposed surface from exceeding 275 degrees F. Check with your property insurance carrier about the required maximum allowable exposed supply duct temperatures.

Dual-Stage, Temperature-Limiting, Heat-Supply Duct Controllers

Back in 1974, I designed the first dual-temperature controller for heat supply ducts on commercial wood dryers. The dual-stage temperature-limiting system operates as follows:

During the early (**first**) stage of drying, when the lumber has a high moisture content, one controller keeps the heat supply duct temperature from exceeding a temperature setting of 400F.

During the later (**second**) stage of drying when the lumber is drier and prone to catch on fire, a second temperature controller keeps the heat supply duct temperature at a temperature setting 100 degrees F lower than the first-stage controller.

The transition from the first stage to the second stage is automatic based on a calculated average heat output from the burner. Once the second stage is engaged, it will lock itself in until the dryer is shut down and then restarted with a cold charge of lumber. The system is also designed such that the dryer operator can lock the second stage controller in for all lumber drying if he chooses to. This is especially useful for re-drying lumber that is slightly wet.

Dual-stage heat-supply duct controllers should be used on all direct-fired wood dryers in which (1) the surface of the heat supply ducts is less than 4′ from the lumber, or (2) wood is used for fuel.

Maximum Allowable Outlet Velocities in Recirculation Blowers

High exit air velocity in blower outlets creates dynamic pressure losses that are very difficult to recover. Because of this, high outlet velocities in blowers should be kept to a minimum to save electricity. Blowers should be engineered to operate at conditions that result in the highest possible efficiency. Small high-speed blowers are not as efficient as large low-speed designs. The type of blower wheel, its speed, operating head, and design efficiency should all be carefully reviewed when selecting a recirculation blower.

The design of blowers and ducting is a task that requires both complex engineering skills and years of experience. Leave this task to engineers who know how to do it the right way. The scope of this subject is too complicated to cover in a book of this nature. The reader should refer to air-handling publications for detailed engineering practices and data to locate the most efficient blower for each application.

Blower Selection, Class Rating, and Certification
Blower construction is rated by AMCA by class, configuration, and application. Go to www.AMCA.org to order the latest publications for blower ratings. This website also has an active list of all blower manufacturers who have AMCA certifications. The AMCA also has numerous publications for engineering blowers for maximum safety, efficiency, and reliability. Never use any blower that has not been certified by AMCA.

Shaft Heat Slingers and Seals
Every recirculation blower should be fitted with a high-temp shaft seal and heat slinger to protect the blower shaft bearing located next to the blower housing.

Main Heat Supply Ducts
The duct that delivers the heated air to the main distribution duct inside the dryer is called the main heat supply duct. This duct should have a cold air velocity no greater than 60% of the recirculation blower's cold exit air velocity. Connections, transitions, and expansion joints should follow accepted engineering design practices. In some cases, turbulizers may be needed inside this duct to improve heat distribution in dryers using a positive blend box design.

All surfaces of the duct located outside the dryer should be insulated and covered with a weather-resistant protective jacket. High-temperature insulation is required. The duct should be fitted with lifting tabs for handling purposes.

Main Interior Heat Distribution Ducts
Main heat distribution ducts located inside dryers should be capable of delivering a uniform distribution of heated air throughout the length of the duct. These ducts should have a cold air velocity no greater than 40% of the recirculation-air blower's cold exit air velocity.

Accepted Methods of Heat-Supply Duct Construction Inside Dryers
The two methods of construction used are the suspended-duct design and the integrated-duct design.

The **suspended-duct** design is a separate structure not connected with any structural component inside the dryer. This design is typically supported by flexible supports connected to the interior metal frame of the dryer. Because this duct will experience numerous cycles of expansion and contraction, it must be designed accordingly. The suspended design is common in low-temp dryers in which the size of the duct is small. Furthermore, small air leaks at joints are of no consequence because all of this duct is located inside the dryer.

The other type is the **integrated-duct** design. This design uses the dryer's fan deck floor panels and the two outer drop-baffles for the upper half of the duct. The lower side of the duct is suspended below the fan deck. Allowances for expansion and contraction of the different parts of the entire duct system must be addressed during the design of the entire assembly. The different parts of the system should be field bolted together with all-steel self-locking nuts and flat washers tightened to sufficient torque to eliminate leaks and then backed off ¼ turn to allow for thermal expansion and contractions. The bottom side of the duct should be insulated if located within 4′ of lumber. Due to the large, required cross-section area of these ducts used in high-temp softwood kilns, this is the predominate design.

Tapered Versus Straight Designs

In all types of heat-distribution ducts, a tapered design is a better choice from an air-handling efficiency and heat-distribution standpoint. Tapered designs perform better at different firing rates of the burner because the distribution of total pressure throughout the length of the duct is stable. Straight designs should be avoided for this reason. Although they should be, tapered ducts are rarely used in large wood dryers because of the additional construction costs.

Fan Deck Heat Outlets

Fan deck heat outlets are the means by which the heated air inside the distribution duct enters the top side of the dryer environment. The heat outlets at the fan deck level should meet the following specifications:

The number of heat outlets should be two or four times the number of propeller fans.

The alignment pattern of the heat outlets and fans should be consistent.

The heat outlets should be of location and sufficient distance from the fan wall to not interfere with the maintenance and inspection of the fans.

The direction of the heat outlet's discharge should be away from the fan wall.

The total cross-sectional area of all the heat outlets should be no less than:

1. 1.5 times the exit area of the recirculation blower (for kilns with no center reheat).
2. .75 times the exit area of the recirculation blower (for kilns with center reheat).

Each heat outlet should include a damper for adjusting the flow of air through the outlet.

Downcomers for Kilns with Center Reheat

Both double-track kilns and back-to-back side-loader package kilns can have downcomers located between the packages of lumber for reheat. These downcomers are vertical slotted ducts designed to inject hot air into the cooled air stream between the packages. These downcomers should meet the following specifications:

The air space between downcomers should not exceed 3′.

The total cross-sectional area of the top of the downcomers should be no less than 2.0 times the exit area of the recirculation blower.
The downcomers should be fitted with vertical slots for the discharge of hot air into the dryer.
The slots should inject hot air perpendicular to the dryer's air stream passing the downcomers.
The slots should be adjustable from a width of ¼″ to ¾″.
Each downcomer should include an adjustable inlet damper that acts as both a scoop and damper.

Design Head Limits and Tests for Heat Distribution Systems
During a cold-flow 70-degree F test of the recirculation blower and complete heating system ducting, the static head measured across the main recirculation blower should be no less than 2 and no more than 4 times the static head across the dryer's fan wall while the kiln is loaded with cold lumber. This final test should be conducted after all the heat outlets and dampers have been adjusted for uniform drying. If designed and adjusted properly, the static heads measured at both ends of the main heat distribution duct should be the same.

Burner/Kiln Safety Controls
All furnace and burner fuel trains and burner management systems must meet the latest applicable design safety codes. Each burner must be designed according to nationally recognized engineering standards. The burner's flame management system must meet all applicable state and national safety codes. Each dryer must be fitted with sufficient high-temperature limits and safety temperature controllers to meet both industry and insurance standards. All burner electrical controls and schematics should be reviewed by an engineer familiar with combustion safety codes before the system is manufactured, installed, or operated.

Before the equipment is installed, the final design should also be reviewed by the property and casualty or boiler and machinery insurance carriers for the plant where the equipment will be installed.

During the commissioning of the system, hard-copy (printed) records should be developed to document the mechanical configuration, electrical schematics, field wiring, safety-interlock adjustments, and logic testing of every safety control loop. These records should be kept in a safe fire-proof storage system for future maintenance and safety inspections.

Documentation of Records
If an outside burner service company is used to service the combustion equipment, both the service contractor and the owner should maintain up-to-date service records of all maintenance and service work, and final adjustments of process controllers and safety-interlock devices. These records will be needed for annual safety inspections and audits.

11.63 Humidity Control Systems for Wood Dryers

When wood is dried, the evaporated water mixes with the stream of air circulated through the packages of lumber. This causes the dry-bulb temperature of the air stream to drop. The amount of temperature drop is dependent on how much water is evaporated out of the lumber. When the air stream exits the lumber, it has to be heated back up to the level it was before it entered the lumber. Once the air is reheated, it now has a higher wet-bulb temperature. To bring the wet-bulb temperature back down to its desired level, fresh air has to be blended in with the air stream.

For every pound of cold outside air brought into the dryer, an equal weight of hot air leaves the dryer at the exhaust vents. The hot air leaving the dryer is mixed with hot water vapor that came from the lumber. This is what a humidity-control venting system does.

If the drying temperature schedule is a very low one, the required quantity of vent-air/pound of water removed from the lumber can be significant. If a high-temperature drying schedule is used, the required amount of vent-air is much less. If the dryer operates above 240 degrees F, the required amount of vent-air can approach zero. In very high-temperature dryers, mostly hot steam exits the venting system. However, because dryer doors rarely seal tight, significant quantities of tramp-air can be sucked in and then forced out through other door leaks and kiln vents. Much of this tramp-air gets heated inside the dryer and eventually leaves the dryer through the vents. If the dryer has a large quantity of tramp-air, the wet-bulb temperature inside the dryer can drop and stay below the set point of the wet-bulb controller.

Venting Terms

Dry-Bulb Temperature

The dry-bulb temperature of an air stream is the temperature that a thermometer would measure if placed in the air stream.

Wet-Bulb Temperature

The wet-bulb temperature of an air stream is the temperature that a thermometer, covered by a wet sock, would read if placed in the air stream.

Relative Humidity

Relative humidity is the ratio of the actual humidity to the maximum possible humidity of air. The term is expressed in percentage. If air is saturated, it has a relative humidity of 100%. If air has no water in it, the relative humidity is 0%.

Humidity Ratio

Humidity ratio is the ratio of water to air on a mass basis.

Water Vapor Pressure

When water vapor mixes with air, the water vapor creates a pressure based on its temperature. The pressure of the water vapor plus the pressure of the air equals 14.7 psig. At 212 degrees F and a relative humidity of 100%, all the air is displaced out and only water vapor is present.

Equilibrium Moisture Content of Wood

There is a predictable relationship between relative humidity and the moisture content at which wood will stabilize at. As the relative humidity increases, the EMC increases. See the USDA Dry Kiln Operator's Handbook for a discussion of EMC.

Humidity Measurement and Control Systems

The term "humidity," as used by wood dryer operators, often implies the wet-bulb temperature of the air stream inside the dryer. Controlling the wet-bulb temperature (thus the humidity) requires both accurate temperature measurement and control.

Wet-bulb temperature sensing is only as accurate as the temperature sensor and the geometry and conditions of the wet sock in which the thermometer is placed. In hardwood drying, the wet-bulb sensor should be able to detect a 1-degree F change in temperature. To accomplish this, the wet sock should be mounted on the thermometer probe such that the following conditions are maintained:

1 The top of the temperature probe should be 1″ above the water level in the wet-bulb well
2 The bottom of the probe should not come within ½″ of the water in the well
3 The wet sock should cover the last 5″ of the probe
4 The lower side of the sock should barely rest on the bottom of the wet-bulb well
5 The sock should not interfere with the flow of water through the well to its overflow
6 The well should have a flow of water to it such that its overflow line has a slight drip
7 The discharge from the overflow line should be checked frequently
8 The air velocity over the probe should be ½ to 1 ½ of the air velocity through the lumber
9 The overflow drain line should carry the water to the outside of the dryer and away from the dryer wall and to a drain line so water does not continually wet the ground near the dryer

Use only commercial wet-bulb socks made of pure cotton cloth. Do not use polyesters, nylon, or rayon blends in the sock. They will give false readings. The sock should be replaced often, depending on the temperature schedule being used. Socks are usually replaced every 2–3 kiln charges. In low-temp drying, the socks can be washed in a washing machine with Tide detergent and reused. If the cotton remains intact and soft after being dried, it is usable. When a sock is changed, it should be first soaked in water and then placed on the temperature probe. Never place a dry sock on a temperature probe, especially in a high-temp dryer.

Wet-bulb assemblies using modern electronic temperature sensing, can be accurate and reliable responding to temperature changes as small as 1 degree F. However, they are like everything else in electronic process controls. They are only as good as their maintenance and calibration.

Operating Control Scheme

Wet-bulb controllers operate on a dead-band principle. The dead band is the amount of temperature differential between the operation of the vents and the humidity spray system. Ideally, for hardwood drying, this dead band should be 1–2 degrees F. For softwood drying using E&C, the dead band can be 2–3 degrees F.

The following describes the operation.

Assume a wet bulb set point of 150-degree F has been chosen for the latest part of a schedule used for drying 4/4 oak, when the lumber is almost dry. The dryer operator adjusts the wet-bulb temperature controller to a set point of 150. The dryer controls immediately start responding to the new set point. If the wet-bulb temperature inside the kiln is at a temperature below 150, such as 140, the humidity spray valve will open, and the vents will close. The sprays will inject water vapor inside the kiln. The wet-bulb temperature inside the dryer will then start increasing. After a period, the wet-bulb temperature will start getting close to 150. Once the wet-bulb temperature reaches 149, the spray valve will shut. When this happens, the wet-bulb temperature inside the dryer will level off to about 150 for a short period of time, when neither the spray valve nor the vents are open. Depending on how well insulated the dryer is and how moist the lumber is, the wet-bulb temperature will react accordingly. Most likely, it will start to fall. Then, when the wet-bulb temperature drops to 149, the spray valve will open. The cycle repeats itself over and over to maintain the 150 F wet-bulb temperature set point.

Now assume that the preceding situation occurred during the early stages of drying when the lumber was wet. In this situation, the amount of water being evaporated from the wood was such that the 150 F set point required venting instead of spraying. In this situation, once the wet bulb reached any level above 151 the vents would open. As the vents open, hot air and moisture are exhausted out of the kiln and fresh air is brought in. In this situation, the vents would activate at the 151 F temperature to maintain the 150 F wet bulb set point. The system is designed such that the spray valve and the vents do not open at the same time. Depending on the actuators used on the spray valve and the vents and the type of temperature controller, the temperature dead-band between spraying and venting can be adjusted down to as low as 2 degrees F. To keep the sprays from coming on when they are not needed, the steam pipe feeding the spray control valve has a gate valve that the kiln operator can close.

Venting Equipment

The first requirement for venting equipment is long-term reliability. Due to their remote location on the roof of the dryer, vents receive very little attention by either dryer operators or maintenance people. They should be designed to operate for years with little to no maintenance. They should be manufactured from materials resistant to corrosion.

All-aluminum construction of vent lids and bases is the norm in the dryer industry. Vent-lid hinge-pins, bushings, and vent linkages should be durable, adjustable, and resistant to corrosion.

Vent linkages can be installed above or under the roof. For track kilns with overhead control rooms, placing the linkages under the roof allows the vent actuator to be installed inside the control room.

Whether using a pneumatic or electric actuator, the vent actuator design must be able to withstand decades of exposure to high humidity and acids from dryers. Never install an unprotected electric actuator on the roof of a dryer. It will fail. All vent operators must respond quickly to the wet-bulb temperature controls to make venting control schemes function properly.

Vent Operating Schemes

During the last century, there have been seven types of venting **schemes** used on wood dryers.

The **open** vent stack is one in which no control damper is used to control dryer humidity. This venting scheme was the first one used on wood dryers and is still used on some types of process dryers, besides wood dryers.

The **conventional** vent controlled-action scheme, used for over a century, lets all the roof vents open and close together. A few wood dryers today still use this method.

The **vent-discharge** scheme only opens the vents on the positive-pressure side of the kiln fans. This design has been used on lumber dryers and other types of process dryers for over 75 years.

The **split-control** scheme uses the preceding vent-discharge scheme, but with a time-delayed and staged operation of the suction-side vents. In this scheme, the wet-bulb controller has to have sufficient controls (either band width, time-delay, or both) to control the operation of the two separate rows of vents. This venting scheme has been used on wood dryers for over 50 years.

The **exit-air** venting scheme is used to reduce psychometric drying energy costs. In many drying systems, negative-air vent blowers have been used to suck the coldest air out of the dryer from the discharge side of the packages of lumber to reduce heat losses caused by the venting process. If this design is used, ducting will be needed inside the dryer to direct the exit-air stream to the inlet of the exhaust blowers. This energy-saving venting scheme has been used in different types of industrial process dryers for over 75 years.

Powered vents are used in both venting heat-recovery heat exchangers, or direct-fired dryers. Some direct-fired kilns use the recirculation blower on the heating system for powering a vent stack. If this design is used, the recirculation blower should be located upstream of the burner to save fuel.

Heat-recovery heat exchanger venting schemes involve a heat-recovery exchanger in the venting process. The device transfers heat energy from the hot air leaving the dryer to the cold air entering the dryer. Some heat recovery designs use a heat wheel to improve the efficiency of the system. These systems have been used in commercial process dryers for over 75 years. See ASHRAE publications for the many heat recovery technologies available today.

There are numerous different technologies and manufacturers of air-to-air heat-recovery systems in existence today. If you decide to purchase a wood dryer heat-recovery system, I recommend you hire a mechanical engineer familiar with heat-transfer technologies before you spend a lot of money on the equipment. There are some unethical salesmen in this business, so be careful.

Tramp Venting Losses

Tramp leaks at the main doors of wood dryers have a degrading effect on the dryer's ability to maintain a high wet-bulb temperature. If leaks exist at the doors, the dryer will be constantly vented driving the wet-bulb temperature down. Condensation on the inside of roofs, walls, doors, and floors will produce the same effect. In direct-fired dryers, the flow of combustion air from the burner into the dryer will also act as a vent driving the wet-bulb temperature down.

Required Venting Capacity

The required capacity of a dryer's venting system is affected by dryer temperature. Both the dry-bulb and wet-bulb temperatures and the depression will determine how much venting capacity is required. Low-temperature schedules require more total venting capacity than high-temperature schedules because the vapor pressure of water drops with temperature. Depending on the design of the dryer, and the temperature schedule used, the total capacity and flow area of the venting system per board foot of lumber in the dryer will vary widely. The static pressure developed across the fan wall has a significant effect on this value.

Temperature Depression Versus EMC in Convection Dryers

The following graph's shaded area demonstrates the relationship between EMC (the moisture content wood will stabilize at) and the temperature depression the wood is exposed to. The following figure was constructed from several published wood drying manuals for a wide range of dry-bulb temperatures found in convection wood dryers. It demonstrates how the surfaces of wood can be over dried if a dryer cannot maintain the required temperature depression.

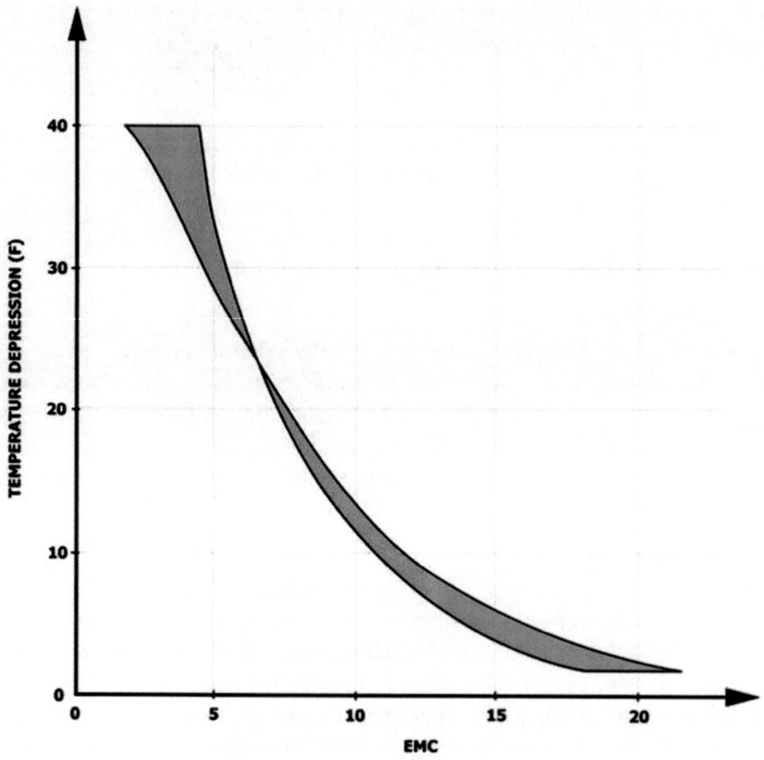

Mass and Energy Principles in Humidification

When the relative humidity inside a dryer needs to be raised, water from an outside source has to both enter the dryer's environment and react with it in such a manner that only the relative humidity is affected. Ideally for wood drying systems, only the wet-bulb temperature is affected during this process and the dry-bulb temperature is not. However, in the real world, when humidification water is introduced into a dryer, both the wet-bulb and the dry-bulb temperatures are affected.

In a perfect humidification system, the humidification water enters the dryer in the form of water vapor that is at the same vapor pressure and temperature as the water vapor inside the dryer just seconds before the humidity controller called for the humidification system to start.

Humidity control in a dryer involves balancing both the flow of moisture and energy during the process. Think of the dryer building as a closed system.

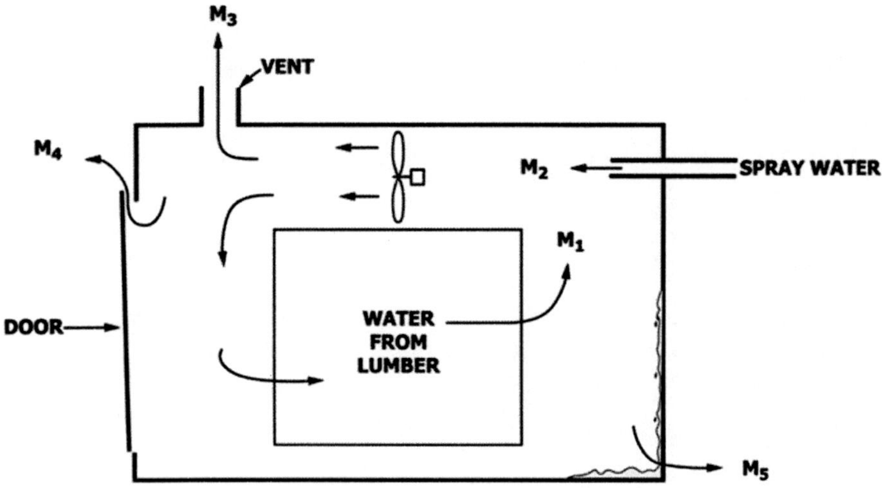

The water vapor mass entering the dryer's internal air circulation comes from:

M1 – The water coming from the lumber
M2 – The water coming from the humidification system

The water vapor mass leaving the dryer's internal air circulation are:

M3 – The water exiting the vents
M4 – The water exiting through leaks in the doors
M5 – The water condensing on the inside of the roof, walls, and doors

For a closed system, the mass balance equations are:

M1 + M2 = the total mass of water entering the dryer's air stream

And

M3 + M4 + M5 = the total mass of water leaving the dryer's air stream

Thus

M1 + M2 = M3 + M4 + M5

Rearranging for the humidification water M2,

M2 = (M3 + M4 + M5) − M1

This equation demonstrates that four variables determine how much humidification water M2 must enter the dryer to maintain a specific wet-bulb temperature during the operation of the dryer.

Since the vents are always closed when the humidification system is being used, M3 = 0.

Thus,

M2 = (M4 + M5) − M1

From this equation, we can see that the combination of M4 and M5 must be greater than M1 for humidification to be required. If the dryer is perfectly insulated, M5 = 0. If no door leaks exist M4 = 0.

This leads us to the finding that: M2 = −M1, or M2 + M1 = 0.

Since we know that during drying, M1 is always greater than 0, it is thus obvious that M2 will reach a value equal to M1, but with an opposite value.

In the real world of lumber drying, what this means is that if the vents are closed and the doors are tight, and there is no condensation on the inside of the dryer, then the water leaving the lumber will exit the kiln through the same route the humidification water entered the dryer. In the real world, this does not happen. What really happens is that the dryer is not perfectly tight, and the water from the lumber ends up venting out of the dryer through leaks at the doors, vents, and building.

Now imagine we had such a dryer in which no humidification system was needed. Although no such commercial (heat-only) wood dryer has ever been built, there are designs that can approach a humidification-free dryer. Such a dryer would have to be highly insulated, free of leaks, and fitted with a pressure relief valve for allowing the evaporated water vapor to escape.

In addition to the mass flow balances are the energy balances. Imagine a dryer system in which every pound of water vapor lost is replaced with an equal pound of water vapor at the same energy level. For water vapor, the energy level is determined by the temperature and pressure of the water vapor. If we had a humidification system that could do this when the humidification water vapor entered the dryer, only the wet-bulb temperature would be affected.

This may seem to be difficult to do, but it is not. Imagine a large wet cotton blanket hanging inside the dryer. Imagine the blanket having a pre-heated stream of water draining down the face of the blanket. The water has been preheated outside the dryer with a heat exchanger that keeps the water at the same temperature of the water vapor inside the kiln air stream. This temperature is the wet-bulb temperature. In such a system, the liquid water enters the dryer at the wet-bulb temperature and evaporates off the blanket at the wet-bulb temperature. This is how water-trough and water-cascade humidification systems work.

Wells for wet-bulb temperature sensors are supposed to be designed this way but are not. No dryer manufacturer that I am aware of preheats their makeup water to the wet-bulb temperature before delivering it to the wet-bulb well. Because of this oversight, the reading of the temperature sensor can have errors as the dry-bulb temperature inside the dryer or the temperature of the feed-water changes.

Steam Humidification Systems

Steam injection is an effective and popular method for raising the humidity inside wood dryers. The best steam spray systems use 2–5 psig steam that comes directly from a low-pressure boiler. Low steam pressures are needed to minimize the energy introduced into the dryer by the steam. If high-pressure steam is used, the energy problem is worse.

Operation

The typical steam spray system includes a main supply gate valve, an air-operated steam control valve, a desuperheater (water-injection nozzle), and a steam header (located inside the dryer). The control valve is an on-off design. The desuperheater is a simple water nozzle through which water is injected into the steam feed pipe downstream of the control valve. The water flow rate is controlled by a needle valve and an electric solenoid valve. The steam header runs the length of the dryer. The topside of the header is drilled with holes from which the steam escapes. The far end of the header is fitted with a drain for the water that collects in the header. The drain is plumbed to the outside of the dryer. Although the system just described is far from being a perfect humidification system, this is the most common method used for raising humidity inside wood dryers.

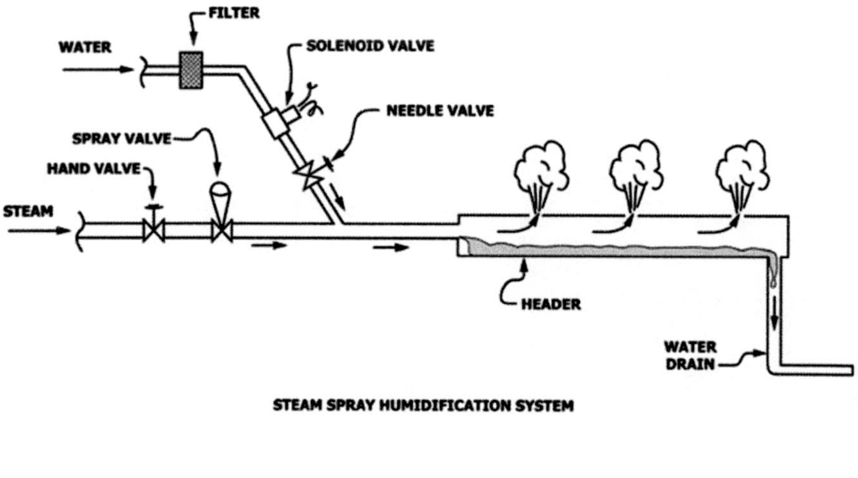

STEAM SPRAY HUMIDIFICATION SYSTEM

Atmospheric-Evaporator Humidification Systems

Atmospheric evaporators are used to raise humidity in all types of process water dryers. In these systems, liquid water is boiled at atmospheric pressure producing a clean low-energy vapor for humidification.

Operation

A steam line (for heating) is submerged in an insulated vessel, partially filled with water to be used for humidification. Water vapor (for humidifying the dryer) is produced when the steam line receives steam. The water vapor leaves the top of the vessel and enters the dryer's air stream. Because the steam vapor is boiled at atmospheric pressure, the amount of energy in the vapor is both predictable and controllable. The steam line is fitted with a control valve and steam trap. The liquid condensate produced inside the steam line passes through the trap and is returned to the main steam boiler that produced the steam.

Another benefit of this type of humidification system is that the main steam boiler does not lose water in the humidification system requiring expensive makeup water pretreatment, chemical treatment, and deaeration. This humidification system makes for a more reliable main boiler operation.

Two versions (the external and the internal designs) of atmospheric-evaporator humidification systems have been used for lumber dryers and treatment steamers.

External Designs

An insulated evaporator assembly is located outside the dryer. A steam vapor line from the evaporator is piped into the dryer. A header with openings distributes the steam vapor through the dryer. The sizes of the vapor line and header outlets are designed such that high pressures cannot be produced inside the water storage tank.

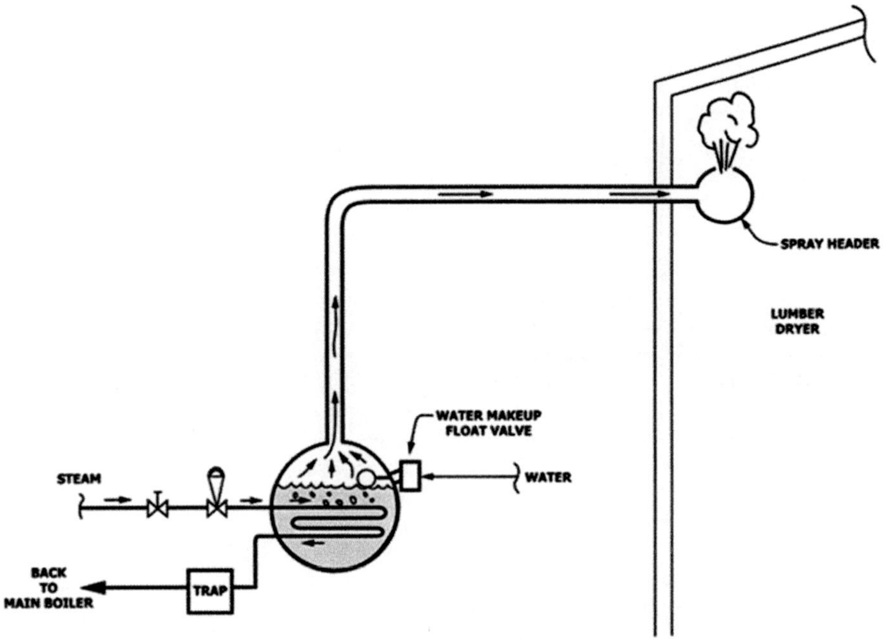

Internal Designs

An insulated evaporator assembly is located inside the dryer. This design involves a long water trough covered with screens and baffles to keep trash from getting into the trough. Another (pipe) design involves a long pipe evaporator fitted with vapor openings located at the top of the pipe. Caps and screens are put over the openings to keep trash from getting into the system.

The same steam-supply valves, condensate traps, and water-level control valves are required as shown in the external design.

The sides and top cover of the internal designs should be insulated if used in slow-drying species of hardwoods. In fast drying species, the insulation is not needed.

Of the two designs, I prefer the external because of its simplicity.

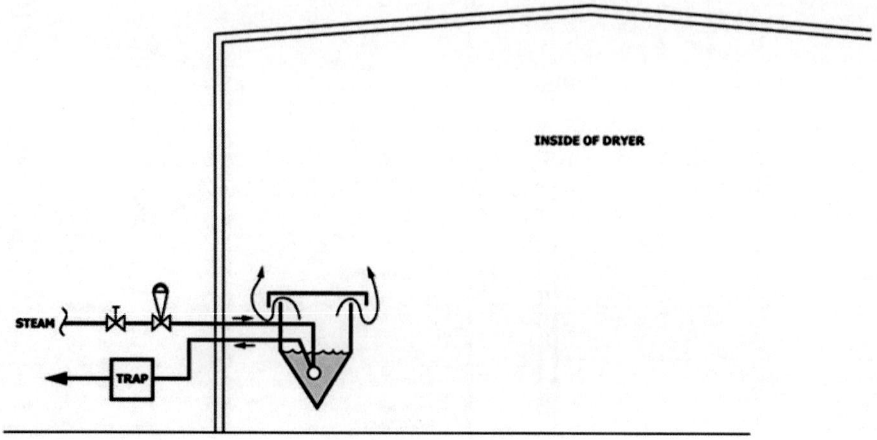

Water-Spray Humidification Systems

Water sprays for humidification involve sonic nozzles, high-pressure atomizing nozzles, and hot-water spray nozzles. To operate properly, the water entering the nozzle must be preheated to the correct temperature for the maximum diameter of the water droplets exiting the nozzle.

If these systems are not designed and maintained properly, large quantities of water droplets will fall to the floor of the dryer and wet the entire inside structure, walls, and floor of the dryer.

These types of humidification systems can cause **massive corrosion damage** to dryers if not working properly. I do not recommend they be used in dryers for corrosive species of wood unless the dryer is of all-aluminum or stainless-steel construction. And even still, there will be significant damage to concrete floors inside the dryer if the floors are not constantly protected with an acid-resistant barrier/coating.

Chapter 12
Veneer and Paper Dryers

12.1 Veneer Operations

Green veneer is controlled peeled from hot-soaked logs, trimmed, inspected for rots and defects, sorted, stacked, and then fed into veneer dryers. The thickness and moisture content of the green veneer will determine the number of green sorts, required green inventory, and drying rates. The target moisture content of the veneer leaving the dryer is dependent on the finished product and gluing process. Too low of moisture content or uneven drying causes excessive veneer warpage, cracks, and splits. Too high of final moisture content causes poor gluing penetration and blow-outs (steam explosions) in hot veneer press operations. If the drying, gluing, and hot presses are not controlled properly, the finished product can suffer costly delamination of glued joints requiring the final product possibly having to be removed and replaced.

Small Veneer Dryers
Small veneer dryers are batch-type “shotgun” high-frequency vacuum dryers in which green veneer is hand stacked onto a tram and then moved into the dryer from only one door. Some of these small dryers include cold internal moisture condensing tubes and collection troughs with evacuation pumps for removing water from the dryer.

Large Veneer Dryers
Large veneer continuous dryers involve long multitray, multizone impulse + convection-type hot-air drying systems. As the green veneer moves through the dryer on powered rolls, it is heated by either long hot air tubes fitted with transverse air discharge slots, or a series of orifice jets configured above the face of the veneer and designed to both control the local drying rate and heat energy distribution across the wide sheets of veneer constantly moving through the dryer.

J E ‘Ed’ Smith, *Industrial Drying Systems*,
https://doi.org/10.1007/978-3-031-31863-4_12

Both the slot and orifice-jet veneer drying systems are excellent candidates for computational fluid dynamics (CFD) modeling to improve drying production rates, energy efficiency, and dry-veneer quality.

Large veneer dryers are fitted with discharge-end veneer-cooling sections and exhaust-air collection and handling systems fitted with environmental-abatement control systems. Large veneer dryers can also be fitted with vent-heat-energy recovery systems to improve energy efficiency and drying production rates. These dryers can also be direct-fired with natural gas or wood, or indirect-fired with steam or hot liquids. Some OEMs offer veneer dryers that use superheated steam for drying to – improve veneer grade by maintaining a high EMC at the surfaces of the veneer.

Fire Hazards Because of the high temperatures most softwood veneer dryers operate at and the high probability of dry-veneer dust and flakes accumulating inside these dryers, fire protection systems are required. Repetitive fires in wood-fueled, direct-fired veneer drying systems are to be expected because of burning wood particles suspended in the discharge streams of wood burning equipment.

There are a limited number of companies that manufacture wood veneer dryers. Some offer complete pre-engineered veneer-handling, grading, sorting, and drying systems.

12.2 Paper Dryers

Paper drying is an old highly developed technology involving contact and convection heat transfer between wide sheets of wet paper mats and large diameter heated rolls in combination with air moisture heating, convection, and moisture removal systems. Because of the involved scale and thus economics associated with large paper dryers, detailed drying performance modeling and engineering calculations already exist with virtually every one of these dryers in existence. Thus, there is no need to waste your time trying to reinvent the wheels on this drying subject. Work only with people with years of experience in these dryers.

Interested readers of this book should contact TAPPI in Atlanta, Georgia, and Taylor & Francis Online, and the Center for Advanced Research in Drying (CARD) in Worchester, Maine, for the latest research being done to develop the performance and energy efficiency of these drying systems.

There are a limited number of companies that engineer, manufacture, test, and service paper dryers.

Chapter 13
Particle Dryers

13.1 Particle Drying Operations

Particle drying operations include press, rotary, flashtube, belt bed, pellet dryers, fluidized bed dryers, suspension burners, high frequency/radio frequency (HF/RF) dryers, infrared dryers, and superheated-steam dryers.

13.2 Press Dryers

Press dryers are mechanical rams driven either by hydraulic cylinders or mechanical rotating, or mechanical oscillating compression devices. All of these systems compress a fixed volume of wood particles configured to develop a pressure sufficient to squeeze out both excess water and a portion of water not enclosed in cell walls of the wood fiber. The higher the pressure created, the more water is removed in the process. After the squeeze event, the bulk material is released and then dumped into a discharge conveyor.

13.3 Rotary Dryers

Rotary dryers are long inclined cylinders rotating on a fixed set of rollers. Wet materials are fed into one end, and internal drum-mounted flights keep the material both in suspension (for drying) and gradually moving through the long dryer. Burner flame-to-material flow configurations can involve parallel flow or counterflow schemes. Modeling these dryers requires both computational fluid dynamics (CFD) technology and field testing for specific types and sizes of materials to be dried.

J E 'Ed' Smith, *Industrial Drying Systems*,
https://doi.org/10.1007/978-3-031-31863-4_13

Wide variations in either size or moisture content of incoming materials will produce wide variations of moisture content of the materials exiting the dryer as well as excessive particle loading of environmental abatement control systems. Some rotary-type (drum) dryers may also use conduction as the primary means of drying. All rotary-type dryers require careful engineering modeling and field testing for specific materials to be dried because of the high capital costs involved in these systems.

13.4 Flashtube Dryers

Flashtube dryers are hot-air suspension dryers. These dryers keep the product particles suspended inside a hot-air tube, and then transported to a discharge port. Retention time for drying is short and dictated by air stream velocity, the length of both horizontal and vertical tube heating sections, hot-air temperature, particle size, particle shape, fiber specific gravity, fiber initial moisture content, and final (target) moisture content. Dryer materials require stringent classification for size and effective density prior to entering these dryers. Because particle carryover is such an ongoing problem with these dryers, many people consider these dryers to be unpredictable hard-to-control drying systems.

13.5 Belt Bed Dryers

Belt bed dryers are moving beds of material supported by a flatbed conveyor system. These dryers can include above radiation heat sources, HF zones, or heated air streams passing through the bed from below. If the heated air streams stay below the lifting velocity of the smallest particle, these dryers can be relatively trouble free. The bed thickness, density, and porosity will impact both drying rates and final moisture content in a fixed system.

13.6 Pellet Dryers

Pellet dryers are typically standard rotary-drum dryers. However, because of the rapidly growing global market for using wood pellets as a heating fuel, research into more energy-efficient methods of drying pellets continues. Contact industry experts in pellet drying to learn where this drying research is headed.

13.7 Fluidized Bed Dryers

Fluidized bed dryers for wood products exist in fluidized bed furnaces for the purpose of reducing moisture before combustion begins. Similar particle heat transfer, liftoff dynamics, and emission loadings exist in other fixed bed (grated) wood furnaces. Go to Taylor & Francis Online at www.tandfonline.com for peer-reviewed published articles about fluidized bed drying.

13.8 Suspension Burner/Dryers

Suspension burner/dryers designed for burning low-moisture content wood dusts involve both radiation and convection turbulence in drying water from wood dusts or milled dry shavings prior to ignition of the wood particles. Both standard suspension and double-vortex combustion technologies are used. The double-vortex high-turbulent burner allows smaller burner combustion chambers to be used when compared to low-turbulent burners. The difference in burner size is dependent on the levels of turbulence and retention time in the main burner zone. All the ash in the wood is discharged from these burner systems and end up in whatever device they are heating. If using these burners with lumber kilns, the ash will collect on the fan blades inside the kilns requiring frequent shutdowns of the kilns for cleaning the ash off the fans. In pine kilns, this task is difficult because of heated pine resins mixing with the ash forming very hard cakes which can be difficult to remove. And, if the cleaning is not done properly, the internal dryer fans will end up out of balance causing failures of bearings. The presence of the hard cakes on the fan blades also degrades the efficiency of the fans causing drying problems. These burner systems also have a long-documented reputation of being "dangerous, dirty and toxic" and causing wood dryer fires. Check with your property and casualty insurance carrier, your attorney, your safety and health manager, and your local fire department before using these heating systems on any wood drying system.

13.9 HF/Radio/Microwave Dryers

HF/radio/microwave dryers are designed for removing water from small wood particle sizes (dust) to wet sheets to moving beds of biomass to moving small solid pieces of wood. The terms high frequency (HF), radio frequency (RF), and microwave are often but erroneously interchanged when discussing these dryers.

13.9.1 The HF/Radio Dryer Technology

The HF/radio dryer technology is over 70 years old, with frequencies ranging from 30 to 400 MHz, and consists of a PLC-controlled generator (a powered oscillating LC circuit connected to a triode with a high-voltage DC power supply), and set of electrodes, used in numerous applications such as adhesives on web paper making systems, bulk materials, and web, sheet, and board conveying systems, are very controllable, and uniform drying throughout the product being dried doesn't cause internal stresses in the product being dried and used in many postdrying low-moisture content drying/tempering systems. The only major downside of these simple drying systems is the cost of electricity to power them and the maintenance of the generator primary components. The final thermal efficiency of these dryers is subject to design and application variables. Some batch wood dryers combine HF technology with vacuum-drying technology to speed up the drying process and collect the water emissions by using heat pump condensing systems fitted with evacuation pumps or drains.

13.9.2 The Microwave Dryer Technology

The microwave dryer technology, because it is a short wavelength technology, only penetrates a few centimeters of depth of a product, causes uneven drying, and cannot be easily scaled up to large production-sized industrial wood drying systems. These low-cost easy-to-manufacture dryers are common in final food preparation and reheat applications.

13.9.3 The Infrared Dryer (IR) Technology

The Infrared Dryer (IR) technology like the HF/RF dryer technology is a very old one. Infrared heating/drying systems are common in the paper industry where it is used in both paper and thin board processes, curing drying after coaters, profiling of uneven moisture content of paper sheets, and preheating for incremental drying processes. Both gas and electric IR dryers exist.

IR drying combined with vacuum drying technology exists to speed up drying systems and save energy. IR systems are also used in both continuous linear tunnel dryers and process tumbling dryers. The final thermal efficiency of these dryers is subject to design and application variables. Field testing is required to determine exact drying/curing rates and energy consumption.

13.10 Superheated-Steam Drying

Superheated-steam drying is an old technology that has the benefit of high energy efficiency when compared to conventional convection drying because of the absence of psychometric energy waste due to the presence of air in the superheated drying stream. This technology also allows fiber moisture contents to be kept near or at fiber saturation point (FSP) because of the balance of vapor pressures in the drying process. The downside of superheated-steam drying is the total (gross) cost of all the required equipment to create, deliver, and maintain a truly superheated-steam drying environment. A slight amount of tramp air entering this type of product drying has a significant detrimental impact on the performance of these systems. Small amounts of (tramp) air entering these systems can cause overdrying and wide variations in fiber moisture contents exiting these systems.

These dryers require modeling and testing of specific products to develop drying schedules (temperature, pressure, turbulence, residence time, and product quality).

Chapter 14
Dry-End Systems

14.1 Introduction

After dried products leave dryers, the dried material must be stored in a protective shelter to protect it from damaging environmental conditions.

Storage capacity, inventory management, and tempering are the three key issues at this point. Sufficient storage capacity must exist to meet the plant's production levels, variances in grade, and material size. In addition, moisture gains or losses or contamination during storage or handling can create significant shifts in product grades.

Earlier in this book, I emphasized the importance of dry storage facilities and their ability to both protect dried products and provide controlled tempering to keep final moisture contents at the desired levels. Those issues apply to all agricultural, food, and dry-wood-fiber storage and inventory control systems, and not just softwood and hardwood lumber systems.

14.2 Example for Wood Dry-End Systems

Once the wood product leaves the dry storage and tempering facility, it then enters the last manufacturing stage, which is the final manufacturing operation.

Material	Final operations
Lumber, timbers	Package breakdown, planer, trimmer, sorter, final storage chemical treatment operations
Poles	Chemical treatment operations
Veneer	Dry trim, patching, glue lines, panel & furniture manufacturing
Wood fuels	Storage bins, metering, combustion, energy conversion

J E 'Ed' Smith, *Industrial Drying Systems*,
https://doi.org/10.1007/978-3-031-31863-4_14

In every wood products manufacturing operation, the success of the product's quality outcome is dependent on the success of the entire drying operation, starting at the logs and going all the way through the dry-end system.

14.3 Final Grading Systems

The last part of a dry-end system is the final grading system close coupled to a dry product moisture content measuring system. This is where management-information system (MIS) reporting of moisture content distributions and the impacts of fiber supply and drying system operations have on the current market value of the specific dry product being manufactured.

This is accomplished by statistical analysis of moisture content, and regression analysis of moisture data producing real time point-batch linear curve bias and slopes, and then generating actual market value production income rates for each product.

The accuracy and calibrations of the moisture content measuring system are crucial to these reporting systems producing valid MIS reports by identifying and quickly correcting problems to upper management and plant dryer operators.

Chapter 15
Agricultural/Food Dryers

15.1 Introduction

Agricultural/food drying systems are classified as any drying system that removes water from any farm-grown food product. The term system includes more than just the dryer proper. It includes all the dedicated material handling systems both upstream and downstream of the dryer.

Examples are grain dryers, corn dryers, soybean dryers, vegetable dryers, and silage dryers.

15.2 Classes of Dryers

The classes of dryer technologies used are batch dryers, jogging batch dryers, and continuous dryers.

15.3 Types of Dryer Energy Systems

The types of dryer energy systems involve air drying, forced air drying, direct-fired or indirect-fired convection drying, conduction, high frequency (HF), microwave, infrared (IR), dehumidification, freeze-drying, superheated steam drying, spray drying, and ram-pressure dryer technologies.

J E 'Ed' Smith, *Industrial Drying Systems*,
https://doi.org/10.1007/978-3-031-31863-4_15

15.4 Types of Product Material-Handling Systems

The types of product material-handling systems include screen dryers, screw dryers, rotary dryers, paddle dryers, sludge dryers, conical dryers, fluidized-bed dryers, press dryers, and centrifugal dryers.

15.5 Dryer Configurations

Dryer system configurations can include single-pass drying, counterflow drying, triple-flow drying, and recirculation-drying schemes to improve final product quality and/or energy-efficiency.

15.6 Target Moisture Contents

The target moisture content of agricultural/food dryers varies from as high as 30% MC to as low as 4% MC depending on the product and its intended use. In some drying systems, only surface water is removed from the product, following a washing cycle, to prevent mold formation during product storage.

15.7 The Terms: Dryers, Dry Kiln, Kilns, Ovens, and Stoves

Although these common terms are often used simultaneously, as well as misused often, they are not technically the same when discussing the removal of water from a product. However, in every one of these systems, water ends up being removed from a product.

Dryers are devices specifically designed to remove water from a product.

Dry kilns are devices specifically designed to remove water from a product and possibly raise the temperature of the product. This popular term is common in the lumber industry.

Kilns are devices specifically designed to heat a product to higher-than-ambient temperatures. This term is popular in the wood, brick, glass, minerals, and ceramics industries.

Ovens are devices specifically designed to heat a product to higher-than-ambient temperatures. This term is often used the same as the following term "**stoves**" which are devices specifically designed to burn a fuel to produce heat inside or on top of the stove for cooking or space heating.

15.8 Dryer System Design Codes

Unlike many other industrial drying systems and although the thermal and HVAC engineering principles are the same, the agricultural/food industry requires stringent health/safety processing and equipment design standards and codes.

Thus, once the readers of this book have mastered the larger wood products sections of this book, they will readily comprehend the same engineering principles apply to drying systems in the agricultural/food industry such as ASAE S248.3 minimum standard for the agriculture industry terms, components, operation, efficiency, and safety in designing installing and operating farm dryer systems.

15.9 List of Organizations and Journals Applicable to the Agricultural/Food Industry

ASABE – The American Society of Agricultural and Biological Engineers publishes a long list of farm equipment minimum design standards. Go to ASABE Standards Published to see all the current standards. For *farm dryer systems*, see the 21-page standard; ASAE S248.3 Construction and Rating of Equipment for Drying Farm Crops.

Also see the following journals for informative peer-reviewed drying articles
Drying Technology: An International Journal, www.tandfonline.com
Drying Technology Journal, www.academic-acceleration.com
International Journal of Food Engineering, www.ijfe.com
International Journal of Heat and Mass Transfer, www.sciencedirect.com
International Journal of Food Engineering, www.sciencedirect.com
Drying Technology, www.researchgate.net
Drying Technology, Scimago Journal, www.scimagojr.com

Organizations connected to the agricultural/food industry – Partial list
AEHAP – Association of Environmental Health Academic Programs
AFDO – Association of Food & Drug Officials
AIFP – International Association of Food Protection
ANAB – ANSI-ASQ National Accreditation Board
ANSI – American National Standards Institute
ASABE – American Society of Agricultural and Biological Engineers
ASPE – American Society of Plumbing Engineers
ASPH – Association of Schools of Public Health
AWWA – American Water Works Association
CDC – Centers for Disease Control & Prevention
CFP – Conference of Food Protection
CNCA – The Certification & Accreditation Administration of the People's Republic of China

CIEH – Chartered Institute of Environmental Health
CIPHI – Canadian Institute of Public Health Inspectors
CODEX – Alimentarius
CSA Group – Canadian Standards Association
DOC – US Department of Commerce
DOE – US Department of Energy
EPA – US Environmental Protection Agency
FDA – US Food and Drug Administration
FDLI – Food and Drug Law Institute
FSSP – Food Safety Services Providers
GFSI – Global Food Safety Initiative
GSA – US General Services Administration
IFPTI – International Food Protection Training Institute
IFT – Institute of Food Technologists
NEHA – Natural Environmental Health Association
NIST – National Institute of Standards and Technology
NRL – US Naval Research Laboratory
NSF – National Sanitary Foundation International, Ann Arbor, MI
OCED – Organization for Economic Co-operation and Development
Office of Energy Efficiency & Renewable Energy, www.energy.gov
OTA – Organic Trade Association
PFA – Phosphate Forum of the Americas
PFSE – Partnership for Food Safety Education
SCC – Standards Council of Canada
USDA – US Department of Agriculture
USCG – US Coast Guard
WHO – World Health Organization

Chapter 16
Process Controls and Automation

16.1 Disclaimer

The format used in this section lists process control loops and includes notes with recommended practices. These practices are included only as a starting point for a project. Because each project has its own nuisances, only qualified people should be used in:

1. Applying applicable safety and code requirements for each loop
2. Evaluating the control dynamics of each loop
3. Implementing the appropriate logic and tuning criteria

All process control systems should be fully documented from the initial design though the final tuning. Design, manufacturing, calibration, and service records should be maintained in a secure location.

16.2 The ISA Organization

The Instrumentation, Systems, and Automation Society (ISA) is the nationally recognized organization for process controls and automation. Their website is www.isa.org. Their mailing address is:

ISA – The Instrumentation, Systems, and Automation Society
67 Alexander Drive
P.O. Box 12277
Research Triangle Park, NC, 27709, USA

J E 'Ed' Smith, *Industrial Drying Systems*,
https://doi.org/10.1007/978-3-031-31863-4_16

The ISA publishes process control books and training and certification materials for the beginner to advanced levels. They also develop process control standards. The phone number for ordering ISA publications is 1-919-549-8411.

16.3 Agricultural, Food, and Wood Dryers, Furnaces, and Boilers

Energy and drying equipment involve the primary control elements – temperature, pressure, and flow. Controlling these variables involves sensors, transmitters, controllers, actuators, and drives. The following technologies are used:

1. Conventional hard-wired discrete electrical devices
2. Programmable logic controllers (PLC)
3. Process computers
4. Pneumatics and hydraulics

The design, maintenance, and tuning of all process control systems MUST comply with all applicable local and national safety standards. Use only qualified design personnel who understand both process control theory and the system they will be used on.

16.3.1 Example of Process Control Notes for Lumber Dryers

Which are like all water dryers in the wood, agricultural, and food industries because of overlapping sciences, technologies, and design safety codes.

IMPORTANT! – The reader of this book MUST research, study, and fully understand all types of industrial dryers to become proficient in any one type of industrial dryer control system.

Internal Fans
Lumber dryer fans operate under the following criteria:

1. Continuous operation during the drying process.

Dryer fans may operate 95% of the available hours in a year.

2. Continuous reversals.

Dryer fans are reversed every 3–18 h depending on the species, board thickness, and drying schedule.

3. Coasting times during reversals.

Fans should coast 10–60 s depending on the fan and drive design.

4. Fan motor amperage control.

Fan motors should be fully loaded with the kiln dry bulb temperature at 70 °F. The dryer should be loaded with lumber during this test.

5. Fan motor speed control.

Fan speed can be reduced by variable frequency drives to save electricity during or at the end of drying schedules.

6. Fan motor torque control.

The hubs of fans can be damaged by excessive stresses created during fan startups and reversals. Soft-start schemes should be used to limit the torque subjected to the fan's hub.

Safety Interlocks
Implosion Prevention

During the startup of a cold dryer, the following sequence should be followed to prevent implosions caused by sudden condensation of moisture in dryer circulated air streams:

1. Dryer internal fans start after switch turned to "on."
2. All roof vents start opening.
3. A safety limit switch trips when the roof vents get fully open.
4. The roof vents stay open for 3 min to let hot steam escape from the dryer.
5. The dryer fans start.
6. Twenty seconds later, the roof vents are put in automatic control.
7. Have your insurance company approve the above control scheme before the dryer is put into operation.

Fire Prevention

Safety interlocks for heating systems involve the following:

1. **All Dryers/Kilns**
   High-temperature limit switches are mounted in the wall of the dryer at the same level as the top of the product load. If a limit switch trips, the dryer shuts down and an alarm sounds.

These same safety switches can also be used to isolate the entire dryer control room from the dryer proper by activating a **fire-rated protection barrier** described below:

Fire-rated protection barriers for industrial dryers.

For (1) high-fire-risk, direct-fired dryers using wood as the fuel, or (2) any industrial dryer operating above 250 °F temperatures, I recommend that a fire-rated barrier wall be located between the dryer building and the dryer control room in which the dryer instrumentation and process controls, the electrical switchgear, and the fuel-burning furnaces and heat recirculation blowers are located. Heat supply and return-air ducts should be fitted with automatic fire dampers and safety disconnect joints designed to prevent smoke, fire, or structural damage to the control room or any ducting or equipment located inside the dryer's control room. The fire-rated wall should be located no less than 8′ from the outer adjacent side wall of the dryer building, or 10′ from the end wall of the dryer. Consult with your insurance carrier for the design and location of these smoke/fire protection barriers. Some property insurance companies or government jurisdictions may require larger fire-separation distances than those listed above. Some insurance companies or government jurisdiction may not require any fire-rated protection barriers, even for high-fire-risk dryers.

2. **Direct-Fired Dryers/Kilns**
   (a) If any fan motor fails, the heating system shuts off.
   (b) Burner purge cycles can only proceed if the internal fans are running, and the roof vents are fully open.
   (c) Heat return ducts are fitted with high-temperature limits.
   (d) Heat supply ducts are fitted with high-temperature limits.
   (e) Heat supply ducts are fitted with automatic dual temperature controllers. One controller is set at a temperature for high heat demands. The other controller is set at a temperature for low heat demands.

3. **Temperature Sensing Devices**

   Liquid/gas-filled systems can be used for temperatures of 0–500 °F.
   Bimetallic sensors can be used for high-temperature safety limits of 250–500 °F.
   Electric resistance RTD can be used for temperatures of 0–300 °F.
   Type K Thermocouples can be used for temperatures of 0–500 °F.

4. **Temperature Controllers**

   Electrical and electronic
   Discrete – used only on small gas burners and heat pumps
   Analog and digital – the predominant design for most modern industrial dryers
   Pneumatic – 3 to 15 psi – the predominant design for older models of dryers

5. **Computer-Based Control Schemes**

   A. **Internal Fan Control**

   Reversing times can be based on:

   Time of day
   Time since startup
   Integrated energy usage
   Temperature drops across the load
   Wood moisture content

   Reversal schemes can be based on:

Motor amperage

Soft starts can be based on:

Maximum allowable fan hub torque
Shaft RPM

B. **Temperature Control**

Storage and selection of drying schedules can be based on:

Species
Board thickness and width
Board moisture content

Dry bulb control schemes can be:

Single-zone designs used for direct-fired dryers
Multiple-zone designs for steam or hot-liquid heated dryers
The effect of zoning on final mc standard deviation
Inlet air temperature schemes can be either

Dual heating schemes
Staged heating schemes

Averaging air temperature schemes
Exit air temperature schemes

Wet bulb control schemes

Single vs. dual vent control schemes
Staged vent operations

C. **High-Temperature Safety Limits and Controllers**

Kiln building safety limits
Direct-fired supply duct safety limits and controllers
Return air duct safety limits

Check with the applicable code for the setting of these safety limits
All high-temperature safety limits should be precalibrated and sealed

D. **Historical Records and Management Information Systems (MIS)**

1. **Data logging**

Used for storage of past drying schedules and moisture content data. Historical process analysis involves the following:

Pattern analysis schemes
Regression analysis
Statistical analysis

2. **MIS reports involve the following:**

Moisture reports involving standard deviations, point-batch linear curve slope, and bias
Market values based on final product moisture content distributions
Energy reports
Drying times
Scheduling information
Maintenance schedules
Equipment repairs and costs

16.3.2 *Process Control Notes for Furnaces and Burners*

A. **Combustion Air Control Loops** involve the following:

Primary airflow control
Secondary airflow control

NOX-limiting control loops may be required
Excess air measurement and trim control
Trimming loops should have upper and lower limits
Modulating controls are required

B. **Wood Flow Control Loops**

Mass impulse flow rate meters are common in wood-handling blowpipe systems.
Variations in moisture content and mass density have a strong effect on mass flow rates.
In-line fuel moisture measurement may be required in some furnaces.
High- and low-flow safety limits are recommended.

C. **Gas and Oil Burner Control Loops**

Consult with the burner manufacturer and your insurance carrier for the type of controls required and the applicable safety codes. See NFPA, ASME, FM, UL, etc.
Safety controls for all furnaces must meet applicable safety standards.
See NFPA, ASME, FM, UL, etc.

Air handling system proving safety interlocks
Fuel train safety interlocks
Purge timers

D. **Burner Management Systems**

Burner management systems involve self-checking designs, pilot controls, low fire start, purge controls, main burner controls, and possibly multinozzle flame detectors.

E. Third-Party Inspections

Depending on local codes and the requirements of project financial institutions, third-party inspections may be required.

Chapter 17
Management of Industrial Drying Systems

I recommend every manager establish his own set of work rules for his employees. The following is a list of rules you can use to design your own.

17.1 Rules for the Workplace

Safety is always #1 in our company.

People are important.
Know your safety risks.
Keep all the risks at bay.
Make statements about safety every day.
Enforce safety rules every minute of the day.

Pride

Take pride in yourself.
Take pride in your job.
Take pride in your equipment.
Take pride in your company.
Take pride in your country.
Cleanliness is next to godliness.

Reach your peak performance

Knowledge is important.

Competency is required.
Continuing education is required.
Training programs make it happen.

J E 'Ed' Smith, *Industrial Drying Systems*,
https://doi.org/10.1007/978-3-031-31863-4_17

Communicate with others.

Share your ideas with others.
Ask others for their ideas and suggestions.

Respect

Do for others what you would want them to do for you.

Fear

Fear is a bad trip.
No healthy person wants to live in fear.
Keep fear out of the workplace.

Discipline
At the end of each day, ask yourself privately what you did right and what you did wrong.

Regroup for tomorrow.
Make it happen.

17.2 Management Models

Management models are derivatives of the human brain's ability to observe, process, and react with information received by our senses in a three-dimensional world. Therefore, management models have the dimensional qualities: top-side-down, bottom-side-up, horizontal, and reactive impulses.

Top-Side-Down Management
An organization that rules with an iron fist can be successful if the person with the iron fist has the knowledge required to make the correct decisions all the time. In an organization with more than one person, the chance of this happening is small. Organizations that adopt this management style have considerable difficulty in an environment where complex machinery exists. When these organizations succeed, it is typically because they have found a niche in a market with very little competition.

Bottom-Side-Up Management
This management style relies on the collective knowledge of the workers at the plant floor level. For this management style to function effectively, the person operating the machine center must have exceptional knowledge of the process. Sometimes, the required level of knowledge exists, and sometimes it doesn't. The success of the outcome is knowledge based.

Horizontal Management
This style of management communicates across horizontal boundaries within the organization chart, spending time and effort on creating procedures and guidelines.

It doesn't react much with upper and lower levels. The success of this type of management style is dependent on a facilitator whose job it is to identify problems, encourage communication, and find solutions. The facilitator must have access to valid information for valid solutions to be implemented.

If facilitators are not present, this management style can evolve into clans debating over which procedures are the best ones to use. Horizontal management styles are common in bureaucratic organizations, government agencies, and government-funded industries. The amount of fraud, corruption, and waste in horizontal-management organizations can exceed 95% of their budgets.

The psychosocial dynamics of horizontal management styles are driven by flaws in human character.

17.3 Reactive Impulses

Reactive impulses are time-based reactions to external data. Two impulses exist in this setting. The closed-door philosophy that blocks data at the door of the decision-maker. The decision-maker doesn't want to be bothered by new data and keeps his door shut to prevent hearing it. The open-door philosophy is used by decision-makers who welcome new data and encourage people to bring new data to the attention of the decision-maker.

17.4 Reward Conditioning

Healthy civilized people want to live successful lives. Unhealthy uncivilized people want to live under the heel of a dictator. Healthy people also want those around them to be healthy. To have a healthy workplace, management must encourage free exchange of ideas and reward those who demonstrate achievements.

17.5 Loss Prevention

Accidents occur in life. They can be caused by simple chance, casual conduct, lack of attention to details, and errors in judgment. Automobile accidents, injuries, and fatalities are examples of such events. To minimize the incidence of accidents, their causes must be understood. To understand the causes, experts should be retained to evaluate the risks and probability for each event to occur. When management gives little to no attention to preventing accidents and failures and refuses to seek advice from experts, the probability of accidents occurring increases.

17.6 Myths and Causes of Accidents

There are many myths about causes of accidents in drying operations.

Myths:

Accidents are an acceptable part of life.
You can live forever with accidents.
Avoiding equipment repair costs saves money in the long run.
Worn out equipment can run just as good as new equipment.
Plant managers are drying experts.
Business owners are drying experts.
Big companies know more about drying systems than small companies.
Big companies are run by geniuses.

Causes:

Faulty subliminal beliefs.
I am the smartest person on the planet.
Accidents do not happen in my plant.
We don't need to hire a safety engineer.
Character Flaws.
Don't worry about safety. Our lawyers will protect us if we are sued.

17.7 Drying Disasters – What to Do About Them

The causes of drying disasters are ignorance, ego traps, and company politics. To recover from a drying disaster, management must have the emotional intelligence and cognitive ability to recognize disasters when they happen and know when and how to ask for help. Once viable solutions are found, management then must both choose and implement effective recovery plans.

17.8 Quality Control

Every drying system should have the following quality control items tracked daily.

Safety meeting notes
Final moisture content of the product
Drying production
Heat energy production
Fuel usage
Dryer loading efficiency
Suggestions from employees

Equipment in need of maintenance
Scheduled outages

Daily reports should go to the plant manager.

Chapter 18
Managing and Investigating an Injury or Property Loss

In many losses, insurance coverage or litigation may become involved. If either or both are involved, the management of **loss scene evidence** must be performed according to nationally recognized legal standards.

If a defective product is suspect, the sales representative for that product must be put on notice the moment that product is suspected to be involved in loss causation.

Failure to notify the product sales representative immediately once that product is suspect could prevent any (later) legal claims against, or financial recovery from the product manufacturer.

And the same legal principle applies to any other party in which they are suspect in the loss causation.

18.1 First Thing to Do

Rescue and provide medical attention to victims.

18.2 Second Thing to Do

Contain the loss.

18.3 Third Thing to Do

Maintain communication with the fire department's officer in charge.

Appoint a person the job of photographing the scene with a camera.

J E 'Ed' Smith, *Industrial Drying Systems*,
https://doi.org/10.1007/978-3-031-31863-4_18

Do not touch, move, or alter physical evidence at the scene.

18.4 Fourth Thing to Do

Contact your insurance company agent and attorney so they can get adjusters and investigators to the scene.

Do not wait several days to notify your insurance agent about the loss.

18.5 Fifth Thing to Do

Establish a recovery plan.

18.6 Fire and Explosion Investigations and Scene Management

NFPA 921 is the nationally recognized standard for the investigations of fires and explosions. This standard was designed specifically for establishing procedures for investigating and preserving evidence that can determine the origin and cause of the loss.

18.7 Product-Defect Loss Investigations

If the loss involves a failure due to a product defect, ASTM E-860 is the standard by which the investigation and evidence documentation is to be done.

18.8 Qualifications of Investigators

Depending on the type of loss, the required qualifications of the investigator will vary. Licensed engineers and fire investigators may work jointly on a loss. Usually, every party involved with the loss will have a representative present during the investigation. Some of the people present during a loss investigation may be observers only, some may be experts, and some may be adjusters or attorneys.

18.9 Preservation of Loss Scene Evidence

This subject is not to be taken lightly. Scene evidence must be preserved and accessible to the investigators doing the investigation. Do not allow people to wander into the scene and disturb evidence until the investigation has been completed. Discuss this with your attorney and insurance adjuster. See ASTM E-860.

18.10 Cooperation During Investigations and Claims Adjustments

Always offer to and cooperate fully with investigators and insurance adjusters.

Comply fully with their requests for information.

Do not commit **insurance fraud** by hiding, misrepresenting facts, or instructing others to. Insurance fraud is a **federal crime**.

Be extremely careful what you say to adjusters and investigators.

Chapter 19
Furnaces and Steam Generators for Industrial Dryers

19.1 Dryer Heat Demand Rates

The first thing to know is the product's heat energy usage/unit of product. Then the dryer system's required production capacity. Special attention should be directed on the effect weather conditions will have on the energy requirements of the entire dryer system.

19.2 One Dryer Versus Multiple Dryers in a Drying System

This is an area in which having accurate data on the product and the type of dryer system is a must have.

19.2.1 If Only One Dryer Is Being Used

If only one dryer is being used, the following rules of thumb can be used, but additional studies should be conducted before selecting a final design.

Batch Dryers Typically, an initial energy demand rate is 2.0 times the average drying cycle usage rate. However, this number 2.0 can possibly be reduced to 1.4 with proper control systems. During the tail end of batch dryer operations, the heat demand rate can approach zero in certain types of products. The specific temperature drying schedules used will impact the total drying cycle's heat energy demand versus time curve. This is especially true in drying refractory products such as hardwood lumber subject to dynamic cracks and splits.

J E 'Ed' Smith, *Industrial Drying Systems*,
https://doi.org/10.1007/978-3-031-31863-4_19

Jogging-Batch Dryers Typically, an initial energy demand rate is 1.7 times the average drying cycle usage rate. However, this number 1.7 can be reduced to 1.4 with proper control systems. Once these dryer systems are up and running at maximum speed, each material flow path (or track) will produce its own heat demand rate that rises and falls in sync with the material stream flow. Because of this, it is advantageous for all jogging-batch dryers with more than one path to stagger the "start" command of the different loaders feeding the dryer. If internal moisture measuring systems are being used, additional control loops will have to be used to allow the moisture control systems to still perform properly.

Continuous Dryers Typically, an initial energy demand rate is 1.7 times the average drying cycle usage rate. However, this number 1.7 can be reduced to 1.4 with proper control systems. Because continuous dryers operate at a continuous drying rate, there are no significant rises or falls in heat energy demand rates unless the dryer automatic controls are changed by an operator.

19.2.2 If Numerous Dryers Are Being Used

If numerous dryers are being used, then studies should be conducted on the impact each dryer has on the total heat demand rate and the ability of automatic controls to both limit and prioritize the numerous dryer operations. Studies should be conducted to determine how much total drying production is possible for a specific (limited) amount of heat energy supply system. These studies will be required in any part of any design-build contract. Also included will be agreed-on adjustment factors for changes in the product entering the dryer. This is a normal practice seen in some but not all drying system equipment contracts.

19.3 Furnaces

The three classes of furnaces used in drying systems involve natural gas, fuel oils, and woody/biomass fuels. There are far too many different types of gas, oil, and biomass combustion systems in existence to list in this book.

When applying any combustion system to dryers, the maximum required continuous firing rate and the minimum continuous firing rate of furnaces must be considered. Every furnace has a rated turndown ratio (TDR) which is the maximum firing rate divided by the minimum firing rate. When using woody/biomass as a fuel, both firing rates may be affected by the heating or cooling of the furnace interior refractory linings. For gas and oil burners, the TDR and flammability limits are determined by shear rates between the fuel and the combustion air stream velocities, and the excess oxygen levels.

19.4 Steam Generators (Saturated Steam)

19.4.1 Safety-Code Compliance

All steam generators are designed, manufactured, and serviced to meet applicable national safety codes. Compliance with these safety codes is mandatory.

Do NOT let unqualified people act as consultants, designers, manufacturers, installers, or service any steam generator. Steam generators can be very dangerous if not designed, manufactured, and maintained by qualified people.

19.4.2 Annual Inspections

Annual inspections are required in most countries. Check with the local authority having jurisdiction over steam generators and comply with their requirements. Company safety managers must be aware of all local and national requirements for safety inspections and keep records of all field inspections in a secure records system.

19.4.3 The Two Classes of Steam Generators

The Two Classes of Steam Generators used in drying systems are fire tube and water tube.

19.4.4 Typical Saturated Steam Generator Operating Pressures and Temperatures Are

For a steam-to-dryer temperature difference of 150°F

Wood dryer schedules	Steam supply temperature (F)	Steam supply pressure (PSIG)
Low-temp. (33–120°F)	200–270	0–27
Conv-temp. (121–180°F)	260–330	21–88
Acc-temp. (181–211°F)	291–361	45–138
High-temp. (212–250°F)	330–400	88–232
Hyper-temp. (251–280°F)	360–430	138–328

Note The above saturated steam supply pressures are typical of many drying systems. However, the reader should understand that designing any product dryer with as little steam boiler pressure as possible increases the overall energy efficiency and long-term reliability of the entire drying operation. Economic studies should always be conducted to determine the total cost and economic return of dryer systems that operate on as low of steam pressure as possible. This is a very common error made by people when matching dryers to steam generators. Doing these economic studies correctly requires designers to understand how convection-type dryers and their internal heat exchangers should be designed, not how they are usually designed. Consult with experts in modern heat exchanger CFD design software to do these studies correctly.

19.4.5 Steaming Capacity

Steaming rates vary from less than 1000 PPH to over 200,000 PPH depending on the type of product manufacturing facility. Additionally, every steam boiler should be fitted with an automatic steam flow overload safety system such that the boiler can be safely pushed to its rated steaming capacity for extended periods of time. Properly designed, these high-flow safety systems can allow maximum product drying capacity for a specific sized boiler and protect the boiler from wild swings in steam loadings. These safety systems require accurate steam flow meters at the boiler. Do not use plate-type orifice flow meters for these safety systems because they are not accurate unless they include pressure-compensation capability.

19.4.6 Avoiding Superheat

Because superheated steam causes heat-transfer dead spots in steam-to-air condensing-type heat exchangers, it is to be avoided in all types of dryer steam heat exchangers.

19.4.7 Boiler Makeup Water Treatment

This is one of the most important and underappreciated part of steam-heated drying systems.

Failure to maintain the chemistry of water in steam systems will lead to costly severe corrosion of steam-to-air heat exchangers, dryer equipment, and boiler equipment.

The entire steam system loop should be designed to achieve **100% condensate return** from drying systems to prevent costly water treatment of boiler makeup water.

Only use experienced steam system designers when planning any steam-heated product drying system. This is especially true when boiler system makeup water is highly alkaline.

In very low-temperature industrial drying systems, always consider the use of closed-loop (contained) condensate return systems that do not require steam traps. In this type of low-pressure steam heating system, the supply steam and condensate return are both inside the same sloped pipe. Such designs will eliminate steam flashing problems, and thus the associated wasted energy and water treatment costs. They are typically seen in 1–5 PSIG steam systems.

Chapter 20
Wood Drying with On-Site Cogeneration Systems

Due to the low temperatures involved in wood drying systems, there exist sufficient temperature differentials to create efficient cogeneration thermal cycles.

Additionally, by using hyper-low-pressure steam-heated drying systems, the numerous health, safety, and fire-hazard problems with direct-fired drying systems are avoided.

EXAMPLE – All dry wood products manufacturing operations should conduct economic and environmental-impact studies to determine the following.

The annual wet tons of log fiber entering the mill.
The annual water tons entering the mill in the logs.
The annual wood fiber tons entering the mill – without the water.
The average specific gravity of the logs.
The average initial moisture content of the logs.
The annual tons of water to be removed due to drying.
The annual heat energy required of the drying system.
The annual wood fuels required of the steam boilers.
The annual electrical energy usage of the entire drying system.
The potential annual tax incentives for energy conservation incentives.
The total drying system capital costs without on-site electricity generation.
The total drying system capital costs with on-site electricity generation.
The added drying system labor and maintenance costs for on-site generation.
The total annual economic value of the electricity generated on-site.
The total annual economic value of the net electricity delivered to the local electric grid.
The return on investment of the proposed co-generation system.

These studies should only be conducted by qualified people with experience in wood drying systems, cogeneration systems, and applicable local laws for cogeneration systems.

J E 'Ed' Smith, *Industrial Drying Systems*,
https://doi.org/10.1007/978-3-031-31863-4_20

IMPORTANT *– Special attention should be focused on the effective H-value of the heat exchangers used in drying systems. This will require CFD modeling expertise. Additionally, all dryer system fans and blowers will require adaptive-speed control systems connected to the drying rates of the individual dryers. And all dryer fan systems must be high-efficiency designs set at the proper pitch setting for maximum propeller aerodynamic efficiency and with as little dynamic head losses as possible. Avoid the use of small diameter high rpm fan designs. The performance of the inner dryer fan system has to* ***match*** *the effective H-value of the inner heat exchangers, something only possible by using CFD technology.*

Chapter 21
Material Handling Systems and Terms

21.1 Introduction

All the many classes and types of material handling systems require secure stable foundations and mechanical supports to operate reliably for long periods of time especially at flow points where one machine center interfaces with other machine centers. This can be a significant problem where highly expansive soils or variable soil water tables or occasional flooding exist.

21.2 Material-Handling Types, Classes, and Terms

The following list of typical terms is common in many drying industries. The number of global companies that manufacture and service these material handling systems are in the millions.

21.2.1 Classes of Material-Handling Systems

Linear – point-to-point delivery system
Shuttling – a back and forth action
Lifting or lowering – fork trucks, straddle carriers, hoists, sky hooks, elevators, etc.
Sliding/falling – one material stream moving against another object or location
Rotating – a circular path in which the handling system is loaded, unloaded, loaded, unloaded, etc.
Tumbling – washers, dryers, classification, separation, mixing, etc.
Intermittent – irregular intervals, not continuous or steady – like automobiles

J E 'Ed' Smith, *Industrial Drying Systems*,
https://doi.org/10.1007/978-3-031-31863-4_21

Jogging – a repetitive action in one direction involving an at-rest period and a stroke distance
Continuous – a smooth steady, nonjogging action
Storage – inventory or accumulation points
Classifiers – separation of products based on size, density, or grade
Centrifuges – separation of materials based on density only
Bottom-supported vs. overhead-supported or lifted

21.2.2 Classes of Materials Handled

Packages, boxes, crates, bulk products, etc.
Boards, small pieces, small packages, small containers, etc.
Lumber, flitches, large timbers, poles, etc.
Sheets, veneer
Finished board products (plywood, OSB, particleboard, fiberboard, etc.)
Particles (biomass, sander dust, furnace particulate emissions, etc.)
Wet sludges (agricultural/food/wood products)
Dry abrasive materials (furnace ash, char, sand, soils, etc.)
Agricultural grains, nuts, fruits, and melons
Market-ready food products manufacturing – stringent inspection and health codes apply

21.2.3 Common Types of Wood, Agriculture, and Food Material Handling Systems

Mechanical conveyors (chains, screws, belts, roller beds, vibrating beds, etc.)
Pneumatic conveyors (low-pressure, high-pressure)
Hydraulic conveyors (sluice systems for abrasive materials, sewage systems, storm water, etc.)
Hydraulic lift stations – common in sewage handling systems, agricultural water handling,
Cyclones and centrifuges
Piping, ducts, and valve systems (cold, hot fluids)
Condensate traps for steam systems
Diverter gates (fixed, powered)
Tipples and fixed slides (static, vibrating, lubricated, etc.)
Inspection stations (rotating, horizontal, fixed, oscillating tables, with mirrors, cameras, recorders, etc.)
Line bars for alignment, positioning, and support for transporting, machining, or sawing operations

Pumps, fans, and blowers (liquids, gases, slurries)
Airlocks (linear, rotary, horizontal, vertical, manual, automatic)
Silos, tanks, bins, and containers

- cargo
- storage
- surge
- metering
- live bottoms
- level measurements and controls

Product loaders and unloaders to and from hot, corrosive, or abrasive process centers
Chop saws used to reduce product size to enhance product handling
Trailers (fixed beds, live beds, tilt dumping, bottom dumping)
Mobile dump trucks (light duty, heavy duty, rough terrain, flat terrain)
Rail systems (flat cars, boxcars, powered, nonpowered, light duty, heavy duty)
Mobile trucks (fork trucks, straddle carriers, rough terrain, flat terrain)
Floor carts (manual, powered)
Skate rolls – low-friction wheels mounted on fixed or movable rails
Lifts & elevators (Hydraulic – 1-post, 2-post; Cable – 2-post, 4-post; Hydraulic/cable combinations)
Elevators – personnel, platform, bucket, etc. – strict design and safety inspection codes apply
Cranes, boom trucks, and skyhooks – design, maintenance, inspection, and operating safety codes apply
Overhead-supported vacuum-gripper devices
Overhead-supported mechanical clamping devices
Overhead-supported magnetic lifting cranes and devices for iron and steel products
Rail systems (1-rail, 2-rail, 3-rail, 4-rail, carts, trams, overhead, custom-design systems)
Roller beds (powered, nonpowered)
Screw beds (level, inclined, 1-screw, 2-screw, multiscrew, single-rotation, counter-rotation, flexible)
Vibrating beds – oscillating spring mounted conveyors
Heated paddle screws – used in the drying of sludges, abrasive, nonabrasive
Continuous press lines – used in panel/board manufacturing
Power transmission (gear boxes, couplings, flat belts, v-belts, serpentine belts, chains, hydraulics, etc.)
Adapted speed controls (electrical usage, inventory control, point- or batch controls)
Biomass handling systems – designs dependent on moisture content and species of fiber being handled

- belt conveyors
- chain conveyors
- chop saws
- disc screens

drum chippers
rotary chippers
debarkers (rotary, drum, knives, scrapers) – highly sensitive to wood species and log diameter
dewatering rolls and presses
drag back chains
hammer mills
live bottom bins
log deck and troughs
smart conveyors
smart feeders
smart containers
storage buildings
trailer receiving stations
trailer loading stations

Types of Continuous Ovens, Furnaces, or Dryers – (rail system, rollers, flat conveyor, chain conveyor, Ferris wheel, carousal, overhead trolley, serpentine, chain-on-edge, spindle, slat type, strip cure, paint plate, pusher ovens, dog beams, washer-drain furnaces, and multitier)

Stackers (boards, packages, boxes, cargo containers)

Solid pilers with freight strapping or wrapping systems

Sorting systems (green-end sorters, dry-end sorters, quality/grade sorters, size sorters, etc.)

Un-stackers (rake-off, tilt hoists, vacuum lift, etc.)

Unscramblers – inclined conveyor with lugs or slots that passes upward through a pile of products and creates a single line of products, common in lumber manufacturing

Over-head rolls – used to hold down products moving on a conveyor

Side-rolls – used to stabilize moving machine carriages or products to prevent side drifting

Robotics – computer-controlled arms for lifting, rotating, stacking, clamping, machining, drilling, welding, etc. – Usually includes AI capability for improving production rates, precision, and energy-efficiency. Robotic systems are now available for handling large heavy items such as heavy timbers, poles, trams, pallets, packages, containers, etc. Detailed time, motion, energy, capital expenses, labor, and safety-risk analysis are required for applying robotics in industrial applications. Fault-tree analysis must be included in all safety-risk studies for all applications of robotics handling systems. Only properly trained design, maintenance, and service people should be allowed to enter a robotics operation safety zone. Safety motion detectors should be used to detect any physical event not part of the normal robotics operation, automatically shut down the system, sound an alarm, and require the operator to restart the system after determining the cause of the event.

21.2.4 *Safety Systems in Material Handling*

Safety limits – sensors specifically designed to prevent injury/damage to equipment or personnel
Safety–safety limits – backup safety limits to primary safety limits
Safety–safety–safety limits – common in military, space, aeronautics, and public elevators
Safety–safety–safety–safety limits – common in nuclear power or weapons systems
Fault-tree analysis – examination of sequential-failure outcomes, usually conducted by third parties. Coordinate with your insurance carrier and attorney when documenting fault-tree analysis work.
Comply with all local and national safety standards and codes.

21.2.5 *Third-Party Inspections and Documentations*

A. All material-handling systems should be inspected by a competent, independent party before the final installed system is put into operation.
B. Coordinate this activity with your insurance provider, attorney, and the local authority having jurisdiction.
C. Maintain records of final adjustments of primary control elements in a secure fire-proof vault.

Chapter 22
Needed R&D Projects in Industrial Drying Systems

The following list includes some of the many R&D projects needed on a global scale to advance industrial drying technology. Expect to see this list grow in future editions of this design guideline. I welcome feedback from people using this book about future R&D topics.

Development of hyperefficient reversible axial fans for convection-type dryers
Development of hyperefficient internal heating coils for indirect-fired convection-type dryers
Investigation of dehumidification drying systems with community heat pump systems
Comparison of co-gen & drying systems vs. wood-fueled direct-fired lumber drying systems which will include environmental emission stack tests of wood-fueled, direct-fired high-temp softwood dryers
Development of CFD-developed tables for plenum designs in convection-type dryers
Development of CFD-developed models for lumber packages being dried in convection-type dryers
Development of a standard platform for lumber dryer digital control systems
Development of regional gross energy data for engineered materials for the coming decades

The below article presented in Appendix D of this book, and published on November 30, 2020, describes research needed in the wood products industry and can serve as a model for the same research needed in the agricultural and food industries.

The Need for Global Energy Efficiency; Studies, Design Guidelines, Standards, and Practices for Industrial Wood-Drying & Wood-Energy Systems in the Age of Global Warming

J E 'Ed' Smith, *Industrial Drying Systems*,
https://doi.org/10.1007/978-3-031-31863-4_22

Chapter 23
Preventing Frauds in Industrial Drying Systems

The following points are common in most books published about financial frauds.

(1) The total costs of financial frauds can vary from 0% to 100% of the total cash flow through a social system.

EXAMPLE – The healthcare system in the United States is a classic example of a social system rampant with frauds. Depending on how the term fraud is defined, and how the accounting is done, the real personal cost of fraud can vary from 45% to over 95% of the total cash flow through the system. In the US healthcare system, the real costs of fraud involve many trillions of dollars each year.

(2) Direct and systemic frauds are significant social problems for all humanity because it conditions (normalizes) both immoral and illegal human conduct by changing people's attitudes about what is considered okay and not okay to do.
(3) Every employee of companies should be diligent and trained to both recognize and confront frauds against their employer.
(4) Government agencies continually underestimate the true amounts of frauds occurring in society including the many government-funded programs because politicians and public servants want to protect their image and careers.

J E 'Ed' Smith, *Industrial Drying Systems*,
https://doi.org/10.1007/978-3-031-31863-4_23

Chapter 24
Industrial Drying Industry Terms, Processes, and Topics

Term	**Definition**
Academy of Sciences	An organization dedicated to promoting the scientific method and publishing peer reviewed science articles and books. Often used to resolve critical social issues without special-interest bias. Most developed countries have a national academy of sciences.
Acceleration	The rate of increasing speed change
Adiabatic	A thermal process with no heat energy losses or gain through the surrounding boundary.
Adiabatic dryer	An infinitely insulated dryer (chamber) with a singular venting discharge stream.
Aerodynamic inspection	This is a term involving inspections of axial fans that must be done before axial fans are installed in any dryer. The fan is mounted on a shaft mounted on secure bearings. The fan assembly hub and blades are then checked for manufacturing defects. Assemble the fan on the shaft, adjust the blade pitch to the amount to be used in the dryer, check the tips of the blades for wobble, final consistent radius, and the swept profile of each blade using two dial indicators. As the fan is slowly rotated on the shaft, each blade outer edge should move the dial indicators the same amount as the edge of the blade passes the dial indicators. Do these measurements on both the front and rear of the fan, and at ½ the radius and at the tip of the blades. These measurements will reveal any manufacturing defects in the hub and blade castings. Once the data is collected, record the data for each fan assembly identification number. The objective of these inspections is to assure that all the blades have the same

J E 'Ed' Smith, *Industrial Drying Systems*,
https://doi.org/10.1007/978-3-031-31863-4_24

	exact pitch to prevent aerodynamic imbalance. Any pitch imbalance over ½° will cause the fan to vibrate when put into service. The higher speed and horsepower put into axial fans will impact the amount of vibration and thus bearing failures significantly. This is a very common problem in axial fans used in wood dryers. This is also the reason all axial fans to be used in any industrial dryer should be purchased and inspected by the buyer months before they are to be used in a dryer, thus allowing the buyer sufficient time to resolve any manufacturing issues with fan manufacturers.
Agriculture (Department)	A government agency that regulates the use of land for crops and animals for food production, and other useful soil-based products such as timber. Every country has a different structure for departments of agriculture. Some include forestry, and some do not.
Airlock	A mechanical device in a flow stream that allows solid materials to pass through without air flow also passing through. The three classes of airlocks are Rotary, Oscillating, and Static pressure.
Airlock (rotary)	A rotating drum fitted with adjustable radial vanes located inside a cylinder designed to allow solid materials to pass through via gravity with minimal air leakage. These devices are common in material air handling systems. Rotary airlocks are also used as rotating personnel doors for entrances to buildings to minimize energy loads on HVAC systems.
Airlock (oscillating)	Manual doors or automatic doors or gates that open and close by sliding either sideways or vertical to allow solid materials, packages, mobile vehicles, or personnel to pass through a point in a flow stream.
Airlock (static pressure)	A pressurized air chamber designed to stop air leakage in a specific direction. The static pressure inside the chamber is maintained slightly above the static pressure of the adjacent system. Differential pressure sensors and controllers are used to automatically adjust the fan or blower servicing the air chamber.
Airlock (dryer door)	Pressurized static pressure air chambers located on both entrance and exit doors of moving-product dryers to prevent toxic dryer compounds from leaking through the doors of the dryer. For lumber dryers, the chamber length should be 2–3 times the spacing of the stacking stickers in the lumber packages. These chambers must be pressurized with sufficient fresh air flow to prevent dryer emissions from passing through the doors of the dryers. These airlocks can also be fitted over negative-draft vent-a-hood systems mounted on dryer doors.

Angle-of-repose	In dry materials handling, the angle of incline at which gravity and friction or attraction forces are equal. The angle of repose is different between slides and piles of materials. Adding lubricants or vibration to the materials will reduce the angle of repose.
Application	Capacity for use
Artificial Intelligence (AI)	A branch of process control engineering involving data collection, statistical regression analysis, predictive analysis, and maximization routines for the purposes of minimizing waste and improving manufacturing process accuracy. Used in repetitive machine applications, such as robotics, and any material flow process where patterns exist. The success of AI systems is highly dependent on the accuracy and maintenance of the sensors collecting data. Consult with a process control engineer familiar with AI for information about specific applications.
Atmospheric	Standard air pressure at sea level
Auto-load-baffle-system	Overhead ceiling baffle and top-of-package loading system for both eliminating bypass air flow over stacked packages and providing a warpage-reduction top load on the packages. These systems can be powered to provide a constant load on the top of the package as the wood dries and shrinks. Common in European multi-zone jogging-batch softwood dryers. Rare in the US.
Auxiliary	Providing help, functioning in a subsidiary capacity (Webster)
Auxiliaries	Multiple subsidiary devices
Average Speed	The arithmetic average of a collection of speed data points
Batch	A quantity (as of bread) baked at one time, a quantity of material for use at one time or produced in one operation. (Webster)
Bidirectional	Movement in two different directions
Blend air	Mixing two or more air streams to achieve changes in temperature and/or humidity
BTU/Board foot	In lumber dryers, the amount of heat energy (BTU) to dry one (sales) board foot of lumber. This number is dependent on the wood species specific gravity, initial moisture content, target moisture content, type drying schedule used, type dryer used, presence of heat energy recovery systems, and weather conditions. For example, in high-temp batch drying of the southern pines, figure an average of 3000 BTU/Board foot (sales). In southern pine continuous kilns with preheating systems, figure 2100–2600 BTU/Board foot (sales). In low-temp drying of both softwoods and hardwoods, the energy usage/board foot will increase. In some cases, the usage can exceed 8000 BTU/Board foot.

BTU/board foot (peak)	In batch lumber dryers, the peak demand is typically 1.75–2.0 times the average hourly heat demand. However, this number is sensitive to the type of dryer and the drying schedule used. In heat-limited drying, this number is 1.0.
Cascade	A succession of stages each of which is closely related to or depends on the output of the previous stage or stages. (Webster)
Cascade drying	A drying scheme in which a result is dependent on previous drying events. Cascade drying schemes are common in vertical package locations in softwood dryers to minimize warpage in the top layers of lumber.
Chaos Theory	In both failure analysis and process controls, the mathematics involved in predicting the probability of failures (chaos) determined by large numbers of critical elements. As the number of elements increases, the probability of failures increases dramatically. This is why the number of critical elements in any system should be kept as low as possible. In complex safety systems, the cost of designing, manufacturing, and inspecting all the individual elements may reach a level unacceptable for the intended process. There are many cases of large digital-control systems being so unreliable that entire plants had to be shut down until major changes were made to keep the systems operating.
Circular tram system	A track-type tram-based product-handling system that operates in a circular fashion, moving a product from one initial location to a final location and returning the trams back to their original starting point. Common in many older softwood lumber drying operations.
Class	A group of the same general status or nature (Webster)
Class of dryer	The three classes are; batch, jogging-batch, and continuous
Color treatment	The use of a dryer system to change the color of a product by select temperature schedules.
Continuous press line	A wood-products panel making process in which materials are fed into a long conveyor system with sequential hot presses.
Cramming	The practice of putting excessive electrical horsepower into axial fans located inside convection dryers to increase drying rates instead of designing the dryer for maximum electrical efficiency. However, designing these fan systems for rare occasional cramming by using variable speed drives is acceptable. Furthermore,

	all convection dryer fan systems should be controlled by adaptive dryer software programs that constantly monitor the required fan system CFM requirements and automatically adjust the speed of the fans. Failure to do this will cause a dryer to waste large amounts of electricity. To put a number on the term "cramming," it is the annual difference in electrical energy usage between the dryer designed to the dryer designed by optimum geometric modeling. In poorly-designed convection dryers, the annual cost of the wasted electricity can be a significant percentage of the total front-end cost of the drying system. Proper CFD and geometric modeling can accurately predict these costs before the dryer system is built.
Code assignment	To put into the form of a code
Code assignments (legal)	Examples are: NFPA, SBC, NEC, EPA, ANSI, ASME, etc.
Conditioning	In drying, a process of controlling fiber moisture content target points
Configurations (Webster)	Structural arrangements of parts
Constant-energy drying	A type of heat-limited drying in which the amount of total heat energy put into a drying cycle is divided equally between drying zones in a multizone jogging batch or continuous drying system. These drying systems require leakage between zones to be kept to a minimum by using tight inner zone baffling barriers. Sounds easy to do but can be difficult with certain types of stacking systems using large numbers of cross-outs.
Continuous	Continuing without interruption
Continuing	To maintain without interruption
Counterflow	Moving in opposite directions
Cooling	The reduction of temperature or heat energy
Cross-shaft fan	For track-type lumber dryers, fan shafts that are perpendicular to the length of stacked lumber
Cv (discharge coefficient)	A measure of flow capacity through an orifice at a specific pressure drop across the orifice. Most manufacturers of process control valves rate their products based on Cv for specific fluids. The flow rate is based on standard fluid dynamics equations. The shape of the orifice impacts the flow capacity.
Cycles – Closed cycle	A system in which all primary process material flow is contained within a boundary.
Cycles – Open cycle	A system in process material flows both enter and exit a boundary.

Decarbonization	A process in which operating systems are void of fixed carbon or the use of carbon compounds for sources of energy.
Deceleration	The rate of decreasing speed change.
Dehumidification	The process of removing water from a flow stream by condensation, absorption, or centrifuge.
Design	The arrangement of elements or details in a product (Webster).
Design documentation	Vendor drawings, parts lists, manufacturing specifications, construction drawings, etc.
Direct fired	Heating systems in which heat or combustion products are injected into the interior of the dryer.
Discontinuous drying	In drying, this term is the same as batch drying.
Door (kiln) – Door (main)	The large doors fitted on kilns for allowing lumber to both enter and exit the kilns.
Door (kiln) – Door carrier	A mechanical device for both lifting and moving large kiln doors.
Door (kiln) – Door safety guards	Fixed structural items designed to prevent main kiln doors from falling.
Door (kiln) – Door (personnel)	Small personnel access doors in walls and main doors of kilns.
Door (kiln) – Door (zone)	Small personnel doors installed in zone separation panels located inside a multizone dryer.
Door (kiln) – Door gasket	Soft flexible strips designed to prevent air/moisture leaks around kiln doors.
Double	Having two features or parts (Webster).
Drum dryer	They come in a wide variety of designs and sizes for specific products to be dried. Modeling and field testing is required for determining drying rates accurately.
Dry bulb temperature	Air temperature measured with a dry sensor device.
Dry sheds (open sides)	A dry product storage shed with only a roof.
Dry storage buildings	A dry product enclosed storage building fitted with access doors or curtains. Some buildings may have both temperature and humidity control systems and including fans.
Dry transfer car	A powered dry lumber package transfer car mounted on rails or tires.
Dryer (proper)	In a drying system, the heat-energy equipment located between the green-end and the dry-end, including the process-control and support equipment located in the dryer control room.

Effective density	A ratio of actual weight to aerodynamic drag on a particle in suspension drying. Particles with very low effective density will quickly pass through some types of low-turbulence suspension drying systems and exit the dryer while still wet.
Electric motor RPM	Electric motor designs are either alternating (ac) or direct current (dc) designs. AC motors are either synchronous or induction-type designs. Synchronous motors run at the motor's supply voltage frequency. Induction motors run at "slip" speeds slightly below the supply-voltage frequency. DC motors run at speeds based on the DC supply voltage to the motor. Because of the increasing cost of electricity, all electric motor and drivetrain speeds should be kept as low as possible to reduce frictional energy losses. This is especially true in air-handling systems. Most modern industrial electric motor speeds are controlled by remote variable-speed drives connected to process optimization/adaptive software. Contact the National Electrical Manufacturers Association (NEMA) for additional information, standards, and technical papers about all types of electrical products.
EMC	Equilibrium Moisture Content that wood reaches in a specific relative-humidity environment.
End Coatings	Moisture barriers applied to the ends of fresh sawed logs, poles, timbers, and lumber to stop moisture loss and thus damaging differential shrinkage at the fresh cut ends of these products.End coatings are used to prevent the many checks and end splits that occur in wet hygroscopic products due to differential shrinkage dynamics. Only used for low-temp drying processes. In thick finished products where appearance is an issue, some type of end coating is required to prevent unsightly end checks from developing after the product is put into use.
Energy (plant)	The total of all types of energy sources used for operating a plant.
Energy (unit cost)	The cost/delivered unit of a specific type of energy.
Energy (drying)	The total of all types of energy (heating, electricity, etc.) for every piece of equipment including all support systems from the green trimmer to the infeed table at the planer mill.
Energy (evaporation)	The energy required to evaporate one pound of water at ambient conditions.

Energy conservation	Methods used to reduce energy consumption. For lumber dryers, typically the use of high-temperature drying schedules, additional process insulation, use of efficient fan designs with adaptive speed controls, use of statistical moisture content control programs, use of predictive equipment inspection and maintenance programs and use of energy recovery systems.
Energy efficiency	For dryers, the ratio of ambient evaporation energy divided by the total energy consumed by the entire drying system, based on a per pound of water removed. Energy efficiency is highly variable to front-end design and plant inspection and maintenance practices.
Energy recovery systems	The use of technologies such as; heat pumps, fluid heat exchangers, sister kilns, recycling technologies, counterflow technologies, etc. to recover energy. See the latest edition of ASHRAE publications; Fundamentals, HVAC Applications, HVAC Systems and Equipment, and Refrigeration for detail engineering principles involved in designing energy recovery systems. Always use experienced thermal engineers when both analyzing the economics and actual designs of energy-recovery systems so the final construction designs meet current safety and efficiency codes.
Entropy & Enthalpy	Two energy terms used in physics and thermodynamics. Entropy is a term that refers to the amount of irreversibility in a thermal process. Enthalpy is the sum of (1) heat energy and (2) pressure times specific volume. Refer to texts on thermodynamics to learn how to use the two different terms in thermal processes.
Equalization	In lumber drying, a process of controlling final EMC set-points to get rid of the wet lumber and add moisture to the dry lumber, done in a stepped fashion.
External fan motors	Kiln fan motors located outside of the heated kiln environment.
Fans (internal)	Axial fans used to circulate heated air through stickered packages of lumber.
Fan reversals	A process of reversing fans in convection lumber kilns to minimize moisture differences. Most lumber kilns require no less than two fan reversals during a drying cycle. The design of the kiln and lumber stacking practices are major factors in the minimum required number of fan reversals. If total energy usage is being monitored; some dryers can perform well with only one fan reversal.
Fault-tree analysis	A systematic analysis of all possible methods by which faults can occur in a complex system including the out-

	come of those faults. A type of risk analysis. Some insurance companies may require fault-tree analysis of industrial processes be done by independent third parties.
Finite-Element-Analysis	A mathematical model (FEA) for slicing a material into numerous sections and conducting energy, stress, and mass flow calculations. My first lumber drying FEA model, back in 1976, revealed that FEA modeling is helpful to understand both batch and continuous drying dynamics.
FEA Hi-Low Limits	Control limits for embedding an FEA model into a conventional drying control schedule. The dryer is controlled by either the conventional drying schedule, or the FEA model. Because limiting fiber stresses to prevent product damage is the objective, the control system will watch the FEA model and decide (by the hi-low limits) if fiber stress is being exceeded. This control scheme requires sufficient sensors that can measure product drying rates and temperatures, and then decide if the FEA model or the conventional drying control scheme is to control the dryer. The dryer operator can also disable either control loop depending on his experience with drying a specific product. Or the FEA hi-low limits can be modified depending on the final amount of product degrading that is occurring. These control systems must be designed by process control engineers with experience in industrial drying control systems.
Flow	To issue or move in a stream. (Webster)
Fork truck	A mobile truck with lifting forks. The key elements are operator safety training, fork length, lifting capacity, lift distance and the type of surfaces the truck will be driven on. For 24/7 operations, proper lighting of the entire plant facility is crucial to safe fork truck operations.
Geometric Modeling	In drying systems, an analytical process, often called configuration modeling, for determining the affect physical product handling dimensions has on the drying production rates and energy efficiency of the entire drying system. For example, in lumber drying, the four primary variables, total-package-height, total-package-width, board-thickness, and sticker-thickness, each have significant impacts on the following variables: dryer production-capacity, electrical-efficiency, final-moisture-distribution, equipment capital-costs, material-handling costs, and economic-payback on investments. Conducting such a study accurately requires the use of CFD software capable of doing both

	fluid heat transfer and fluid flow calculations to find an optimum economic solution. During these studies, numerous CFD runs will have to be made using different combinations of the four primary variables. This same capability can also be used to find optimum energy-efficiency solutions which are usually different from optimum economic solutions.
Green moisture sorting	A process in which green wood fiber is sorted by moisture content prior to or during green machining or drying operations. This practice is used in certain species of lumber and veneer manufacturing to minimize product degrading.
Green sorters	In virtually all modern lumber mills today, green sorters are highly engineered systems designed to separate different lumber products by; species, grade, board thickness, board width, board length, and in some mills flitches and rejects. After boards are produced by the sawing operation and then measured by sensors, a computer directs those boards to specific receiving bays where an inventory count is maintained. Then, when a sorter bay is full of a specific product, it no longer receives any more boards. The number and capacity of sorter bays are determined by the number of, and the hourly and daily volume of products produced by the sawmill, and the width and height of the stacked package used in the drying system. Also, the green sorter is a means of temporary product inventory and control until enough boards are in a bay or bays to make up a full stacked package of lumber. Products produced in a specific mill's sawing operation are also highly variable to the size and types of logs entering the mill and can cluster for long periods of time making some types of sorters run out of available bays. The computer system managing the green sorter must be able to inform the sawmill operators of what is occurring in real time.
Green buggies	A cart used for temporary inventory control and for transporting green lumber to the green stacking system.
Green chain	A cross transfer chain conveyor for green lumber.
Green stacker	An electromechanical machine for breaking down rough bundles of lumber and stacking the lumber into stickered packages for drying.
Green transfer car	An electromechanical platform running on rails for transporting green packages of lumber.
Greenwashing	The production of misleading or false information claiming benefits to the environment.
Heat recovery	The recovery of energy to reduce fuel consumption.

Heat limited drying	A dryer system in which the drying rate is determined by a fixed rate at which heat energy is put into the dryer. If designed properly, and sufficient dryer capacity exists, this simple method of drying can minimize product degrading due to drying. Both direct fired and indirect fired dryers can be designed to create heat limited drying.
HP	Horsepower – 550 ft.-lbf./second
Heat pump	A closed electromechanical fluid pump, condensing and expansion system designed to transfer heat energy from one environment to a second environment.
High-temperature limit	An electrical device designed to sense temperature in a process and shut off specific parts of a closed system to prevent fires or equipment damages.
Humidity (in air streams)	See the ASHRAE handbook for detailed data and engineering/explanations of the following: Absolute humidity Relative humidity Humidification Dehumidification
HVAC	The term applies to heating, ventilation, and air conditioning.
HVAC/R	The letter R in the term HVAC/R refers to the "repair" of HVAC systems. Check with the local authority having jurisdiction to determine how much of an existing HVAC system requires licensing. Repair activities do not include designing.
Hype in drying	A common psychological disorder seen especially in lumber drying. Extravagant claims of drying equipment capability, reduction of product degrading, energy consumption, longevity, etc. Hype is used to convey an "image of total success" to people outside the sphere of actual drying reality, done for political and emotional reasons.
Hyper-low-pressure steam	A low level of saturated steam pressure for use in industrial dryers in which the long-term environmental benefits dictate the design of the steam system at the dryer. This task can only be achieved by careful CFD analysis of the dryer's design and its steam system heat exchangers. In some industrial dryer applications, the operating pressure of the steam boiler may be modulated down to extremely low operating pressures. Consult with the boiler manufacturer before installing pressure-modulating control loops on any boiler. Furthermore, all the steam traps in the entire steam system have to be able to handle the expected condensate flow at low steam pressures.

ISO dryers	Industrial dryers that meet ISO International Standards Organization requirements.
Idle-energy rate	The energy consumption rate of a thermal or material handling system with little to no product loading. For example, adaptive controls should be used to automatically lower the speeds of blowers or fans to a safe low level while product loading is near zero, for the purpose of reducing electrical energy usage. Controls should include anticipation capability such that the system reaches the minimum required speeds before product loadings start to increase.
Indirect fired	A heating system in which heat exchangers are located inside the kiln structure to prevent products of combustion from entering and contaminating the lumber being dried, as well as damaging kiln equipment located inside the kiln building.
Industry (Webster)	A department or branch of a craft, art, business, or manufacturer.
Industry (examples)	Agriculture, food, wood, clothing, minerals, biomass, chemicals, masonry, paint, etc.
Internal fan motors	Fan motors located inside a dryer.
Jerk	A quick sharp sudden movement (Webster).
jerk (motion)	The rate at which acceleration increases or decreases (braking).
Jog	To give a slight shake or push to, to go at a slow monotonous pace, to run or ride at a slow pace. (Webster).
Jogging batch dryer	A dryer or kiln in which the materials to be processed are fed into and removed from dryer system by a series of starting and stopping of the material flow stream.
Jogging process	An interruptible, repetitive (possibly jerking) method for moving a process material.
Kiln	A heated enclosure for processing a material by burning, firing, or drying.
Kiln (batch)	The most common type of lumber kiln, designed to dry one batch of lumber at a time.
Kiln (firewood)	A kiln specifically designed for drying green firewood stacked on a tram or in a pile.
Kiln (track)	A common type of kiln used mostly for softwoods. Stacked packages of lumber are placed on trams that run on a set of rails that traverse the kiln structure.
Kiln (side-loader)	A common type of kiln used mostly for hardwoods. Stacked packages of lumber are placed inside the kiln with fork trucks.
Kiln (package)	The same as a side-loader kiln.
Kiln (sidewinder)	Typically a small low-cost kiln with all equipment located at ground level. Although now rare in the United States, sidewinder kilns are common in other countries.

Kiln (predryer)	A kiln designed specifically for predrying lumber. Typically, large buildings designed to accommodate large numbers of packages of mostly hardwood lumber.
Kiln (dehumidification)	A low temperature, highly insulated kiln fitted with a heat pump for condensing moisture out of the kiln airstream. Also, typically a wooden structure.
Kiln (steamer)	Often called steaming chambers, used for stress relief, MC control, and setting color. These chambers must be built of corrosion resistant materials, such as aluminum or stainless steel.
Kiln (sterilization)	A process specifically for destroying infestations of insects. The process can be either a kiln or hot-water dip tank in which the center of the wood is heated to a sufficient temperature and time to kill off insects.
Kiln (continuous)	A kiln in which lumber is dried as the lumber moves through the kiln via a continuous, nonstop, or nonjogging, driving process. Continuous kilns are different from jogging-batch kilns.
Kiln (jogging-batch)	Kilns that transfers lumber through the kiln via repetitive pusher or puller devices. The cycle times for jogging the lumber loads in these kilns vary widely depending on species, board thickness, and package length. Jog cycle times in jogging batch kilns can vary from minutes to hours to days.
Kiln (counterflow)	Any multiple-track-type kiln in which the direction of lumber flow on the tracks is in opposite directions. The counterflow double-track is the most common type. This process is also referred to as multidirectional or bidirectional flow.
Kiln (dual path)	A popular (but technically misleading) marketing term for a counterflow kiln.
Kiln (dual pass)	The same meaning as dual path.
Kiln (zig zag)	A continuous kiln with alternating air flow through the packages, common in Europe.
Kiln (E&C)	A kiln structure, often referred to as a chamber, designed for E&C purposes only.
Kiln (lumber)	Any kiln designed specifically for drying lumber.
Kiln (pole)	Any kiln designed specifically for drying poles.
Kiln (roller bed)	A kiln that uses fixed roller beds for moving lumber packages before, in, and downstream of the kiln. Typically used in small low-temp drying operations.
Kiln (screen)	A kiln that uses a metal screen for transporting materials through the kiln. A common practice in the food processing industry and particle drying.

Kiln (shotgun)	A track kiln with only doors at one end. The lumber enters and leaves the kiln via one door.
Kiln (sister)	A second kiln designed to complement a primary kiln, for improving grade control and/or reducing energy consumption via condensation of primary venting streams.
Kiln (treated wood)	Any kiln designed specifically for drying treated lumber.
Kiln door height	The physical distance between top-of-rail and bottom of the door lintel beam in a track kiln. The distance between the floor and the bottom of the door lintel beam in a side-loader kiln.
Kiln tracks	The steel rails that kiln trams run on. These rails must be supported properly and aligned carefully to prevent wear and tear of kiln carts and trams that ride on them in addition to preventing spillage of lumber from packages. Most all problems with kiln carts and trams are caused by settlement and heaving of supporting soils located under kiln tracks. Furthermore, if the soils are highly reactive to moisture content, changes in weather conditions can cause rails to move up and down due to seasonal changes. This is the reason a soils engineer must be retained before kiln tracks are installed. In some geographic locations, the cost of installing stable kiln rail systems can be a significant portion of the total cost of the entire drying system.
Kiln cart	A short, assembled metal structure with wheels for supporting packages of lumber. Kiln cart lengths can vary from 4′ to 16′.
Kiln truck	A pair of opposing steel channels with pairs of axles and steel wheels. Kiln trucks usually vary in length from 4′ to 8′.
Kiln tram	An assembled steel structure with wheels for use in track kilns. The length of trams varies depending on the material (lumber or poles) being carried by the tram. Some trams have posts on both sides to prevent loads from spilling off the trams. The actual length of trams must match the actual (effective) length of products resting on the tram.
Kiln truck cross-outs	Typically a 4 × 4 steel or wooden beam for loading lumber packages on kiln trucks.
Kiln charge	The actual board footage of lumber inside a kiln.
KW	Kilowatt of electricity.
KWH	Kilowatt-hours of electricity.
Line-shaft fan	A type of fan system in which the fan shaft is parallel to the lumber in a kiln.
Load	The board footage of lumber inside a kiln.
Load width	The physical outer width of lumber packages loaded inside a track kiln. The physical outer length of lumber packages loaded inside a side-loader (package) kiln.

Load height	The total physical height of lumber packages loaded inside a kiln.
Load depth	The total physical depth of lumber packages loaded inside a side-loader kiln.
Load length	The total physical length of lumber packages loaded inside a track kiln.
Loader (dryer)	A machine for loading the three classes of dryers (batch, jogging-batch, and continuous). Batch dryer loaders are fork trucks, rail systems, or roll cases. Jogging-batch dryers use fork trucks, automatic pusher devices for rails, belts, or roll cases. Continuous dryer loaders are powered rail systems, belts, or roll cases.
Loading efficiency	The statistical average percentage of maximum possible capacity of lumber inside a kiln.
Matrix	Something within or from which something else originates, develops, or takes form.
Material	The substance to be burned, fired, or dried in a kiln.
Material (examples)	Crops, food, lumber, cotton, lime, bark, salt, brick, particles, sheets, boards, packages.
Mixed package	A wood products stacked package containing different thicknesses, widths, or lengths of wood products, and a practice that causes significant drying problems.
Model	Being a miniature representation of something, for study or analysis (Webster).
Models, kiln	Test kilns, computer models, design models, etc.
Mean Speed	The arithmetic mean (average) of a collection of speed data points.
Moisture content trimming	A process in which the moisture content exiting a dryer or entering a final grading system is trimmed either up or down by adjusting previous drying centers.
Moisture Tracking System	A process in which the statistical averages and standard deviations of fiber moisture content, specific gravity, and density is tracked by species, tree diameter, age, geographic harvesting location, time of year, and weather at each of the following processing centers: At the forest harvesting site. Entering the log storage yard Entering the sawing, peeling, chipping facility Entering the green fiber storage area Entering the dryer Exiting the dryer and before entering the dry fiber storage area Exiting the dry fiber storage area and entering the dry-end moisture measuring system and including statistical reporting of averages and standard deviations of product moisture content

	Entering the final product process (grading, trimming, gluing, etc.) Changes in moisture content during product shipment Changes in moisture content while in the final warehouse or sales outlet Changes in moisture content of the product during construction activities Changes in moisture content while the product is in use
Multizone	More than one temperature or humidity control zone.
Operations	In lumber drying systems, the entire system design, equipment, maintenance, and support systems, operating procedures, including safety, environmental, and risk issues.
Optimum Steam Pressure	For steam-heated convection drying systems, the optimum economic steam supply pressure (OESSP). This value (PSI) is determined by a series of calculations that evaluates; primary drying variables, required dryer production rate, dryer dry-bulb and wet-bulb temperatures approaching the product, heat-release rates of the dryer steam coils, air pressure drop through the steam coils, dryer fan system performance curves, boiler water chemistry, supply steam quality, cost of electricity, steam boiler design, steam piping and condensate handling system designs, and system safety code requirements. The OESSP is different for different classes and types of convection drying systems and whether or not a CO-GEN system supplies the steam to the dryers. CFD modeling is required to determine the OESSP accurately. In some drying systems, the steam boiler pressure may be automatically modulated up and down to maintain OESSP. However, not all steam boiler systems can accommodate wide changes in operating pressures.
Owner's Rep.	An independent technical/engineering provider and auditor for new construction projects to make sure projects comply with contract legal and code requirements. The owner's rep works closely with the construction manager but their roles are different.
Package	A stacked assembly of wood products with or without stickers. If no stickers exist, the package is solid, and if stickers exist, the package is stickered.
Package vector	The end-to-end direction of a stacked lumber package with stickers.
Package angle	The angle between the direction of package flow and the direction of air flow.
Package angle of 90°	The package angle found in most convection kilns.
Package angle of 0°	The angle found in counterflow (air to package) continuous kilns.

Parade drying	A process control method used mostly in multizoned jogging batch and continuous drying systems by which dry-bulb and wet-bulb temperature conditions in different dryer zones are changed based on different wood products entering the dryer, like in a parade. Dryers using parade drying routines require stored drying temperature schedules for each one of the different products entering the dryer. These systems also require in-dryer moisture sensing systems to provide feedback data into the individual dryer zones. Parade drying schemes are mostly used with thin narrow easy-to-dry light softwoods. These schemes do not work with refractory hardwoods. Parade drying schemes can also be used in batch drying operations but require time-consuming preplanning of mixed size products. Some people have developed computer models for batch parade drying in attempts to get more production through batch dryers. The actual cost versus benefit of batch parade drying systems is debatable.
Parallel	Lying or moving in the same direction but always the same distance apart (Webster).
Parallel motion	In lumber kilns, often referred to as unidirectional process flow.
Peak Speed	The maximum speed of a movement.
Pest treatment	The use of a drying or heating system to kill harmful agents in a product.
P-I-D Controller	A three-element process controller that includes proportional-integral-derivative capability. Depending on whether the drying process is batch or continuous, the required capability of process controllers varies. In some drying processes, very small temperature proportional bands are required (such as in the drying of hardwoods). In some steam boilers, a three-element drum pressure controller is required. Always document the final settings of all process controllers after final tuning has been completed. Keep these records in a secure place for future troubleshooting of process controllers. Do not let unqualified people adjust these settings. Improper adjustments of process controllers can both damage products, or equipment, and possibly cause personal injuries or death. Some property insurance companies may require records be kept on final settings of safety-critical process controllers.
Point-batch MC analysis	A comparative linear method of analysis of dry-end moisture content data to determine the root causes of product degrading. This technique can distinguish between drying systems and fiber causation in product degrading.

Precision stacking	The stacking of product packages into tightly controlled top, edges and sizes, required for successful energy-efficient drying in convection dryers.
Process	A series of actions or operations directed toward a particular result (Webster).
Process (kiln) (Webster).	For burning, firing, heating, or drying materials
Product (to be dried)	A physical material and size to be dried, equalized, conditioned, or treated in a drying system. Product specifications include the species, grade, thickness, width, length, or diameter prior to entering a dryer. In some drying systems, products may also be classified by initial moisture content (groups) before entering a dryer. In some drying systems, product finished color may also be classified or specified before entering a dryer.
Product Grade Bias Curves	A set of curves that demonstrates the effect of fiber shrinkage, allowable offsets, and market value of the product plotted against the moisture content of a product. These curves are useful in predicting the root causes of product degrading due to drying processes and fiber supply.
Product temperature limit	The upper temperature limit a finished product can be subjected to for a specified length of time.
Profile (moisture)	The distribution of moisture inside an ag/food/wood product. The moisture profile starts to change once a product is harvested due to moisture moving toward an environment of either lower vapor pressure or lower total absolute pressure. If the drying rate is excessive the product may develop degrading. This profile is different between low-temp and high-temp drying. The moisture profile and product degrading are highly sensitive to the temperature of the product during drying. Moisture profiles can be estimated by FEA modeling.
Progressive (Webster)	Adjective – moving forward, proceeding onward, advancing, as progressive motion or course.
Progressive (kiln)	A common term used by European kiln manufacturers for jogging-batch and continuous kilns.
Quad	Having four features or parts.
Rebound (degrade)	Product degrade that occurs after a product exits a drying system or a specific dryer.
Recovery (product)	A measure of the actual volume of final product as a percentage of actual green input volume.
Repetitive Relaxation (RR)	A type of drying process (schedule) in which a product is dried in an alternating intermittent manner caus-

	ing product fibers to be stressed then relaxed then stressed then relaxed repeatedly with the intended purpose of reducing product degrading due to dynamic stresses caused by excessive drying rates. The RR drying theory is based on the claim that a specific size product can be dried in the same dryer residence time with less dynamic degrading than those degrades when using proven conventional drying schedules. RR drying is the process of repeatedly starting and stopping drying energy being put into the product being dried. It is possible to test RR drying claims using finite-element-analysis software once sufficient data has been collected from conventional drying schedules.
Rotation	Direction of spin about a central point.
RPM	Revolutions per minute.
Rake off un-stacker	An electromechanical machine that rakes layers of lumber off the top of packages.
Saturated Moisture Content	The maximum moisture content that can exist in wood fiber for a certain specific gravity.
Sawing Solution	A geometric-based decision to how a product is sawed into smaller pieces for maximum economic value. When sawing anisotropic materials, such as wood, sawing solutions can impact how the smaller pieces are dried.
Scanner	An electronic device that scans products for different manufacturing operations. Usually configured for optimizing routines, grade detection, defects, metals, sales value, etc.
Scanner (density)	An electronic system that scans products for density.
Scanner (grade)	An electronic device that scans products for defects. Common in lumber manufacturing.
Scanner (moisture)	An electronic device that scans products for moisture content.
Scientific Method	A system of observations, measurements, and experiments based on formulations, testing, and modifications of hypotheses. (Webster)
Single-zone	One dry bulb temperature control zone.
Sluice system	Inclined flowing streams of water used to convey abrasive ash from wood-fired steam boilers to an ash-holding-settlement pond. Also used in furnace stack pollution abatement systems. These systems require sufficient stream velocities to prevent ash from settling in the water streams. The sluice-water recirculation pumps must be designed to handle dirty abrasive water.

Spreadsheet	An accounting program for a computer, a method of organizing data.
Steaming (chamber)	A steam chamber used for fiber equalization, stress relief, bend forming, or pest sterilization. Refer to high-temperature EMC charts for using steaming chambers.
Subtype	Divisions of a type.
Sub-subtype	Divisions of a subtype.
Sister Loads	Two or more adjacent (sister) loads of materials to be dried. Start normal drying in one load and use the air and moisture vented from one load to pre-heat the other sister load. Moisture that condenses on the sister load is collected and pumped to a storage tank. Use this water for humidity sprays in any kiln. Cycle this process of drying and condensing between adjacent loads.
Sprays (water)	Pressure injection of water inside a dryer for either fire protection or increasing humidity.
Stacking stick	A narrow plastic, metal or wooden stick placed between layers of green lumber to allow air flow through a package of lumber to be dried. Some stacking sticks are serrated to minimize contact and staining of the product being dried.
Stacking stick machine	A mechanical system that automatically places stacking sticks in a package.
Sticker vector	The end-to-end direction of a stacking stick.
System	A group of interacting, or interrelated entities that form a unified whole (Webster).
Temperature (kiln) – Dry-bulb	Use of a dry sensing element for measuring air stream temperature.
Temperature (kiln) – Wet-bulb	Use of a wet-sock sensing element for measuring air stream temperature.
Temperature depression	In a convection-type dryer, the difference between dry-bulb and wet-bulb temperature. Material drying rates are dictated by the temperature depression and the mass air flow rates through a sticker opening. The available energy rate for drying is a multiple of temperature depression times the mass air flow rate through a sticker opening. In dryers designed to dry thick refractory species, such as white oak, temperature depressions can be as low as 1 °F during the early stages of drying. Thus, it is paramount that the accuracy of all temperature sensors be sufficient to prevent excessive stresses in some species being dried. Also, in some situa-

	tions, the air flow through sticker openings must be reduced to prevent damage to some types of wood.
Temperature rise in drying	In dryers, the rise in dry-bulb temperature due to moisture condensing on a dry product caused by the heat of adsorption. Common during E&C schedules and product preheating systems.
Tempering	A term in drying that refers to controlling small variances in final moisture content.
Third-party	A neutral and independent party retained to evaluate processes or systems.
Transition algorithm	A process control algorithm for changing dryer controls from one method of measuring moisture content (during a drying cycle) to a second method of measuring moisture content.
Trim (control valve)	The relationship between valve stem travel and flow capacity. Valve trims are either on-off, linear or equal percentage. Valve trim Cv is a measure of flow capacity.
Type	Something distinguishable as a variety (Webster). Subtype: A subset of a type. Sub-subtype: A subset of a subset of a type.
Tumbler dryer	A dryer that uses a continuous tumbling action to mix products for drying. Common in convection and conduction drying systems.
Tunnel (Webster).	An elongated passage (as a tube or conduit)
Tunnel kiln (dryer)	All track, roller-bed, and screen kilns are tunnel dryers. Sometimes referred to a track-type dryer with building insulation panels located inside the main structure of the dryer.
Turbine Kiln Fan	A line-shaft type axial fan driven by a steam turbine, the exhaust of which is used to heat the kiln. The benefits are no electricity is used by the kiln fans, thus making the energy efficiency of the drying system much higher. Easy to do with long line-shaft type convection kilns.
Turn-down ratio	The ratio of high to low stable firing rates of a combustion or heating system.
Tilt hoists	An electromechanical tilting device for breaking down packages of lumber into layers.
Type	For dryers, a predryer, kiln, E&C chamber, storage, tempering, etc.
Unidirectional	Moving in the same (one) direction.

Unscrambler	For lumber, an inclined chain with lugs designed to remove individual boards from a bundle.
Vacuum drying	A method of drying in which the product is subjected to low atmospheric pressures to cause entrained moisture to effectively boil out of the product.
Vent assembly	Metal dampers located on a kiln to exhaust hot moist air.
Vent heat recovery	Any process by which the heat energy contained in hot moist vent air streams is recovered.
Vent (negative –air)	In psychometrics, the optimum location for removing high-humidity vent air. In lumber kilns, this location is at the discharge (negative-air-pressure) side of the lumber packages.
Voodoo Engineering	Slang term for any engineering not based on valid scientific principles. A common practice seen in many unregulated industries.
Wet bulb temperature	See ASHRAE publications for detailed description of wet-bulb sensor dynamics.
Wood products	Any commercial product obtained from wood. Lumber, veneer, timber, poles, plywood, OSB, fiberboard, particleboard, craft, paper, biomass, shavings, bark, etc.
Wood waste	Portions of flow streams leaving a wood products plant considered of no commercial value.
Zigzag	A turbulent material flow stream with many twists, turns, and alterations used to improve product mixing during a mechanical, thermal, or drying process. Common in food manufacturing and some (rare) convection grain and wood drying processes.
Zone (kiln)	Either a single zone or multizoned temperature control scheme.
Zone leakage	The amount of air heat energy and moisture leaking from one zone to an adjacent zone.
Zone partition wall	A baffled wall located between temperature zones, often fitted with flexible seals to allow lumber packages to move freely between zones.

For additional terms used in the forest products industry, see ***Terms of the Trade: A Handbook for the Forest Products Industry***, by Random Lengths Publications and **Thesaurus of Forest Products Terms (FPRS)**.

Appendixes

Appendix A – Industry Codes, and Standards

This section introduces the reader to codes and standards for designing safe, environmentally compliant energy, and drying systems.

Codes, Standards, Guidelines, Practices, and Ethics

Term	Definitions derived from *The Merriam Webster Dictionary*
Codes	a systematic statement from a body of law (principles and rules)
Standards	something as a rule for measuring or as models to be followed
Guidelines	an indication or outline of policy or conduct
Practices	actual performance or application
Ethics	a discipline dealing with good and evil and moral duty

Codes can be adopted at different levels of authority (government). They can be adopted at the federal, state, or local levels of government. If a code is adopted as law, the adopting agency will be the "authority having jurisdiction" and responsible for both inspections and enforcement. Codes are developed by committees made up of experts from different facets, perspectives, and interests. Codes are reviewed by committees during their development. Although all safety and energy codes should be free of special-interest influences, not all are.

Standards are not enforced by government agencies because they are not laws unless they are referenced in a code. They are viewed as models of which practitioners are expected to follow. Standards are developed by committees like the formation of codes.

J E 'Ed' Smith, *Industrial Drying Systems*,
https://doi.org/10.1007/978-3-031-31863-4

Guidelines are summary outlines which practitioners should be aware of. These are less scrutinized documents, some of which may be published without peer reviews.

Practices are what a practitioner does in the performance of their profession.

Ethics are prescribed classes of moral human conduct. Professional people are both expected and required to be morally responsible, complying with published standards for their profession. Their conduct should meet the ethical standards set forth by their profession. Linear models and tests define what constitutes ethical conduct.

The following ethics **examples** are for professional engineers in many of the states in the United States. All engineers should become familiar with the applicable ethics models in the country or state where the project exists.

However, many theoretical ethics models have received considerable criticism from some engineers as being too broad, fear-based, and thus impractical in competitive marketplaces. And, in countries with adversarial legal systems, the interpretations of these models are, in my opinion, so vague they cannot hold up in any court of law.

Professional Responsibility Models for Professional Engineers

The following "typical" linear ethics models for professional engineers have been around for many decades in the United States. And because the United States did not license engineers at the national level many decades ago, the same following models are constantly being challenged by engineers and thus are constantly changing.

If you look at the numerous global ethics models for professional engineers today, they probably will vary significantly from the following older models, the origin of which I have never been able to locate during my many decades of working in the engineering profession. Most likely, these ethics models were in the past, and today, written by lawyers not engineers.

The Good Works model – The ultimate effort by an engineer by investing time and consideration far beyond what is not only required, but even beyond what would be reasonably expected.

The Reasonable Care model – The more demanding, requires that an engineer is expected to consider factors, most often related to safety and quality, that are not explicitly addressed in standards or codes.

The Malpractice model – The least demanding, requires that an engineer need only perform at a level that meets standards of the profession and applicable laws or codes for the specific situation. This model is often called the **minimalist** model.

Professional Responsibility Tests for Professional Engineers

The Golden Rule Test – Asks whether we (the engineer) would be willing to exchange places with those affected by the engineer's actions.

The Utility Test – Asks what action (by the engineer) would provide the most well-being for the most people, even if some individuals are harmed by the engineer's actions.

The Rights Test – Asks whether the engineer's actions violate the rights of others, including the right to free and informed consent, to life, and to health.

If professional engineers find themselves involved in litigation matters, the above legal concepts may be used in a court of law to argue for or against either moral or legal negligence, or both.

Associations, Societies, and Institutes

Many of the current published codes, standards, and guidelines were originally conceived by one organization. Today, it is the norm for committees to be made up of several groups, each which have their own task development committees. These groups can have different needs and roles, some of which may be reflective of special interests.

Federal, state, county, and municipal government agencies and organizations
Industry associations (lobbying groups, minimum standards groups, etc.)
Industry societies (typically for research, education, and communication)
Institutes (authoritative organizations assembled for a cause; technical, business, social, etc.)

The following list of organizations was prepared from equipment designs and manufacturing issues in the agriculture/food/forest products and ancillary industries. Each of these is on the Internet. Most have search engines for their publications, or a customer service person that can assist in locating a specific code or standard. Internet search engines can locate outlets for codes and standards by entering key words.

search – boiler code ASME
search – gas burner code NFPA
search – UL standards

Global Engineering Documents, www.global.ihs.com phone # 1-800-854-7179 – This organization is a distributor for engineering codes and standards. Also go to www.techsmart.com/publisher/list for a list of organizations involved in codes, standards, and guidelines.

Legal Requirements for Equipment Codes and Standards

Because of the many codes and standards that apply to energy and drying systems, I suggest you contact the following before installing new equipment or upgrading existing equipment:

1. An attorney in the state where the project will be located.
2. Your business insurance carriers.
3. A licensed engineer in the state where the project will be located.

Always check for the latest published code. If litigation is involved in a loss, get a copy of the published code in effect just before the loss incident occurred.

Always let plant engineers, maintenance managers, safety managers, attorneys, insurance carriers, and financial entities know which codes were in effect when the drying project was originally put into operation, including who conducted third-party inspections if any.

For questions about equipment design codes, standards, safety, health, and environmental codes, standards, and development activities, contact the appropriate following organizations.

Global Organizations – Partial List Includes Organizations with Multinational Memberships

United Nations (UN), www.un.org/en

OECD – Organization for Economic Cooperation and Development, www.oecd.org

IEA World Energy Outlook, www.iea.org

CICERO – Center for International Climate and Environmental Research located in Oslo, Norway

AGAI, www.agai-outlook.org

ASHRAE – American Society of Heating, Refrigeration and Air Conditioning Engineers, www.ashrae.org

AHRI – American Heating and Refrigeration Institute, www.ahri.org

ASABE – American Society of Agricultural and Biological Engineers, www.asabe.org

ASME – American Society of Mechanical Engineers, www.asme.org

ICBO – International Council of Building Officials, www.icbo.org

ICC – International Code Council, www.iccsafe.org

IEC – International Electrotechnical Commission, www.iec.ch,

IEEE – Institute of Electric and Electronic Engineers, www.ieee.org

ISO – International Organization for Standards, www.iso.org – For additional information how the ISO works for ag/food/wood products industries, go to www.asabe.org and go to publications and standards/standards development/international standards.

IUCN – International Union for Conservation of Nature

Chemical Engineering Journal, www.chemengoline.com – See article "Solids Drying: Basics and Applications" for different classes and types of industrial dryers.

WATER, www.water.org

WCK – World Central Kitchen, www.wck.org

WHO – World Health Organization, www.who.it

National and Regional Organizations – Partial List

Agri-Food & Biosciences Institute, www.afbini.gov.uk

ABMA – American Boiler Manufacturers Association, www.abma.com

ACGIH – American Conference of Governmental Industrial Hygienists, www.acgih.org

ACI – American Concrete Institute, www.aci-int.org

AEE – Association of Energy Engineers, www.aeecenter.org

AEM – Association of Equipment Manufacturers, www.aem.org

AGA – American Gas Association, www.aga.org

AIHA – American Industrial Hygiene Association, www.aiha.org

AIChE – American Institute of Chemical Engineers, www.aiche.org

AISC – American Institute for Steel Construction, www.aisc.org

AMCA – Air Movement & Control Association International, www.amca.org

ANSI – American National Standards Institute, www.ansi.org

APA – Engineered Wood Association, www.apawood.org

API – American Petroleum Institute, www.api.org

ASABE – American Society of Agricultural and Biological Engineers, www.asabe.org

ASCE – American Society of Civil Engineers, www.asce.org

ASHRAE – American Society of Heating, Refrigeration and Air Conditioning Engineers, www.ashrae.org

AHRI – American Heating and Refrigeration Institute, www.ahri.org

ASME – American Society of Mechanical Engineers, www.asme.org

ASSE – American Society of Safety Engineers, www.asse.org

ASTM – American Society Testing and Materials, www.astm.org

ATSDR – Agency for Toxic Substances and Disease Registry, www.atsdr.cdc.gov

AWC – American Wood Council, www.awc.org

BCSP – Board of Certified Safety Professionals, www.bcsp.com

BOCA – Building Officials and Code Administrators International, Inc., www.bocai.org

Clean Energies Group, Queen's University, Belfast, www.pure.qub.ac.uk

CAGI – Compressed Air and Gas Institute, www.cagi.org

CARB – California Air Resources Board, www.arb.ca.gov

CCPS – Center for Chemical Process Safety, www.aiche.org

CDC – Centers for Disease Control and Prevention, www.cdc.gov

Chemical Engineering Journal, www.chemengoline.com – See article "Solids Drying: Basics and Applications" for different classes and types of industrial dryers. – JScholar Publishers is located in Frisco, Texas.
CIBO – Council of Industrial Boiler Owners, www.cibo.org
CSA – Canadian Standards Association, www.csa-international.org Clean Air Revival, Inc., www.burningissues.org/fact-sheet.htm
CTI – Cooling Technology Institute, www.cti.org
DOE – United States Department of Energy, www.doe.gov
DOL – Department of Labor, www.dol.gov
EJMA – Expansion Joint Manufacturers Association, www.ejma.org
EPA – United States Environmental Protection Administration, www.epa.gov
Environmental Research Institute – University College, Cork, www.ucc.ie/en/eri
FDA – United States Food and Drug Administration, www.fda.gov
FCI – Fluids Control Institute, www.fluidcontrolsinstitute.org
FM Global – Factory Mutual Global, www.fmglobal.com
FPS – Forest Products Society, www.forestprod.org
FPL – Forest Products Laboratory, www.fpl.fs.fed.us
GAMA – Gas Appliance Manufacturers Association, www.gamanet.org
GTI – Gas Technology Institute, www.gastechnology.org
HEI – Heat Exchange Institute, www.heatexchange.org
HI – Hydraulic Institute, www.hydraulicinstitute.com
HYDI – Hydronics Institute, www.gamanet.org
ICBO – International Council of Building Officials, www.icbo.org
ICC – International Code Council, www.iccsafe.org
IEC – International Electrotechnical Commission, www.iec.ch
IEEE – Institute of Electric and Electronic Engineers, www.ieee.org
IRI – Industrial Risk Insurers, www.industrialrisk.com
ISA – Instrumentation, Systems, and Automation Society, www.isa.org
Living Planet Index – www.livingplanetindex.org
ISO – International Organization for Standards, www.iso.org
MSS – Manufacturers Standardization Society – Valve and Fittings Industry, www.mss-hq.com
NAECA – National Appliance Energy Conservation Act, www.epa.gov
NBBPVIA – National Board of Boiler and Pressure Vessel Inspectors, www.nationalboard.net
NBIC – National Boiler Inspection Code, www.nationalboard.net
NCEES – National Council of Engineering Examiners and Surveyors, www.ncees.org
NEC – National Electrical Code, www.nfpa.org
NEMA – National Electric Manufacturers Association, www.nema.org
NESF – National Electrical Safety Foundation, www.nesf.org
NFPA – National Fire Protection Association, www.nfpa.org
NGC – National Gas Code, www.aga.org
NIH – National Institute of Health, www.nih.gov
NIOSH – National Institute for Occupational Safety and Health, www.cdc.gov/niosh

NPC – National Plumbing Code, www.bocai.org
NSC – National Safety Council, www.nsc.org
NSF – National Sanitation Foundation, www.nsf.org
NSIE – National Safety Information Exchange, www.nsie.org
OSHA – Occupational Safety and Health Administration, www.osha.gov
SAE – Society of Automotive Engineers, www.sae.org
SFPE – Society of Fire Protection Engineers, www.sfpe.org
SBCCI – Southern Building Code, www.sboci.org
SPIB – Southern Pine Inspection Bureau, www.spib.org
TAPPI – Technical Association of Pulp and Paper Industries – www.tappi.org
TECHSMART, www.techsmart.com/publications/list for standards and codes
TEMA – Tubular Exchanger Manufacturer's Association,www.tema.org
TPI – Timber Products Inspection, www.tpinspection.com
UL – Underwriters' Laboratory, Inc., www.ul.com
ULC – Canadian Underwriters' Laboratory, www.ulc.com
USFA – United States Fire Administration, www.usfa.fema.gov
USCFR – United States Code of Federal Regulations, www.gpoaccess.gov/cfr
WMMA – Wood Machinery Manufacturers of America, www.wmma.org

Organizations Connected to the Ag/Food Industry

AEHAP – Association of Environmental Health Academic Programs
AEM – Association of Equipment Manufacturers
AFDO – Association of Food & Drug Officials
AIFP – International Association of Food Protection
ANAB-ANSI-ASQ National Accreditation Board
ANSI – American National Standards Institute
ASABE – American Society of Agricultural and Biological Engineers
ASPE – American Society of Plumbing Engineers
ASPH – Association of Schools of Public Health
AWWA – American Water Works Association
CDC – Centers for Disease Control & Prevention
CFP – Conference of Food Protection
CNCA – The Certification & Accreditation Administration of the People's Republic of China
CIEH – Chartered Institute of Environmental Health
CIPHI – Canadian Institute of Public Health Inspectors
CODEX Alimentarius
CSA Group – Canadian Standards Association
DOC – US Department of Commerce
DOE – US Department of Energy
EPA – US Environmental Protection Agency
FDA – US Food and Drug Administration
FDLI – Food and Drug Law Institute

FSSP – Food Safety Services Providers
GFSI – Global Food Safety Initiative
GSA – US General Services Administration
IFPTI – International Food Protection Training Institute
IFT – Institute of Food Technologists
IUCN – International Union for Conservation of Nature
NEHA – Natural Environmental Health Association
NIST – National Institute of Standards and Technology
NRL – US Naval Research Laboratory
NSF – National Sanitary Foundation International, Ann Arbor, MI
OECD – Organization for Economic Co-operation and Development
Office of Energy Efficiency & Renewable Energy, www.energy.gov
OTA – Organic Trade Association
PFA – Phosphate Forum of the Americas
PFSE – Partnership for Food Safety Education
SCC – Standards Council of Canada
USDA – US Department of Agriculture
USCG – US Coast Guard
WATER – www.water.org
WCK – World Central Kitchen
WFPUSA – World Food Program USA
WHO – The World Health Organization

Recommended NFPA Safety Standards and Codes for Drying Systems, Partial List

The NFPA publishes a quarterly NFPA Catalog that lists their many safety publications.

The following NFPA codes can be ordered at 1-800-344-3555 or www.nfpacatalog.org.

Always determine what safety standards and codes are applicable to the region and country in which the drying system will be located in.

National Electrical Code
NEC Handbook
NFPA 70E – Electrical Safety in the Workplace
NFPA 13 – Installation of Sprinkler Systems
NFPA 30 – Flammable and Combustibles Liquids Codes
NFPA 31 – Installation of Oil-Burning Equipment
NFPA 54 – National Fuel Gas Code
NFPA 58 – LP-Gas Code
NFPA 68 – Venting of Deflagrations
NFPA 69 – Explosion Prevention Systems
NFPA 79 – Electrical Standard for Industrial Machinery

NFPA 86 – Ovens and Furnaces
NFPA 90A – Installation of Air-Conditioning and Ventilation Systems
NFPA 90B – Installation of Warm Air Heating and Air-Conditioning Systems

Appendix B – OEM Directory

The following lists do not include all the existing companies and websites for the following product groups. Additional lists of OEMs can be found by contacting wood/ag/food industry trade associations and groups in the country and geographic location where the drying system is to be located.

A. Drying System Equipment, Parts, and Service

A. W. STILES CONTRACTORS, www.awscontractors.com
AEROVENT, www.aerovent.com
AGRIDRYERS, www.agridryers.com
AGRIEXPO, www.agriexpo.online
ALIBABA, www.alibaba.com
ALAN ROSS MACHINERY CORP., www.rossmach.com
ALVAN FLANKA, grain dryers
AMERICAN WOOD DRYERS, INC., www.drykilns.com
AMF TECHNOLOGIES, www.amftechnologies.com
ANDRITZ, www.andritz.com
ARTI INTERNATIONAL, INC., www.artiusa.com
AUTOMATED MACHINE SYSTEMS, INC., www.amssystems.com
AWMV INDUSTRIAL PRODUCTS, www.awmv.com
BADGER INDUSTRIES, www.badgerind.com
BAILSCO BLADES & CASTINGS, INC., www.bailsco.com
BEN JONES MACHINERY, www.benjones.com
BETTER BUILT DRY KILNS, www.betterbuiltdrykilns.com
BES BOLLMAN BV, www.bes-bollman.nl
BIOMASS ENGINEERING and EQUIPMENT, www.biomassengineeringequipment.com
BIOLEXIS ABRA, manufacturers multifuel gasifier stoves
BOLDESIGNS INC., www.boldesignsinc.com
BRANDT GROUP of COMPANIES, www.brant.ca
CATHILD INC., www.cathild-inc.com
CFESA, www.sani-servant.com
CHINA – Chinese kiln/dryer manufacturers – A very large number of companies manufacture, install, and service ag/food/wood-drying systems with electronic controls. Go to www.made-in-china.com to locate companies that manufacture ag/food/wood-drying equipment.

One Chinese company that manufactures HF/Vacuum wood dryers is Shijiazhuang Meinianshun Technology Co. Ltd., www.m.made-in-china.com/company-mnswooddryer

One Chinese company that manufactures lumber kilns is Liyang Forwood Drying Equipment Co., Ltd., www.forwooddry.com

CHRISTIANSON ENTERPRISES, www.lumberdrykiln.com
CLARK FORKLIFTS, www.clark-forklift.com
COEN CO., INC., www.coen.com
COMBILIFT FORKLIFTS & TRUCKS, www.combilift.com
CONNECT2INDIA, www.CONNECT2INDIA.com
CUSTOM DRY KILN COMPANY LTD, www.drykiln.com
DELTECH KILN AND FURNACE DESIGN, LLC – www.bidgroup.ca
DURAN-RAUTE, www.raute.com
DURASTICK, www.durastick.com
DUSKE ENGINEERING COMPANY, www.duskeengineering.com
EBAC INDUSTRIAL PRODUCTS, INC., www.ebacusa.com
ENERGY UNLIMITED, INC., www.energyunlimited.com
EURODRYERS LIMITED, www.eurodryers.com
FOETH, www.foeth.com
FRANK CONTROLS LTD., www.frankcontrols.com
FRATELLI PEDROTTI, www.pedrottisrl.it
GEM ALLIED GROUP, www.gemalliedgroup.com
GERLINGER CARRIER, www.gerlingercarrier.com
GRENZEBACH AKI CORP., www.akidry.com
GSI COMMERCIAL TOWER DRYERS, www.grainsystems.com
GTS ENERGY, INC., www.gtsenergy.com
HEAT ENERGY SYSTEMS, LLC, www.heatenergysys.com
HEAT TEK, www.heattek.com
HEATWAVE USA, www.heatwave.com
HYSTER FORKLIFTS, www.hyster.com
IQS, www.iqsdirectory.com
IMRIE DRY KILN COMPANY, INC., www.deltanet2.com/imrie
JPW Industries, www.jpwindustries.com
KAB ENTERPRISES, www.kilnvent.com
KATOM RESTAURANT SYSTEMS, www.katom.com
KILN DRYING SYSTEMS & COMPONENTS, www.kdskilns.com
KILN-DIRECT, www.kiln-direct.com
KILN VENT, www.kilnvent.com
KOMLINE-SANDERSON, www.komline.com
LANLY, www.lanly.com
LIGNA MACHINERY INC., www.lignamachinery.com
LIGNOMAT USA, INC., www.lignomat.com
LINCOLN – food dryers & ovens – could not find a website address
NORTHLAND KILNS, INC., www.northlandkilns.com
NORTHWEST DRYER & MACHINERY CO., www.northwestdryer.com

MACHINIO, www.machimio.com
Mc LAREN INDUSTRIES, www.mclarenindustries.com
MEC COMPANY, www.m-e-c.com
MECMAR GROUP, www.mecmargroup.com
MET-CHEM Inc., www.metchem.com
MICROWAVE TECHNOLOGY, www.sairem.com
MUHLBOCK KILNS, www.muehlboeck.com
MULTI-WING, www.multi-wing.com
NOVA DRY KILNS, www.novadrykiln.com
NYLE CORPORATION, www.nyle.com
OPICO, www.opico.co.uk
P C SPECIALTIES, www.pcspecialties.com
PERMATHERM, www.permatherm.com
RESOLVED ANALYTICS, www.resolvedanalytics.com
SHIVVERS GRAIN DRYERS, www.shivvers.com
SIGMAE, www.sigmaequipment.com
SII DRY KILNS, www.siidrykilns.com
SMITHCO MANUFACTURING, INC., www.smithcomfg.com
STILES MACHINERY, www.stilesmachinery.com
TAYLOR MACHINE WORKS, www.taylorforklifts.com
TEXAS REFINERY CORP., www.texasrefinerycoatings.com
THE ONIX CORP., www.theonixcorp.com
TIMBER PRO KILN, www.timber-pro.com
TURATTI GROUP, www.turatti.com
TUFF-STIK, www.redwoodplastics.com
UNIVERSAL DOOR CARRIER, INC., www.universaldoor.com
US NATURAL RESOURCES, www.usnr.com
UNIVERSAL DRYING SYSTEMS, www.universaldryingsystems.com
UZELAC DRYING SYSTEMS, www.uzelacind.com
VACUTHERM, www.vacutherm.com
VALUTEC WOOD DRYERS, www.valutec.com
VULCAN DRYING SYSTEMS, www.vulcandryingsystems.com
WELLONS INC., www.wellons.com
WDE-MASPEL, www.wde-maspel.com
WINGFAN, www.wingfan.com
WOOD DRYING EQUIPMENT, www.wooddryingequipment.com
WOODLANDPARTS, www.woodlandinc.com
WOOD-MIZER PRODUCTS, INC., www.woodmizer.com

B. Moisture Sensors, Instruments, & Control Systems

The following list of companies is far from complete because of the many ways moisture measurement and control technologies can be integrated into modern process control systems.

ACCUDRY, www.accudry.com
ACCUMASTER, www.lowes.com
DELMHORST INSTRUMENT CO., www.delmhorst.com
DOVEY CORP., www.doveyco.com
DRYING TECHNOLOGY, INC., www.moisturecontrols.com
ENERCORP INSTRUMENTS LTD, www.enercorp.com
ESI ENVIRONMENTAL SERVICES, INC, m.facebook.com/environmental services
EXTECH, www.extech.com
FINNA SENSORS, www.finnasensors.com
GRAYSTONE ENERGY SYSTEMS INC, www.graystoneenergy.com
KILN SCOUT, www.finnasensors.com
LIGNOMAT USA LTD., www.lignomat.com
LUCIDYNE TECHNOLOGIES, www.lucidyne.com
MERLIN, www.merlinhvac.com
MICROTEC, www.microtechservices.com
MOISTTECH CORP., www.moisttech.com
NOVATECH CONTROLS INC., www.novatech.com.au
PROTIMETER, www.protimeter.com
SCALE-TRON LTD, www.scaletron.com
TRI-MET INSTRUMENTS, LTD, phone 204-774-8889
VAB SOLUTION INC., www.vab-solutions.com
VOGEL, www.vogel-germany.de
WAGNER ELECTRONICS INC., www.wagnerelectronics.com

C. Steam Boilers, Hot Liquid Systems and Furnaces

The American Boiler Manufacturers Association (ABMA) has the complete list of boiler manufacturers in the United States, www.abma.org.

The following is a partial list of manufacturers:

Auto Flame Engineering, LTD, www.autoflame.com
Babcock and Wilcox Company, www.babcock.com
Burnham Commercial, www.burnham.com
Cleaver-Brooks, Inc., www.cleaver-brooks.com
COE Manufacturing Company, www.caenewnes.com
Coen Company, Inc., www.coen.com
Deltak, LLC, www.deltak.com
Detroit-Stoker Company, www.detroitstoker.com
Ducon Technologies, Inc., www.ducon.com
EASCO Boiler Corporation, www.easco.com
Eclipse, Inc., www.eclipsenet.com
Elite Combustion Solutions, www.ecombustion.com
Energy Unlimited, Inc., www.energyunlimited.com

English Boiler and Tube, www.englishboiler.com
Foster Wheeler, www.fosterwheeler.com
Gordon-Piatt Energy Group, www.gordon-piatt.com
GTS Energy, Inc., www.gtsenergy.com
Hurst Boiler and Welding, www.hurstboiler.com
Indeck-Keystone, www.indeck-keystone.com
John Zink, www.johnzink.com
Johnson Boiler Company, www.johnstonboiler.com
KDS, www.kdskilns.com
Kipper & Sons Fabricators, Inc., 1-253-856-2625
Maxon Corporation Industrial Combustion, www.maxoncorp.com
McConnell Sales & Engineering Corporation, www.mcconnellsales.com
Nebraska Boiler Company, www.neboiler.com
North American Manufacturing Company, Ltd., www.namfg.com
Power Flame Inc., www.powerflame.com
Pyronics Inc., www.pyronics.com
Riley-Stoker Corporation, www.rileystoker.com
Sellers Engineering Company, www.sellersengineerng.com
Superior Boiler Works Inc., www.superiorboiler.com
The Boileroom Group, www.boileroomgroup.com
USNR, www.usnr.com
Webster Combustion, www.webstercombustion.com
Wellons Inc., www.wellons.com
Weishaupt North America, www.weishaupt-corp.com
York-Shipley Global, www.aesystech.com

D. Process Controls and Instrumentation

See the ISA Directory of Automation and Control at www.isadirectory.org. The number of companies in this category is far too large to list. Many manufacturers of process control equipment have dedicated design-built engineering firms and can recommend a firm in your area. When designing process control systems for industrial thermal systems, always conduct a fault-tree analysis before installing the system. Records of final process control parts lists, configurations, software, and tuning criteria must be stored in a safe fire-proof filing system.

Appendix C – Recommended Reading and Information Resources

From Basic Drying Research to Final Industrial Drying Systems

Systems of Measurements – Different Countries Use Different Measurements of Unit

Depending on which country the industrial drying system is located, the system of measurement (units) will vary. See Wikipedia "System of Measurements" for a complete history and current practices of the global systems of measurements such as Imperial and US Existing, Metric, and SI systems. Numerous conversion tools and tables exist on the Internet for converting measurements.

Internet Resources for Industrial Drying Technologies

Numerous sources of information about drying technologies, energy conversion processes, and research articles can be found on the Internet. The key to finding what you are looking for is to enter the exact correct technical term into any search engine. However, because the Internet is not regulated for the accuracy of its content, any information seen on the Internet must be viewed with caution. This is especially true when seeking reliable data on energy efficiency.

Universities, Schools, Institutes, Science, Engineering, and Industrial Organizations

An endless amount of information is available on the Internet some of which is peer-reviewed, some of which is not because some organizations represent special interests.

Industrial Search Engines

Frasers is Canada's industrial search engine and has a long list of the different types of industrial dryers including the names of OEMs. Go to www.frasersdirectory.com and enter the word "dryers," and a list of 81 different types of dryers will appear. Then, refine a specific type of dryer search by entering additional words.

Videos of Wood, Agricultural, and Food Industrial Dryers in Operation

Many OEMs now have informative videos on the Internet that show their products in operation.

ASHRAE standards index – www.techsmart.com/publications/list when discussing standards and codes

ASME Publications – Go to www.asme.org/catalog

Industrial Energy Systems, by R.E. Putman
Process Piping, The Complete Guide to ASME B31.3, Second Edition, by C. Becht, IV Heat Exchanger Engineering Techniques, by M.J. Nee

ASABE Publications – Go to www.asabe.org

Go to the ASABE Online Technical Library for the latest issue of "Applied Engineering in Agriculture."

Health Effects of Wood Drying and Burning

US EPA
US Forest Products Laboratory, www.fpl.fs.fed.us.gov
Clean Air Revival, Inc., www.burningissues.org/fact-sheet.htm

Pollution Control Technology

Pollution Equipment News, www.rimbach.com

Fire Protection

National Fire Protection Association, www.nfpa.org
Society of Fire Protection Engineers, www.sfpe.org

Licensed Engineers www.ncees.org/licensure/licensing_boards

Click on a state to locate engineering boards for that state. Each state, and country, has different laws for the legal definition of what engineering is, and the licensing requirements for engineers. Also expect to see a wide variance into what is legally defined to be and not to be engineering. Also expect to see many local engineering boards offer different opinions about what their laws state about what is and what is not considered to be engineering.

Lumber Drying, Lumber Dryers, and Wood Technology

Forest Products Research Society publications
Dry Kiln Operator's Manual Handbook #188

Air Drying of Lumber – A Guide to Industry Practices Handbook #402
Drying Eastern Hardwood Lumber Handbook #528
Quality Drying: The Key to Profitable Manufacturing – 2002
Lumber Drying Sourcebook: 40 Years of Practical Experience – 1998
Dry Kiln Schedules for Commercial Woods, Temperate and Tropical – 1993
Wood Handbook: Wood as an Engineering Material – 1974
Utilization of the Southern Pines – Peter Koch – 1972
Effective Predryer Operations
Bibliography for the Drying of Forest Products Materials
Glossary of Terms Related to the Drying of Wood

Other Lumber Drying Book Sources
The Drying of Wood, C. Arthur Hart, School of Forestry, N.C. State University at Raleigh
Kiln Drying of Western Canadian Lumber, Bramhall and Wellwood
Kiln Drying of Western Softwoods, Edwin Knight
Principles of Lumber Drying and Practical Advice for Dry Kiln Operators
Solar Cycle Lumber Kilns
Vacuum Kiln Drying for Wood Workers

Drying Books Involving Advanced and/or Modern Drying Research

The Book "**Modern Drying Technologies**" – An excellent five-volume handbook of current advanced drying modeling, research, and applications. This is a must-have set of five books for any person serious about advancing all types of industrial drying systems.

NOTE: The terms advanced and modern are often both misunderstood or misused if people do not understand the scope of what industrial technologies currently exists and what research is ongoing. This is one of many reasons this industrial drying handbook was written and why I highly recommend this series of books.

Drying Books Available from Springer Series in Wood Science

Authors; Roger B Keey, Timothy A. G. Langrish, John C. F. Walker
Kiln-Drying of Lumber
The Structure of Wood
Wood–Water Relationships
Evaporation and Humidification
Wood-Drying Kinetics
Moisture Diffusion
Multiple-Mechanism Models
Lumber Quality
Stress and Strain Behavior

Airflow and Convection
Kiln Operation
Pretreatment of Green Lumber
Less-Common Drying Methods

Drying Books Available from Taylor and Francis Publishing

"Drying Technology" covers a large list of drying topics for different industries.

American Wood Products Associations and Organizations, Partial List

Directory of Wood Products Industry Associations, American Wood Council
APA – The Engineered Wood Association
American Forest and Paper Association
Forest Products Laboratory
Forest Products Research Society
International Society of Wood Science and Technology
National Hardwood Lumber Association
National Lumber and Building Materials Dealers Association
Northeastern Lumber Manufacturers Association
Softwood Lumber Board
Southern Forest Products Association
Southeastern Lumber Manufacturers Association
Timberline Magazine Editor
Western Wood Products Association

Food Science and Technology

Food Process Engineering and Technology – A volume in Food Science and Technology, 2009, Author: Zeki Berk – A very good resource about food drying systems.

Food Manufacturing Trends at the Global Scale with Significant Health, Safety, and Social Risks

See the Internet documentary: "**Food, Inc.**" produced by Robert Kenner, and the book: "**Fast Food Nation**" by Eric Schlosser about the enormous political and economic power that large multinational food companies now have, and the associated costs and health and safety risks to society. Many scientists and experts now predict the current "modern" industrial food system will create levels of health care costs sufficient to bankrupt the United States.

Combustion Technology, Environmental, and Safety Codes and Standards

North American Combustion Handbook, North American Manufacturing Company, Ltd.
The NFPA Fire Code
The OSHA codes
The EPA codes

Process Controls and Automation

ISA Directory of Automation and Control, www.isadirectory.org
PC Systems Handbook for Scientists and Engineers, www.cyberresearch.com
ASABE also has a section "Information Technology, Sensors, & Control Systems" in their publication "Applied Engineering in Agriculture".

Insurance Resources

www.americafirst-ins.com/corporate/resources/resources_safe.cfm has a list of sites for safety information, standards and codes.

Business Ethics, Moral Attitudes, Psychopathy, and Social Responsibility

There are numerous well-researched, science-based videos on the Internet, and YouTube that discuss these issues. I suggest the following videos as a starting point to learn about the science and research behind moral reasoning and the high costs to society when people do bad things:

"The Neuroscience of Real-Life Monsters: Psychopaths, CEOs and Politicians" by Octhuio Choi, MD, PhD, and **"Living in the Future's Past"** by Jeff Bridges.

Scientific Frauds, Misinformation, and Disinformation

There are numerous well-documented conduct (frauds) of self-serving people, including many scientists, making claims of benefits to humanity when in fact no such science-based benefits exist. This crime is rampant in the age of the Internet, and any system funded by government agencies. Human greed is the driving vector behind all scientific frauds.

Go to www.backreaction.blogspot.com to watch German physicist Sabine Hossenfelder discuss the long list of current frauds now occurring in science. Many trillions of global dollars are wasted annually on frauds in science.

Appendix D – Needed R&D Projects in Industrial Drying

The Need for Global-Energy-Efficiency; Studies, Design-Guidelines, Standards, and Practices for Industrial Wood-Drying & Wood-Energy Systems in The Age of Global Warming

By
Ed Smith, PE, MSME
J E Smith Company (JES) – The Lufkin R&D Center
PO Box 428, Terrell, Texas 75160
Email: ed.jes4278@outlook.com, Phone: 903-257-5053
November 30, 2020

Table of Contents

Project Title
The Need for Global-Energy-Efficiency; Studies, Design-Guidelines, Standards, and Practices for Industrial Wood-Drying & Wood-Energy Systems in the Age of Global Warming

Objective of the Study
Improve Energy-Efficiency in Lumber-Drying Operations by
Measuring energy efficiency of existing lumber-drying systems
Measuring energy efficiency of existing lumber dry kilns
Examine Current Drying System Configurations, Design Guidelines and Standards

Mill Data to be Collected
Determine the number, location, and type drying systems to be examined
Document the wood species being examined
Document the wood product being produced
Document green lumber dimensions
Document the mill's geographic elevation and average ambient temperature
Document total electrical energy usage/pound of water removed during drying
Document total heat energy usage/pound of water removed during drying
Document existing drying system configurations
Document existing kiln equipment configurations
Document existing drying temperature schedules
Document current lumber drying rates
Document initial average lumber moisture content
Document final average lumber moisture content
Document average wood specific gravity

Methodology for Collecting Field Data

Prepare project to do list after receiving funding
Choose a project manager for the United States
Choose project managers in each participating country
Get global agreements on project scope and field data to be collected
Get global agreements on drying energy terms, definitions, and formulas
Prepare field data collection forms (in English)
Prepare instructions for collecting field data (in English)
Begin collecting field data, including a start and completion deadline

Publication of Final Report of Findings by
Forest Products Laboratory & the Lufkin R&D Center
Complete analysis of data
Prepare draft report of findings
Conduct reviews of draft report of findings
Publish the report of findings

Project Funding
Government entities only to avoid conflicts of interests

Background and Discussion

Introduction

Although the author is experienced with industrial wood-drying processes and equipment, and wood-energy utilization processes and equipment (see the attached newsletter about the Lufkin R&D Center which includes the author's industrial CV), the following initial survey was focused on seeing what new "modern" lumber drying systems in the United States looked like.

2019–2020 US Survey

Between January 2019 and November 2020, the author conducted a survey of US industrial lumber drying and wood-energy utilization systems. The mills were a mix of hardwood and softwood operations across the country. This survey included mill site visits and interviews with mill managers, mill operations managers, lumber kiln operators, sales and service people employed with lumber kiln manufacturing companies, wood-furnace manufacturers, steam boiler manufacturers, and engineers and process-control techs working in the forest-products industry. The purpose of this work was to both see recently installed equipment and gather information about wood-drying operations.

Kiln Types and Survey Findings

Hardwood Mills

Other than hardwood mills gradually upgrading or replacing their existing kilns with new all-aluminum buildings to solve the inherent problems of acid-laden wood sap corroding steel kiln structures, the newer kilns were fitted with improved in-kiln temperature and lumber moisture sensing probes and programmable temperature and humidity control systems with remote MIS capability.

Of the hardwood kilns now in the United States, most are the same indirect-fired (heated) side-loader batch type that has been the industry standard for the last 75 years. Mixed in with these kilns were conventional predryers, forced-air dryers, low-temperature heat-pump dehumidification dryers, a few vacuum + RF kilns, and some dedicated steaming chambers for pest or color treatments.

Thus, other than digital control-reporting technologies and changes in drying and treatment schedules, there is not any significant drying technology news to report in the hardwood kiln industry.

Intrinsic Low Energy Efficiencies in Hardwood Lumber Drying

Also true about hardwood kilns, other than the dehumidification dryers, is their extremely low energy efficiency (high BTU consumption/pound of water removed in drying) due to the slow drying schedules, large building losses, spray-humidification processes, equalization and conditioning activities, and the inherent wasteful nature of low-temperature venting psychometrics. And, the further north one travels, the worse the energy problem is.

Softwood Mills

Batch-Type Lumber Kilns

Prior to about 2015, virtually all softwood kilns in the United States were a mix of side-loader batch kilns for mostly small lumber operations and wood-treating plants, and track-type batch kilns for both low-temp and the prevalent high-temp operations found in the larger lumber mills.

Most softwood batch kilns during the last 40 years have gone through minimal changes in basic designs.

However, most of these kilns have adopted digital control technology. Most of the old analog kiln temperature control systems have been replaced with digital controls with remote MIS capability. Some convection kilns now have adaptive-controlled, variable-frequency drives for their internal fans.

Progressive Lumber Kilns, Jogging-Batch Lumber Kilns, and (True) Continuous Lumber Kilns

These three kiln designs are based on the thermodynamic continuity principle that predicts higher drying production rates and energy efficiency can be achieved in continuous operations when compared to batch operations. But there are limits to what these kilns can do and how they should be applied in industrial lumber manufacturing operations.

This is also an area where **misinformation** abounds about what is the **correct name** to call these kilns.

The following discussion has been included to explain what has occurred in the US wood-products industry during the last century.

The **progressive lumber kilns** were common in the United States during the early part of the century and are still in use in Europe. These early single-track-type kilns were all poorly engineered steam-heated or natural draft types that created enormous amounts of lumber degrade due to their inability to prevent over drying. Add to this problem was that many of the older sawmills produced a mix of softwood and hardwood products with vastly different drying challenges. This situation led to disastrous levels of degrading occurring during the drying process. Thus, the progressive lumber kilns did not last long and were eventually replaced by air-drying yards and batch-type kilns. There are no known progressive kilns in the United States today. Furthermore, the reason they were called progressive was the lack of heating capacity in these kilns that caused the temperature inside the kilns to increase the further the drying packages moved through the long kiln buildings. Thus, since the kiln temperatures increased in a progressive fashion, so did the belief that this was helping to keep lumber degrade to a minimum.

The **jogging-batch lumber kilns** (erroneously referred to as continuous kilns) appeared in large numbers in the United States around 2015. The first of these double-track kilns used the counter-flowing of lumber packages through the kiln building to allow lumber preheating and equalization of moisture to occur at the same time. The technical term for this material flow path is **bidirectional.** The bidirectional double-track jogging-batch kiln is also commonly called a **dual-path** kiln. Then, several years later the **unidirectional double-track jogging-batch lumber kiln** appeared in the United States. In this kiln, the trams carrying the lumber packages through the kiln both traveled in the same direction. Both types of these kilns are configured to provide lumber preheating (vent-energy recovery) and lumber moisture content equalization capability. And too, because these kilns should only be used for thin narrow easy-to-dry softwood lumber, they are not designed for or intended to provide any significant amount of **conditioning** for limiting wood fiber stresses. However, both types of these kilns are more heat-energy-efficient simply because their designs promote counter-flow, vent-energy-recovery schemes. Furthermore, there are numerous ways of configuring the energy-recovery schemes in these and the following (true) continuous kilns.

The **(true) continuous lumber kiln** has been used in many configurations in Europe for over fifty years. They are true continuous kilns because of one reason.

All of the trams or roll cases carrying the lumber packages through the kiln move in a slow continuous fashion instead of the lumber package jogging-jerking fashion seen in the US kilns.

There is also another reason for this unique distinction from jogging-batch kilns. All jogging-batch kilns employ a certain amount of **tram resting time** between each jog cycle for loading green packages onto, or unloading dry packages off the trams. The **period** of the jogging cycle is the total elapsed time between each start of the jogs. This period can vary from minutes for thin softwoods to days depending on kiln design, species, package lengths and thicknesses of the product being dried. If heat energy for the kiln is limited or reduced, the jogging period will increase accordingly. Properly designed kiln control systems should include the ability to monitor energy-usage rate/pound of water being removed and automatically adjust the jogging period. If both board thickness and board width are also monitored by the control system, the accuracy of the jogging process can be improved further.

Unlike jogging-batch kilns, true continuous kilns have to be both fed packages of stickered lumber and unloaded in a continuous fashion such that the smooth continuous speed of the kiln's package handling equipment is not affected.

True continuous kilns can also employ preheating (vent-energy-recovery) and wood fiber equalization configuration schemes as well as reverse, counter-flow or zig zag air-flow capability, multizone capability, and guillotine zone gates, baffles, and dampers.

Sidewinder Lumber Kilns

Sidewinder lumber kilns are different from conventional batch-type kilns such as the side-loader or track-type kilns. Sidewinders are batch kilns that have all of their internal fans and heating equipment located at ground level. These kilns are usually loaded by trams running on tracks but some side-loader configurations have been used.

Sidewinders have been used successfully in many hardwood and softwood drying operations around the world. They are very low cost, easy to maintain, very durable with some lasting over 50 years, but limited in holding and thus drying capacity. Thus, sidewinders are only found in small drying operations.

Although the number of sidewinder kilns in the United States is not known by the author, these kilns are common in Europe and Africa because of their low construction cost, reliability, and simplicity.

Types of Industrial Lumber Drying Systems

Successful lumber drying is a **system** that requires the following steps:

Managing log inventories to minimize wide variances in average log moisture content

Precision sawing of green board thickness (a critical variable)
Trimming green sawed boards for defects and length
Transporting green trimmed boards to a green sorter
Sorting green boards by; species, thickness, widths, lengths, and grade into separate bundles
Transporting the sorted green bundles of lumber to an inventory-storage area
Managing the sorted green bundle inventory area
Transporting sorted green bundles of lumber to a green stacker
Stacking the green lumber into packages with drying stickers
Transporting the sorted green stacked packages to a green inventory area
Managing the sorted green package inventory area
Transporting the green lumber packages to the infeed area of the kilns
Loading the kilns with green stacked stickered packages
Drying the green lumber with appropriate conditions for a target moisture content
Adjusting kiln operations to control final lumber moisture content
Unloading the dried stacked stickered lumber packages from the kiln
Transporting the dried lumber packages to the dry inventory storage area
Managing the sorted (weather-protected) dried lumber package inventory area
Transporting the dried stacked stickered lumber packages to an unstacker
Separating the dried lumber boards from the kiln stickers
Transporting layers of boards to the infeed of a dry planer operation
Collecting the kiln stacking stickers into bundles
Transporting the kiln sticker bundles back to the green lumber stacker
Feeding the kiln sticker bundles into the green stacker operation
Measuring individual board moisture content at the dry planer facility
Reporting dry-end board moisture content to kiln operators and mill management
Adjusting drying system operations to minimize variances in dry-end board moisture content
Adoption of re-dry handling systems for certain operations

The above lumber, sticker, and package handling system involves roll-cases, cross-transfer chains, fork trucks, package straddle carriers, hoisting equipment, conveyor gates, diversion tipples, green package transfer cars, buggies, kiln carts and trams, dry package transfer cars, lumber slides, and belted conveyors.

The lumber drying heat energy sources involve electricity, natural gas, fuel oils, hogged green fuels, green sawdust, dry shavings, heat pumps, process heat exchangers, solar, and wind. The kilns can be either direct- or indirect-fired configurations. Energy saving-recovery systems may or may not exist.

Modern lumber manufacturing operations may employ warehouse-inventory-management software to both track and report to mill management the flow rates and sorted lumber inventory levels throughout the entire drying system. Technologies exist to fully automate many areas of lumber drying systems.

Virtually every lumber manufacturing facility on the planet has a different quantity and combination of the above handling, drying, and automation methods. The

final configuration is **the** lumber drying system as explained above. The lumber kilns are only **one part** of a lumber drying system.

Global Energy Consumption and Challenges in all Methods of Drying Lumber

This section involves discussions about required heat- and electrical energy.

Required Heat Energy for Drying Lumber

One – When a pound of water evaporates from a bowl sitting on a table at sea level and 70 F, that pound of water requires about 1000 BTUs of heat energy.

Two – When a pound of water evaporates from a wet board sitting on that same table at sea level and 70°F, the required energy is a bit more due to molecular attractive forces inside the cells of the wood fiber. It takes a little more energy to break the bond.

Three – When a pound of water evaporates from that same board resting in a lumber air-drying yard, the same amount of energy is needed as the previous case (two). However, there is additional energy involved in getting that pound of water dried from that board. It required additional energy due to the handling and storage activities of that board after it left the green trimmer located at the sawmill and another additional amount of energy for handling and storage activities of that board to get it to the infeed point at the dry-end planer machine. Thus, the total drying energy/pound of water removed has increased above the original 1000 BTU. Although most of the energy was provided by Mother Nature delivering free air around the surfaces of the board, some additional **gross energy** was incurred due to board handling and storage activities. In this example, it was both the fuel for the fork truck engine and all of the additional gross-energy costs for manufacturing, shipping, and maintaining that fork truck as well as all associated gross energy costs for the training and work activities provided by the fork truck drivers and the people who maintain the fork trucks.

In lumber manufacturing operations, the collective additional gross energy can be a substantial portion of the total energy requirements for lumber drying. Then add to this figure the total gross energy associated with the heat-energy supply system used to provide heat energy to a lumber kiln and the total energy number gets much larger.

This is the reason that **competent-neutral-objective** energy studies of any industrial process must examine the entire picture and not erroneously focus on just one part of that process.

Thus, researchers must state in exact details the scope of the research and what other contributing factors in the process were both included and excluded during any study about energy.

Failure to follow scientific methods will lead to faulty conclusions, spin, and the organized spreading of misinformation in both media and all branches of society.

These very common activities create significant problems for the entire human race now confronted with the task of combating global warming.

The following is a partial list of wood-drying terms, definitions, and formulas:

Gross energy – The sum total of all forms of energy consumed by a process including all of its support systems.

Drying system gross energy – The sum total of all forms of energy consumed in a drying system.

Drying system – The sum total of all equipment, processes and activities to both dry a product and support the operational needs of the drying process. For lumber manufacturing, this starts at the green trimmer located at the sawmill and follows the flow path of the lumber until it reaches the infeed table at the dry end planer operation.

Drying system gross unit energy rating – The gross energy usage based on total energy consumed / pound of water removed in the drying process.

Drying kiln – An enclosed structure in which the drying of lumber occurs.

Drying kiln unit energy rating – The energy usage based on the total energy consumed **inside** the kiln / pound of water removed by the kiln.

Gross drying system carbon footprint – The total amount of carbon dioxide emissions produced due to drying system gross energy produced.

Total Drying Unit Energy (TDUE) The sum total of all energy consumed/pound of water removed in drying by an industrial wood-drying system.

Benchmark Drying Energy (BDE)

In this research proposal, we will use 1000 BTUs/pound of water removed. However, depending on geographic location, atmospheric pressure determined by mill elevation, and the average ambient temperature, BDE will change. Site-specific BDE must be reported for each drying system examined.

Unit Drying Energy Losses (UDEL) – Based on BTU/Pound of Water Removed

For heating the wood fiber during drying
For thawing out ice in frozen wood
For heating the liquid water in the wood
For heating air for wet bulb control (the venting process)
For superheating the water vapor leaving the kiln
For humidification spraying (both steam and water)
For heat-energy losses through buildings, doors, and foundations
For net electrical energy losses due to fans, blowers, and conveyors, etc.
For all additional energy losses due to heating and utility support systems
These losses can be a significant portion of the total energy losses

Steam boiler losses
Steam handling losses
Condensate system losses
Hot-oil system losses
Hot-water system losses
Furnace/combustion system losses

Thus,

$$\mathbf{TDUE = BDE + UDEL}$$

And wood-drying system Rated Energy Efficiency **(REE)** is defined as:

$$\begin{aligned}\mathbf{REE} &= \mathbf{(BDE / TDUE) \times 100\%} \\ &= \mathbf{(1000 / TDUE) \times 100\%}\end{aligned}$$

Note The above calculations are all based on energy studies based on the pounds of water removed in drying systems. Some former drying energy studies have used data based on /board feet dried. This is a flawed method because the initial (green) moisture content and specific gravity of lumber can vary significantly over time in industrial drying systems.

The **REE** of industrial wood-drying systems involves numerous factors dependent on mill production rates and the design configurations used for the entire drying system. The examples shown below are for demonstration purposes only. Once this research project is completed, the range of REEs of mill operations around the globe will be published.

Examples:	**REE**
For water evaporating from a bowl sitting on a table	!00%
For air drying of lumber including handling energy	85–98%
For drying lumber in a high temp direct-fired kiln	60–70%
For drying lumber in a high temp steam-heated kiln	40–50%
For drying lumber in an accelerated steam-heated kiln	35–45%
For drying refractory hardwoods in steam-heated kilns	15–20%
For drying refractory hardwoods in extreme cold weather	10–15%

Benefits of Having a Simple Term for Lumber Drying System Efficiency Studies

Most people do not have experience with either wood-drying science or lumber drying systems.

However, most people can relate to watching a pound of water evaporating in a heated pan.

The simple term **REE** gives these people an idea of how efficient different methods of drying lumber are.

Once governments and industry around the planet have access to reliable data on how efficient or inefficient their lumber drying systems are, attention will then be focused on which parts of industrial drying systems should receive additional research funding for improving energy efficiency.

Significant Issues in Existing Industrial Lumber Drying Systems

One – The absence of globally-supported research, minimum energy-efficiency guidelines, or design standards for wood-drying systems and or stand-alone dryers in the age of global warming.

Two – The absence of ANSI minimum equipment or process design efficiency guidelines or standards for wood-drying systems in the age of global warming. There are **six critical** engineering design issues in need of multigovernment-funded research for reducing electrical energy consumption in industrial convection-type lumber drying systems on a global scale.

- One – An advanced high-efficiency reversible axial propeller kiln fan design with application software.
- Two – A high-efficiency kiln heat exchanger for indirect-fired applications with application software.
- Three – High-efficiency convection-kiln-plenum design software.
- Four – Kiln drying-energy-configuration design software.
- Five – Lumber-package air-flow-wood-energy-dynamics design software.
- Six – A universal kiln-control software "package" for all types of convection kilns.

Three – The adoption of short-term economic approaches by equipment manufacturers and consumers when both designing and purchasing new equipment, and without any significant consideration of the impact global-warming will have on future industrial processes.

Examples

In one "modern" large southern pine lumber manufacturing facility, it was found that the electrical consumption (of the internal air-circulation fans)/pound of water removed during drying was over four times what it should have been.

Human Health vs. Benefit: There exist today established industry-wide significant worker safety and health issues in numerous softwood lumber mills using direct-fired, wood-fueled, wood-drying kilns.

--------**End of Proposal**--------

Biography of the Author

Executive Biography

The majority of Ed Smith's 60+ years of progressive technical and engineering experience were gained through onsite real-world problem-solving coupled with basic research and development of equipment in the forest products industry, lumber manufacturing, plywood manufacturing, and biomass and fossil-fueled cogeneration and large electrical-power utility systems.

Ed's tenacity, strong work ethic, and exemplary operational skills are keys to his ability to lead vision development, complex industrial process development, and system implementations across diverse business segments, engineering, manufacturing, and operation teams in centralized and decentralized organizations.

A leader known for establishing positive solutions, Ed can quickly connect the dots in complex technical challenges with a wide range of changing variables. From large green-field projects to simple equipment upgrades, Ed has the ability to lead his clients in the right direction. His ability to create innovative solutions to difficult technical challenges is his specialty.

He is also a role model in advancing the science of industrial wood drying. Ed led the industry when in the early 1970s he designed continuous lumber kilns now common in the softwood industry, developed high-performance cross-shaft kiln fans now used around the globe, developed hyperperformance southern-pine lumber kilns, and developed numerous computer models for simulating the complex mass-energy processes involved in all types of convection-type lumber drying.

Ed holds a Bachelor of Science in Mechanical Engineering from Mississippi State University, a Master of Science in Mechanical Engineering from Memphis State University, and an Associate Degree in Criminal Justice from El Centro College in Dallas, Texas, and is a licensed professional engineer in Texas.

Ed Smith, MSME
J E Smith Company (JES), PO Box 428, Terrell, Texas 75160

J E 'Ed' Smith, *Industrial Drying Systems*,
https://doi.org/10.1007/978-3-031-31863-4

Phone: 903-257-5053, Email: ed.jes4278@outlook.com

Education: B.S. in Mechanical Engineering
Mississippi State University, 1968
M.S. in Mechanical Engineering
Memphis State University, 1978
A.A.S. in Criminal Justice
Dallas County Community College, Dallas, Texas, 2001

Experience Overview

Work experience involves decades of research and development, design, manufacturing, and construction management. Projects include wood-products manufacturing facilities, cogeneration systems, energy systems, wood-drying systems, small and large construction projects, accident prevention and reconstruction, fire-and-explosion prevention, safety audits and training, and environmental audits. Forensic projects include vehicle-accident reconstruction, origin-and-cause studies, steam boiler failures and studies, machinery failures, electrical accidents, industrial roof examinations, industrial accidents and fatalities, personal-injury-death civil investigations, arson investigations, and industrial-sabotage investigations. **Systemic-Fraud Studies:** Since 1995, he conducted personal research and modeling in the psychosocial factors causing systemic fraud in the United States. Most of this work was fraud committed in the insurance business.

Articles, Publications, Presentations, and TV Appearances

The Coming Deaths of Science, Reason, and Truth: Case Studies in Systemic-Fraud Dynamics, published 2023

Lumber Drying Systems Configurations, Terms, Energy, Safety and Worker-Health and the Scientific Method in the Age of Global Warming, published 2021

The Need for Global-Energy-Efficiency: Studies, Design-Guidelines, and Practices for Industrial Wood-Drying and Wood-Energy Systems in The Age of Global Warming, published 2020

East Texas Investigation of Propaganda and Human-Control Systems Developed by; Psychologists, Mathematicians, and Computer Scientists for Social Media Platforms, published 2020

The Engineering Licensing Racket, published 2020

2020 Batch and Continuous Lumber Kiln Design, Systems and Analysis Study Manual, published 2020

200 Trillion Dollars Stolen and Plenty More to Come: A Personal Survival Manual in the Age of Greed, pub. 2020

The Old Ford Tractor: An Engineer Traveled to the Future to Save the World, and then he took a trip to Washington., published 2020.

The National Academy of Sciences Bows Down: The Book World-Renowned Scientists Wanted Witten but Those Scientists Could Not, published in 2019.

Proposal to Congress for Registering Engineers at the National Level, published 2019

331 Million Reasons: Why scientists and engineers should run our country instead of special-interest groups, published in 2019.

The Pros, Cons and History of Batch and Continuous Lumber Dry Kilns, published in 2014
Fundamentals of Wood Drying & Energy Conversion Technologies, book published in 2005
Dry Kiln Design Manual, book published in 2005
Insurance Claims and Appliance Fire Investigations, AHAM Product Liability Seminar, 2005
Mothers Against Drunk Driving, Spring, 2004, Vol. #20, Issue #1 – "The Effects of Alcohol on Personality, Mental Cognition and Driving Ability While Behind the Wheel"
McCuistion TV, The Multi-Billion Dollar Cost of Insurance Fraud, 2003
McCuistion TV, Insurance Costs, Regulations and Consumer Education, 2003
The Trillion Dollar Insurance Crook: A Texas Forensic Engineer Exposes Massive Fraud, Collusion and Waste in American Insurance, book published in 2003
A Feasibility Study for a Central Property Database for the State of Texas, 2001
Attitudes About Crime in Dallas County, Texas, 2000
Electrical Damage – A Course for Texas Insurance Adjusters, 1996
Energy Efficiency in Lumber Drying – National Hardwood Magazine, 1976

Work Experience (1960–2022)

Accident/Loss Scene Evidence Collection: Experience in the collection of evidence at accident and loss scenes. This work involved vehicle accidents, industrial accidents, fires, explosions, worker injuries and fatalities, machinery failures, and numerous types of insurance claims. Experience in interviewing witnesses, photography, scene measurements, and physical evidence collection and storage to comply with ASTM E-860 and/or NFPA 921 standards.

Aerospace/aviation: Nine months experience on the Boeing 747 structures project and 18 months experience on the Saturn V static test firing project. Experience in resolving aircraft structural claims due to collisions with ground-service vehicles.

Automotive and Trucking: Experience in the reconstruction of hundreds of vehicle accidents and fires involving site data gathering, site photography, interviews with witnesses, examination of the vehicles, review of police and fire reports, and the use of Newtonian physics, mathematics and EDVAP computer models.

Experience in the repairs of G.M., Ford and Chrysler automobile engines, transmissions, suspensions, differentials, fuel systems, electrical systems, and brake systems. During the last 40 years, have built high-performance Ford, Oldsmobile and Chevrolet racing engines, transmissions, differentials, and suspension systems.

Building Products: Experience in the design, manufacturing and application of building construction products and the investigation of insurance claims due to wind and storm damage, roof damage, water damage, humidity damage, decay, insect damage, vibration damage, foundation distress and fire losses.

Combustion of Coal Oil, Gas and Wood. Designed coal, oil, gas and wood combustion systems for lumber kilns, veneer dryers, steam boilers, incinerators, and gasifiers.

Ten months design experience on a magnetohydrodynamics nozzle-mix, membrane-walled, pulverized-coal combustor, and associated fuel support systems.

This work included safety and environmental code requirements and the installation and troubleshooting of these systems. Designed and tested a prototype wood gasifier in 1979.

Construction Management: Since 1971, experience in managing numerous small and large construction projects involving purchasing, contract preparation, pert charts, cost tracking and control, compliance with OSHA regulations, and safety audits.

Electrical: Experience in the design and troubleshooting of electric motor control systems up to 500 H.P. motor sizes and experienced in ladder logic relay and circuit designs. Also have taken two G.E. series six PLC maintenance and systems courses.

Evidence Collection: Experience in the collection and storage of evidence involved in civil and criminal matters. Experienced in investigative techniques, interviews, photography, scene management and evidence preservation.

Environmental: Experience in designing and troubleshooting equipment (PCB incinerators, bag houses, precipitators, wet scrubbers, ash collectors and furnace controls) for steam boilers, incinerators, drying, and waste handling systems.

Fluid Mechanics: Decades of design experience and troubleshooting piping and valve systems, pumping systems and control systems for steam, hot and cold water, hot oil, municipal waste/sewage, hydraulics, and air handling systems.

Hydronics: Experience with a 300-horsepower experimental water tunnel (Mississippi State University). Experience in designing hot-water heating systems for lumber kilns and HVAC systems. This work involved the sizing of heat exchangers, control valves, pumps and lines and calculations of heat balances. Experience in reconstructing scalding accidents and reviewing code compliance of hot water heating systems.

H.V.A.C. Systems: Design, installation and troubleshooting systems up to 60-ton refrigeration and 60,000,000 btu/hr heating sizes. This work involved calculations of heat balances, duct sizes, analysis and selection of air moving equipment, design of electrical controls, design of heating units, cooling towers and units. Experience in failure analysis of HVAC compressors.

Insurance Fraud: Training and experienced in the modeling of the many faces of fraud in the insurance business. Since 1995, have conducted personal research into the causative factors and different faces of fraud in the American insurance institution. Has assisted SIU investigators, fire investigators, state and federal agencies, adjusters, and individuals in understanding how simple and complex fraud is committed in the insurance business.

Lumber Handling Systems: Twenty years of experience in designing, installing, and troubleshooting lumber handling systems. This work involved; stackers, sorters, breakdown units, hoists, lifts, conveyors, trimmers, saws, chippers, waste handling systems, kilns, planer mills, rough-mill equipment and finished storage areas.

Process Measurement & Controls: Design of analog and digital systems for; temperature, pressure, mass flow, flow limiters, and stack gases for energy and drying systems. Included statistical feedback controls. Six months project design work on Fisher Provox distributed control systems. Attended configuration schools by Fisher and PLC schools by G.E, and boiler control school by ISA.

Property Conservation: Co-authored the L.U.A. Fire Prevention Standards for Lumber Kilns and International Paper Company's Wood Products Property Conservation Manual.

Prepared fire-protection design guidelines and specifications for 54 of Texas Utilities fossil-fueled, power generating units.

Experienced in auditing industrial plants for OSHA and insurance requirements.

Roofing systems: Designed and inspected roofing systems (gabled, single-sloped, and flat built-up types) for lumber kilns, lumber predryers, and commercial structures starting in 1973. This work included the design, installation and inspection of steel, aluminum-paneled, built-up, shingled and single-ply roofing systems including their drainage systems. Assisted insurance adjusters in the assessment of claims of hail, wind and storm damage and the cause of roof collapse due to inadequate water drainage. Designed storm-water-handling gutter and drainage systems to improve drainage, roof life and prevent roof collapses.

Safety in the Workplace: Experience in the auditing of industrial plants for insurance coverage, their safety programs and fire and explosion prevention programs for conformance to corporate, insurance and OSHA requirements.

Steam Production & Utilization: Experience with boiler systems up to 450,000 pound/hour steam production. Included design through troubleshooting installed systems and the testing of installed systems to meet EPA stack emissions. Projects included small low-pressure packaged boilers up to large high-pressure utility cogeneration systems. Experience with gas-fired, oil-fired, coal-fired, wood-waste-fired and co-generation systems. Experience with most boiler stack emission equipment and the required EPA testing. Designed and performed carbon monoxide parametric studies on two bark-fired boilers at the I.P.Co.

Structural: Aeronautical: Nine months experience in designing structures for the Boeing 747.

Steel buildings: Designed and set design standards for steel pre-engineered, prefabricated kiln and predryer steel buildings per the American Institute of Steel Construction Manual and applicable building codes.

Concrete, Wood and Masonry: Designed and set design standards for kilns and predryer buildings and their foundations. Designed and set design standards for numerous lumber-handling and sawmill equipment systems foundations. Designed concrete foundations for the steel, wood, block, and concrete structures listed above. Designed high-temperature kiln concrete gas ducting systems including blend chambers, expansion joints, connectors, reinforcements, burner attachments and heat riser distribution systems. This work included the evaluation of steel and masonry structures and their underground ducting systems and foundations for repair following foundation settlements, fire losses, age deterioration and for upgrades. Designed concrete foundations for lumber kiln tram tracks and conveying systems and loading pads. Experience in the inspection of and testing of concrete silos following explosions and fires.

Lumber-Handling Machinery: Designed and set standards for numerous lumber-handling machines that included stress, strain, fatigue and deflection analysis of their fixed and moving elements.

Lumber-Drying and Kiln Technology

Ed Smith has many decades of experience in research, modeling, designing, manufacturing, installing, and troubleshooting all types of wood-drying systems. Ed's very first exposure to lumber drying kilns was in 1955 at a southern pine mill in Brookhaven, Mississippi. See the following.

Master's Thesis Subject – Moisture Movement in Lumber Drying. A finite-element computer program that predicted optimum drying rates for diffusion drying of lumber.

Participated and taught in numerous lumber drying and boiler schools throughout the forest products industry.

Designed the first 15-hour drying cycle commercial southern pine lumber kiln in the United States (Dierks, Arkansas Weyerhaeuser Plant, 1977).

Developed the first successful direct-fired, high-temperature pine kiln using dry wood waste as fuel (Energex program, 1973).

Designed and developed the fuel-to-air ratio and burner control systems currently being used on vortex wood dust burning systems now in use throughout the forest products industry.

Designed the first negative-air C.R.T. lumber kiln for the Weyerhaeuser Company, 1976.

Coauthored the National Lumber Kiln Fire & Safety Standards used by the Lumberman's Underwriter's Alliance, 1975.

Designed the first high-air-velocity, white-pine commercial lumber drying system in the United States, 1976.

Designed lumber kiln vent heat recovery and lumber preheat systems, 1974

Designed and developed the first high-efficiency cross-shaft fan system now used in high-temperature lumber kilns, 1974.

Designed four types of continuous lumber kilns fitted with E&C chambers for drying stacked softwoods.

One of these is the counterflow continuous kiln now common in the softwood lumber industry, 1974.

Participated in the design of an ocean-bound, hot-oil-heated, mahogany drying system for drying in route between the United States and South America.

Developed a computer model of a lumber kiln in which individual board moisture content and drying rates could be predicted from kiln parameters such as: air velocity, sticker thickness, board thickness, dry bulb and wet bulb temperature, initial moisture content, and specific gravity.

Developed a computer model of a lumber kiln in which the performance of fin pipe heating systems could be predicted.

Developed a computer model of a lumber kiln in which the performance of fan systems could be predicted.

Designed flash/energy recovery systems for steam heated drying systems.

JES Organization
Ed Smith, President
J E Smith Company (JES)
PO Box 428, Terrell, TX 75160
Cell: 903-257-5053
Email: ed.jes4278@outlook.com

JES Research and Development
Energy and Drying Technologies
Fan and Pump Technologies

JES Industrial Services
Equipment Replacement, Repair, Parts, and Service
Equipment Installation
Construction Services

- Project Management
- Vendor and Owner's Oversight

Property Loss Remediation Services

JES Consulting
Energy and Drying Technology Improvements
Industrial Drying Systems
Steam Boilers
Steam, Condensate, and Hot Liquids Systems
Combustion and Incineration Systems
Process Control Systems
Property Conservation and Safety

- Fire and Explosion Protection
- Equipment Safety Code Reviews
- ANSI Matrices

Property and Casualty Losses and Insurance Claims

- Catastrophe Recovery Management
- Forensic Investigations/Failure Analysis/Materials Testing
- Accident Reconstruction
- Data Recovery and Storage
- Observer Services

JES Dynamics
Seminars, Training and Surveys for Private and Government Entities
Primary focus on property conservation and fraud detection

Smith Books
Publishing (study books, training manuals, safety, and security manuals)

Index

J E 'Ed' Smith, *Industrial Drying Systems*,
https://doi.org/10.1007/978-3-031-31863-4

F

G

H

I

J

K

L

M

N

O